Student Study Guide with Solutions

Precalculus

FIFTH EDITION

J. Douglas Faires

James DeFranza

Prepared by

J. Douglas Faires

James DeFranza

BROOKS/COLE
CENGAGE Learning™

Australia • Brazil • Japan • Korea • Mexico • Singapore • Spain • United Kingdom • United States

ISBN-13: 978-1-111-42736-8
ISBN-10: 1-111-42736-4

Brooks/Cole
20 Davis Drive
Belmont, CA 94002-3098
USA

Cengage Learning is a leading provider of customized learning solutions with office locations around the globe, including Singapore, the United Kingdom, Australia, Mexico, Brazil, and Japan. Locate your local office at: **www.cengage.com/global**

Cengage Learning products are represented in Canada by Nelson Education, Ltd.

To learn more about Brooks/Cole, visit **www.cengage.com/brookscole**

Purchase any of our products at your local college store or at our preferred online store **www.cengagebrain.com**

Printed in the United States of America
1 2 3 4 5 6 7 14 13 12 11 10

PREFACE

This Student Study Guide accompanies the text PreCalculus, Fifth Edition, by Faires and De-Franza. The PreCalculus book introduces algebra review topics as they are needed in the context of precalculus. This permits students to proceed with the main elements of the course with minimal delay. Some students, however, need or prefer to have a more intensive mathematical review earlier at the beginning of the course. This Guide provides those students with access to review material, additional worked-out examples, and detailed step-by-step solutions to exercises to make the transition to precalculus more comfortable.

The Guide contains extensive algebra review in the early sections, and additional algebra, geometry, and trigonometry topics throughout. There are numerous examples and supplemental explanations in the Guide, as well as the detailed solutions to all the odd numbered exercises. In addition, solutions are given for all the exercises in the Chapter Tests that presented in the book.

Our approach to PreCalculus places a heavy emphasis on graphing techniques to help students visualize and master important topics. The Guide also emphasizes a visual and intuitive approach to learning mathematics and will give you a good indication of the material that is essential to master to be successful in PreCalculus and in Calculus. We recommend that students studying calculus keep this Guide readily available as a reference for the background material that is assumed standard in a calculus course.

Two short examinations that students can use to test their readiness for Precalculus and for Calculus can be downloaded in Adobe PDF from the book web site

http://www.math.ysu.edu/~faires/PreCalculus

We suggest that you work one of the examinations at the start of the PreCalculus course and the second at the end of the course. If you score approximately 16 or higher on the 40-question examination you should be prepared to take a PreCalculus course based on our book. A score of approximately 28 or higher indicates that you have the background needed for a University Calculus sequence.

If you have suggestions for improvements in future editions of the PreCalculus book or this Guide, we would be most grateful for your comments. Additional information about the book and any updates will be placed at the book web site.

We greatly appreciate the work of Jena Baun, a second-year student at Youngstown State University who performed much of the detail work in preparing this Guide, and Krista Foster, a graduate student at Penn State University, who checked the accuracy of the solutions.

Doug Faires   Youngstown State University
faires@math.ysu.edu

Jim DeFranza   St. Lawrence University
jdefranza@stlawu.edu

iii

Table of Contents

CHAPTER 1: Functions

1.1   Introduction

All branches of mathematics are layered vertically, with the most basic material at the bottom of the stack and the most sophisticated at the top. In order to master the most difficult material you must first become versed in the most basic and work up slowly toward the top. This is certainly the case with Algebra, PreCalculus, and Calculus. In this chapter we review topics from elementary algebra that will be used freely in the material that follows as part of PreCalculus. The first topic is the most basic of all, the number system that is the foundation for most of our work in *PreCalculus*.

1.2   The Real Line

The collection of *real numbers* $\mathbb{R}$ consists of several different kinds of numbers. The *natural numbers* $\mathbb{N}$ consist of all positive whole numbers, 1, 2, 3, 4,..., and the *integers* $\mathbb{Z}$ consist of the positive and negative whole numbers and 0, that is, $\ldots -3, -2, -1, 0, 1, 2, 3, \ldots$. The *rational numbers* $\mathbb{Q}$ include all fractions of the form p/q, where p and q are integers with $q \neq 0$. Finally, the *irrational numbers* consist of all numbers that cannot be represented as a rational number, for example, $\sqrt{2}$ or π.

An important way to distinguish between rational and irrational numbers is through the decimal expansions of these numbers. Rational numbers have decimal expansions that have a repeating pattern, and irrational numbers are those with no repeating pattern.

EXAMPLE 1

Classify each number as a natural, integer, rational, or irrational number.
(a) 192.237   (b) $-2.132132\overline{132}\ldots$ (c) 2.01001000100001...   (d) -3

Solution   (a) It is rational since $192.237 = 192 + \frac{237}{1000} = \frac{192237}{1000}$.

(b) It is rational since the decimal expansion has the pattern 132 repeated indefinitely. It also isn't hard to find the rational number that is represented by $-2.132\overline{132}$. The trick is to move the decimal point to the right one repeated block. First set

$$x = 2.132\overline{132}.$$

Then

$$1000x = 2132.132\overline{132}$$
$$1000x - x = 2130.\overline{000000}$$
$$999x = 2130$$
$$x = \frac{2130}{999}$$

1

$$-2.132\overline{132} = -\frac{2130}{999}.$$

(c) This is irrational since the number does not have any finite sequence of digits that repeats indefinitely. Each successive block of zeros contains one additional zero.

(d) This is an integer, but is not a natural number.   □

     A letter used to represent an arbitrary real number is called a *variable*, and a letter used to represent a fixed value, one that does not change, is called a *constant*. An *algebraic expression* is any combination of variables and constants formed using a finite number of the operations of addition, subtraction, multiplication, division, raising to a power, and taking roots. Examples are,

$$ax^2, \quad \frac{x^3 + \sqrt{2x-1}}{x^5 + x^3 - 1}, \quad \text{and} \quad x^2yz^4 - 3xyz^2 + 2x^3y - 6.$$

The first and third examples are called polynomials and will be considered in more detail later.

     The basic rules for combining real numbers are important to recognize when manipulating algebraic expressions.

- **Commutative Laws:** $x + y = y + x$ and $xy = yx$

- **Associative Laws:** $(x + y) + z = x + (y + z)$ and $(xy)z = x(yz)$

- **Distributive Law:** $x(y + z) = xy + xz$

EXAMPLE 2

Simplify the algebraic expression using the properties for combining real numbers.

(a) $2(x - 3) - 3(2 + 5x)$ (b) $(x + 1)(2y - 3)$

Solution   (a) First use the Distributive Law to multiply the 2 and the -3 through the parentheses, being careful to multiply both terms in the second parentheses by -3. Then combining like terms gives

$$2(x - 3) - 3(2 + 5x) = 2x - 6 - 6 - 15x = -13x - 12.$$

(b) Similarly

$$(x + 1)(2y - 3) = (x + 1)(2y) - (x + 1)(3)$$
$$= 2xy + 2y - 3x - 3.$$

We used the Distributive Law twice, with the end result that each term in the first parentheses is multiplied by each term in the second.   □

Set and Interval Notation

A *set* is a collection of objects called *elements*. Sets are denoted using capital letters, and elements are usually denoted using lower case letters. There are three standard ways of representing a set:

- Listing all the elements, separated by commas, in brackets { }.

- Listing a few of the elements that give the pattern followed by . . . .

- Describing a property or properties that all elements satisfy in the form, $\{x \mid x$ satisfies the property $P\}$.

The set with no elements, called the empty set, is denoted ϕ. The two basic operations on sets are **union** ($\cup$) and **intersection** ($\cap$). If A and B are sets, then $A \cup B$ is the set of all elements from either A or B, and the intersection $A \cap B$ is the set of all elements common to both sets.

EXAMPLE 3

Determine the intersection and union of the given sets.

(a) $A = \{-5, -2, 0, 1, 2, 3, 6, 8, 13\}$, $B = \{-3, -2, -1, 5, 7, 9, 11, 13, 15\}$
(b) $A = \{x \mid -2 < x < 10\}$, $B = \{x \mid x \geq 3\}$

Solution (a) The union of the two sets simply lists all the unique elements that are members of A or B, so

$$A \cup B = \{-5, -3, -2, -1, 0, 1, 2, 3, 5, 6, 7, 8, 9, 11, 13, 15\}.$$

The intersection consists of the elements common to both sets, so

$$A \cap B = \{-2, 13\}.$$

(b) The union is the set of all real numbers to the right of -2, not including -2, so $A \cup B = \{x \mid x > -2\}$. Since 3 lies between -2 and 10, the intersection is $A \cap B = \{x \mid 3 \leq x < 10\}$.   □

An **interval** is a set of real numbers between two given real numbers where the endpoints may or may not be included. When an endpoint is not included in the interval, a round parenthesis, (or), is used. When the endpoint is included, a square bracket, [or], is used. Since ∞ is only a symbol, not a real number, only a round parenthesis can be placed next to ∞ and $-\infty$.

EXAMPLE 4

Rewrite the sets using interval notation. Sketch the intervals on a real line.

(a) $\{x \mid -1 < x < 2\}$ (b) $\{x \mid 2 \leq x < 5\}$ (c) $\{x \mid -2 < x \leq 0\}$
(d) $\{x \mid -5 \leq x \leq -2\}$ (e) $\{x \mid x < 3\}$ (f) $\{x \mid x \leq -1\}$
(g) $\{x \mid x > -2\}$ (h) $\{x \mid x \geq 0\}$

Solution

(a) $\{x \mid -1 < x < 2\} = (-1, 2)$

(b) $\{x \mid 2 \leq x < 5\} = [2, 5)$

(c) $\{x \mid -2 < x \leq 0\} = (-2, 0]$

(d) $\{x \mid -5 \leq x \leq -2\} = [-5, -2]$

(e) $\{x \mid x < 3\} = (-\infty, 3)$

(f) $\{x \mid x \leq -1\} = (-\infty, -1]$

(g) $\{x \mid x > -2\} = (-2, \infty)$

(h) $\{x \mid x \geq 0\} = [0, \infty)$

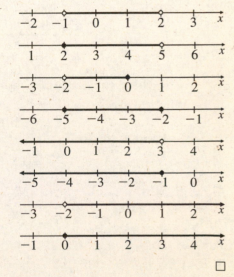

□

The length of an interval is the distance between two points on the real line. The *absolute value* of a real number is the distance from the number to the origin, 0. The absolute value of a number x is defined by

$$|x| = \begin{cases} x, & \text{if } x \geq 0 \\ -x, & \text{if } x < 0. \end{cases}$$

EXAMPLE 5

Determine the exact value of the absolute value.

(a) $|3.5|$ (b) $|-3|$ (c) $\left|\sqrt{2} - 1\right|$ (d) $\left|1 - \sqrt{3}\right|$

Solution   (a) Since $3.5 > 0$, the absolute value is itself and $|3.5| = 3.5$.

(b) Since $-3 < 0$, we have $|-3| = -(-3) = 3$.

(c) Since $\sqrt{2} > 1$, we have $\sqrt{2} - 1 > 0$, and $\left|\sqrt{2} - 1\right| = \sqrt{2} - 1$.

(d) Since $\sqrt{3} > 1$, we have $1 - \sqrt{3} < 0$, and $\left|1 - \sqrt{3}\right| = -(1 - \sqrt{3}) = \sqrt{3} - 1$.   □

The distance between two real numbers a and b is $|a - b| = |b - a|$.

EXAMPLE 6

Determine the distance between the two real numbers.

(a) $2, 12.7$ (b) $-3, 5$

Solution   (a) $|2 - 12.7| = |-10.7| = 10.7$

(b) $|-3 - 5| = |-8| = 8$, or $|5 - (-3)| = |8| = 8$ □

Notice in (b) that the order of subtraction does not matter as long as the computations are done inside the absolute value.

Exponents and Radicals

Listed are the basic definitions and properties.

$$x^n = \underbrace{x \cdot x \cdot x \cdots x}_{n\text{-times}} \qquad x^{-n} = \frac{1}{x^n} \qquad x^{1/n} = \sqrt[n]{x}$$

$$x^m x^n = x^{m+n} \qquad (xy)^n = x^n y^n \qquad (x^m)^n = x^{mn}$$

$$\frac{x^m}{x^n} = x^{m-n} \qquad \left(\frac{x}{y}\right)^n = \frac{x^n}{y^n}$$

$$\sqrt[n]{xy} = \sqrt[n]{x}\sqrt[n]{y} \qquad \sqrt[n]{\frac{x}{y}} = \frac{\sqrt[n]{x}}{\sqrt[n]{y}} \qquad x^{m/n} = \sqrt[n]{x^m} = \left(\sqrt[n]{x}\right)^m$$

EXAMPLE 7

Simplify the expression.

(a) x^3x^4 (b) x^2x^{-9} (c) $\dfrac{x^5}{x^2}$ (d) $\dfrac{x^2}{x^7}$

Solution   (a) $x^3x^4 = x^{3+4} = x^7$

(b) $x^2x^{-9} = x^{2-9} = x^{-7} = \dfrac{1}{x^7}$

(c) $\dfrac{x^5}{x^2} = x^{5-2} = x^3$

(d) $\dfrac{x^2}{x^7} = x^{2-7} = x^{-5} = \dfrac{1}{x^5}$

A shortcut in problems like (c) and (d) is to subtract the smaller exponent from the larger, placing the result in the numerator or denominator depending on which has the term with the larger exponent.   □

EXAMPLE 8

Simplify the expression.

(a) $(2x - 1)(x + 2)$ (b) $(2x^2 + 3xy)(x - 2xy^2)$

Solution   (a) Distributing the terms gives

$$(2x - 1)(x + 2) = 2x^2 + 4x - x - 2 = 2x^2 + 3x - 2.$$

(b) We have

$$(2x^2 + 3xy)(x - 2xy^2) = 2x^3 - 4x^3y^2 + 3x^2y - 6x^2y^3.$$ □

EXAMPLE 9

Simplify the expression.

(a) $(2x^3y^2)(x^4y^3)^3$ (b) $\left(\dfrac{x^2y}{xy}\right)^3 \left(\dfrac{xy^5}{x^6y^2}\right)^2$ (c) $\left(\dfrac{2x^{-2}y^3}{3x^4y}\right)^{-2}$

Solution   (a) First bring the outer exponent of 3 inside the second parentheses by multiplying each inside exponent by 3, and then combine the exponents on like terms. This gives

$$(2x^3y^2)(x^4y^3)^3 = (2x^3y^2)(x^{12}y^9) = 2x^{15}y^{11}.$$

In parts (b) and (c), it is easist to simplify inside the parentheses first and then apply the outside exponent, as we did in the solution.

(b) $\left(\dfrac{x^2y}{xy}\right)^3 \left(\dfrac{xy^5}{x^6y^2}\right)^2 = (x)^3 \left(\dfrac{y^3}{x^5}\right)^2 = x^3 \dfrac{y^6}{x^{10}} = \dfrac{y^6}{x^7}$

(c) $\left(\dfrac{2x^{-2}y^3}{3x^4y}\right)^{-2} = \left(\dfrac{2y^2}{3x^6}\right)^{-2} = \dfrac{2^{-2}y^{-4}}{3^{-2}x^{-12}} = \dfrac{3^2x^{12}}{2^2y^4} = \dfrac{9x^{12}}{4y^4}$ □

Rationalizing a fraction involving radicals refers to eliminating all radicals from the denominator of the expression by multiplying by a suitable fraction equaling 1. This was done originally to simplify calculations that were performed by hand. It is much easier to divide radical approximations by integers, for example, than it is to divide integers by radical approximations. While this is no longer an important application because of the wide-spread availability of calculators, it is still a useful procedure.

EXAMPLE 10

Rationalize each of the fractions involving radicals.

(a) $\dfrac{1}{\sqrt{3}}$ (b) $\sqrt{\dfrac{3}{5}}$ (c) $\dfrac{\sqrt{3}-1}{\sqrt{3}+1}$

Solution (a) $\dfrac{1}{\sqrt{3}} = \dfrac{1}{\sqrt{3}}\dfrac{\sqrt{3}}{\sqrt{3}} = \dfrac{\sqrt{3}}{3}$

(b) $\sqrt{\dfrac{3}{5}} = \dfrac{\sqrt{3}}{\sqrt{5}} = \dfrac{\sqrt{3}}{\sqrt{5}}\dfrac{\sqrt{5}}{\sqrt{5}} = \dfrac{\sqrt{3}\sqrt{5}}{5} = \dfrac{\sqrt{15}}{5}$

(c) $\dfrac{\sqrt{3}-1}{\sqrt{3}+1} = \dfrac{\sqrt{3}-1}{\sqrt{3}+1}\dfrac{\sqrt{3}-1}{\sqrt{3}-1} = \dfrac{3-2\sqrt{3}+1}{3-1} = \dfrac{4-2\sqrt{3}}{2} = 2-\sqrt{3}$

In (c), we are using the algebraic result that $(x+y)(x-y) = x^2 - y^2$, where, in this case, $x = \sqrt{3}$ and $y = 1$. $\square$

EXAMPLE 11

Simplify the expression.

(a) $\sqrt[3]{8x^6 y^5}$ (b) $\left(\dfrac{9x^{2/3}y^4}{4x^9}\right)^{3/2}$

Solution (a) $\sqrt[3]{8x^6 y^5} = \left(8x^6 y^5\right)^{1/3} = 8^{1/3}\left(x^6\right)^{1/3}\left(y^5\right)^{1/3} = 2x^2 y^{5/3}$

(b) $\left(\dfrac{9x^{2/3}y^4}{4x^9}\right)^{3/2} = \dfrac{(\sqrt{9})^3 xy^6}{(\sqrt{4})^3 x^{27/2}} = \dfrac{27xy^6}{8x^{27/2}} = \dfrac{27y^6}{8x^{25/2}}$

In part (b), it is a easier to first expand using the exponent 3/2 and simplify. $\square$

Special Products and Factoring

The following formulas are useful for quickly multiplying certain algebraic expressions which arise frequently, and are especially useful when going in the reverse direction of factoring an algebraic expression.

$$(x+y)^2 = x^2 + 2xy + y^2, \quad (x-y)^2 = x^2 - 2xy + y^2, \quad (x-y)(x+y) = x^2 - y^2,$$
$$(x+y)^3 = x^3 + 3x^2 y + 3xy^2 + y^3, \quad (x-y)^3 = x^3 - 3x^2 y + 3xy^2 - y^3,$$
$$x^3 + y^3 = (x+y)(x^2 - xy + y^2), \quad x^3 - y^3 = (x-y)(x^2 + xy + y^2)$$

EXAMPLE 12

Perform the indicated multiplication.

$(a)(2x - 2y)(2x + 2y)$ 　　　　　 $(b)\left(x^2 + y^2\right)^2$

$(c)\left(3x^3 - y\right)^3$ 　　　　　 $(d)\left(x - \sqrt{x}\right)\left(x + \sqrt{x}\right)$

Solution In each part we will apply the appropriate special product formula.

(a) $(2x - 2y)(2x + 2y) = (2x)^2 - (2y)^2 = 4x^2 - 4y^2$

(b) $\left(x^2 + y^2\right)^2 = \left(x^2\right)^2 + 2(x^2)(y^2) + \left(y^2\right)^2 = x^4 + 2x^2 y^2 + y^4$

(c) Since

$$(3x^3 - y)^3 = (3x^3)^3 - 3(3x^3)^2 y + 3(3x^3)y^2 - y^3$$

we have

$$(3x^3 - y)^3 = 27x^9 - 27x^6 y + 9x^3 y^2 - y^3.$$

(d) $(x - \sqrt{x})(x + \sqrt{x}) = x^2 - (\sqrt{x})^2 = x^2 - x$

□

The Distributive Property is used to expand algebraic expressions. *Factoring* is used to perform the reverse operation of writing a complicated expression in terms of simpler ones. For example,

$$x^2 - 2x - 3 = (x - 3)(x + 1).$$

Expanding the right side to produce the left side uses the Distributive Property. Rewriting the left side to produce the right side involves factoring.

EXAMPLE 13

Factor each expression.
(a) $4x^2 - 9y^2$ (b) $x^2 + 4x + 4$ (c) $4x^3 - 20x$ (d) $x^4 + 2x^2 + 1$

Solution Each problem fits one of the special algebraic formulas.
(a) To obtain an expression that fits the formula for the difference of two squares, note that $4x^2 = (2x)^2$ and $9y^2 = (3y)^2$. So

$$4x^2 - 9y^2 = (2x)^2 - (3y)^2 = (2x - 3y)(2x + 3y).$$

(b) $x^2 + 4x + 4 = (x + 2)^2$
(c) Always look first for common terms that can be factored out from the expression. In this case we have

$$4x^3 - 20x = 4x(x^2 - 5) = 4x\left(x^2 - (\sqrt{5})^2\right) = 4x\left(x - \sqrt{5}\right)\left(x + \sqrt{5}\right).$$

(d) There are no common factors, but treating the term x^2 as the x and the 1 as the y in the formula for $(x + y)^2$, we have

$$x^4 + 2x^2 + 1 = (x^2)^2 + 2(x^2) + 1 = (x^2 + 1)^2.$$

□

Problems that do not fit one of the special formulas are more difficult to factor. In the next example we will factor a general quadratic, one of the form $ax^2 + bx + c$.

EXAMPLE 14

Factor each expression.
(a) $x^2 - x - 2$ (b) $6x^2 + 7x + 2$ (c) $2x^3 - 3x^2 - 2x$

Solution (a) The only factor of $a = 1$ is 1, and the only factors of $c = 2$ are 1 and 2. So the only choices for a possible factoring are

$$x^2 - x - 2 = (x \pm 1)(x \pm 2).$$

To get the constant term of -2, the signs of 1 and 2 must differ. Since the middle term is $-x$ rather then x, the minus sign must be with the larger factor 2. So

$$x^2 - x - 2 = (x + 1)(x - 2).$$

(b) The factors of $a = 6$ are $1, 2, 3, 6$, and the only factors of $c = 2$ are 1 and 2. Since the constant term is positive, and the middle term is also positive, the signs inside the factors will also be both positive. If both signs were negative, we would have a positive constant term but a negative middle term. The possibilities are

$$(6x + 1)(x + 2) = 6x^2 + 13x + 2$$

$$(6x + 2)(x + 1) = 6x^2 + 8x + 2$$

$$(3x + 1)(2x + 2) = 6x^2 + 8x + 2$$

$$(3x + 2)(2x + 1) = 6x^2 + 7x + 2$$

and only the last gives the correct result.

(c) $2x^3 - 3x^2 - 2x = x(2x^2 - 3x - 2) = x(2x + 1)(x - 2)$ □

EXAMPLE 15

Simplify each expression.

(a) $\dfrac{x^2 + 5x + 6}{x^2 - 2x - 8}$ (b) $\dfrac{x^2 + 2x + 1}{x^2 - x - 2} \cdot \dfrac{2x^2 - 5x + 2}{x^2 + 4x + 3}$ (c) $\dfrac{2x^2 + 8x}{x^2 - 3x + 2} \div \dfrac{x^2 + 6x + 8}{x - 1}$

Solution (a) When simplifying fractional expressions, try to first factor as much as possible and then cancel any like terms from the numerator and denominator.

$$\frac{x^2 + 5x + 6}{x^2 - 2x - 8} = \frac{(x + 2)(x + 3)}{(x + 2)(x - 4)} = \frac{x + 3}{x - 4}, \, provided \, x \neq -2.$$

(b) The rule

$$\frac{a}{b} \cdot \frac{c}{d} = \frac{ac}{bd}$$

for multiplying two number fractions is also used for algebraic expressions.

$$\frac{x^2 + 2x + 1}{x^2 - x - 2} \cdot \frac{2x^2 - 5x + 2}{x^2 + 4x + 3} = \frac{(x + 1)^2}{(x + 1)(x - 2)} \cdot \frac{(2x - 1)(x - 2)}{(x + 3)(x + 1)}$$

$$= \frac{(x + 1)^2(2x - 1)(x - 2)}{(x + 1)^2(x + 3)(x - 2)}$$

$$= \frac{2x - 1}{x + 3}, \, provided \, x \neq 2 \, and \, x \neq -3.$$

(c) The rule

$$\frac{\frac{a}{b}}{\frac{c}{d}} = \frac{a}{b} \cdot \frac{d}{c}$$

for dividing two number fractions is also used for algebraic expressions.

$$\frac{2x^2 + 8x}{x^2 - 3x + 2} \div \frac{x^2 + 6x + 8}{x - 1} = \frac{2x^2 + 8x}{x^2 - 3x + 2} \cdot \frac{x - 1}{x^2 + 6x + 8}$$

$$= \frac{2x(x + 4)}{(x - 1)(x - 2)} \cdot \frac{(x - 1)}{(x + 2)(x + 4)}$$

$$= \frac{2x}{(x - 2)(x + 2)}$$

$$= \frac{2x}{x^2 - 4}, \, provided \, x \neq 1.$$ □

EXAMPLE 16

Simplify each expression.

(a) $\dfrac{1}{x - 2} + \dfrac{4}{x + 1}$ (b) $\dfrac{1}{\sqrt{x}} - \dfrac{1}{\sqrt{x + a}}$

Solution (a) To add the fractions, we need to find the least common denominator. The rule

$$\frac{a}{b} + \frac{c}{d} = \frac{ad + cb}{bd}$$

for adding number fractions is used.

$$\frac{1}{x-2} + \frac{4}{x+1} = \frac{(x+1) + 4(x-2)}{(x-2)(x+1)}$$
$$= \frac{5x-7}{x^2 - x - 2}$$

(b) Taking the common denominator,

$$\frac{1}{\sqrt{x}} - \frac{1}{\sqrt{x+a}} = \frac{\sqrt{x+a} - \sqrt{x}}{\sqrt{x}\sqrt{x+a}}.$$

Expressions like these can also be rationalized to eliminate some of the radicals. In this case we will eliminate the radical in the numerator. That is,

$$\frac{1}{\sqrt{x}} - \frac{1}{\sqrt{x+a}} = \frac{\sqrt{x+a} - \sqrt{x}}{\sqrt{x}\sqrt{x+a}}$$
$$= \frac{\sqrt{x+a} - \sqrt{x}}{\sqrt{x}\sqrt{x+a}} \cdot \frac{\sqrt{x+a} + \sqrt{x}}{\sqrt{x+a} + \sqrt{x}}$$
$$= \frac{(x+a) - x}{\sqrt{x}\sqrt{x+a}(\sqrt{x+a} + \sqrt{x})}$$
$$= \frac{a}{\sqrt{x}\sqrt{x+a}(\sqrt{x+a} + \sqrt{x})}. \qquad \square$$

Solving Equations

Solving equations arise as subproblems to many problems encountered in a PreCalculus or Calculus course. To solve an equation means to find all real numbers that satisfy the given condition. Equations are solved by using the basic rules of arithmetic.

EXAMPLE 17

Find all values of x that satisfy the equation.

(a) $5x - 9 = 6$ (b) $\frac{2}{3}x + 3 = 2$ (c) $\frac{2x}{x+1} + \frac{1}{2} = 3$

Solution (a) To solve the equation, first add 9 to *both* sides of the equation and then divide by 5.

$$5x - 9 = 6$$
$$5x - 9 + 9 = 6 + 9$$
$$5x = 15$$
$$\frac{5x}{5} = \frac{15}{5}$$

$$x = 3$$

(b) First subtract 3 from both sides of the equation, and then multiply by 3/2.

$$\frac{2}{3}x + 3 = 2$$

$$\frac{2}{3}x = -1$$

$$\frac{3}{2}\left(\frac{2}{3}x\right) = \frac{3}{2}(-1)$$

$$x = -\frac{3}{2}$$

(c) First subtract 1/2 from both sides of the equation, and then multiply both sides by $(x + 1)$ assuming $x \neq -1$.

$$\frac{2x}{x + 1} + \frac{1}{2} = 3$$

$$\frac{2x}{x + 1} = \frac{5}{2}$$

$$(x + 1)\frac{2x}{x + 1} = \frac{5}{2}(x + 1)$$

$$2x = \frac{5(x + 1)}{2}$$

When solving equations, it is easier to first eliminate fractions. Multiplying both sides of the last equation by 2 gives,

$$4x = 5x + 5$$

$$-x = 5$$

$$x = -5.$$ □

EXAMPLE 18
Solve each equation.
(a) $2x^2 + 5x - 3 = 0$ (b) $x^2 + x - 2 = 4$ (c) $x^4 + x^2 - 2 = 0$

Solution (a) To solve the equation for x, we first factor the left side. This gives

$$0 = 2x^2 + 5x - 3 = (2x - 1)(x + 3).$$

The product of two real numbers can be 0 only if at least one of the numbers is 0. So

$$2x - 1 = 0 \quad \text{or} \quad x + 3 = 0$$

and

$$x = \frac{1}{2} \quad \text{or} \quad x = -3.$$

(b) First rewrite the equation so 0 is on the right side, then factor.

$$x^2 + x - 2 = 4$$
$$x^2 + x - 6 = 0$$
$$(x + 3)(x - 2) = 0$$
$$x = -3, x = 2$$

(c) If we let $u = x^2$, then $x^4 + x^2 - 2 = u^2 + u - 2 = (u + 2)(u - 1)$. The original equation can then be factored as

$$0 = x^4 + x^2 - 2 = (x^2 + 2)(x^2 - 1) = (x^2 + 2)(x - 1)(x + 1).$$

Set each factor to 0

$$x^2 + 2 = 0 \quad \text{or} \quad x^2 - 1 = 0. \qquad \square$$

Since $x^2 + 2 = 0$ has no real solutions, the only solutions are when $x = 1$ or $x = -1$.

If the solutions to an equation of the form $ax^2 + bx + c = 0$ are rational numbers, with $a \neq 0$, then the expression can be factored and the solutions found as we did above.

In general, the solutions to the **quadratic equation** $ax^2 + bx + c = 0$, with $a \neq 0$, can be found using the **quadratic formula.** The solutions are

$$x = \frac{-b + \sqrt{b^2 - 4ac}}{2a} \quad \text{and} \quad x = \frac{-b - \sqrt{b^2 - 4ac}}{2a},$$

which are written compactly as

$$x = \frac{-b \pm \sqrt{b^2 - 4ac}}{2a}.$$

EXAMPLE 19

Find all solutions to $x^2 + 4x + 2 = 0$ and factor the expression $x^2 + 4x + 2$.

Solution Using the Quadratic formula with $a = 1$, $b = 4$, and $c = 2$, gives

$$x = \frac{-4 \pm \sqrt{16 - 4(1)(2)}}{2}$$
$$= \frac{-4 \pm \sqrt{8}}{2} = \frac{-4 \pm 2\sqrt{2}}{2}$$
$$= -2 \pm \sqrt{2}.$$

Once the solutions (called the *roots*) are known, the expression can be factored completely as

$$x^2 + 4x + 2 = \left(x - \left(-2 + \sqrt{2}\right)\right)\left(x - \left(-2 - \sqrt{2}\right)\right). \qquad \square$$

In the previous example, the solutions are irrational numbers, and we used the Quadratic formula to solve the equation. We cannot find the factors of the expression using a trial and error approach, as was done when the solutions were rational numbers.

EXAMPLE 20

Find all solutions to the equation.

(a) $\sqrt{2x-1}=3$ (b) $\sqrt{x-2}+x=2$

Solution (a) First square both sides and then isolate x, giving

$$\sqrt{2x-1}=3$$
$$2x-1=9$$
$$2x=10$$
$$x=5.$$

Because we squared the equation, we need to check the answer to be sure that we have not introduced an *extraneous* solution. That is, a solution to the final equation that was not a solution to the original equation. In this case we did not since $\sqrt{2(5)-1}=\sqrt{9}=3$.

(b) In problems like this, it is easiest to first isolate the radical term and then square both sides.

$$\sqrt{x-2}+x=2$$
$$\sqrt{x-2}=2-x$$
$$x-2=(2-x)^2$$
$$x-2=4-4x+x^2$$
$$x^2-5x+6=0$$
$$(x-2)(x-3)=0$$
$$x=2, x=3$$

It is important that you check the answers, since squaring both sides of an equation may introduce extraneous solutions. Checking, when $x=3$

$$\sqrt{3-2}+3=4\neq 2,$$

but when $x=2$

$$\sqrt{2-2}+2=2.$$

So $x=2$ *is a solution*, but $x=3$ *is not* a solution. □

EXAMPLE 21

Find all values of x that satisfy the equation $\left|\dfrac{2x-1}{x+3}\right|=4$.

Solution The solution to an equation of the form

$$|\square|=a$$

can always be solved by solving the two equations

$$\square=a \quad \text{and} \quad \square=-a.$$

Separating the absolute value equation into two equations gives

$$\left| \frac{2x-1}{x+3} \right| = 4$$

$$\frac{2x-1}{x+3} = 4 \quad \text{or} \quad \frac{2x-1}{x+3} = -4$$

$$2x - 1 = 4x + 12 \quad \text{or} \quad 2x - 1 = -4x - 12$$

$$2x = -13 \quad \text{or} \quad 6x = -11$$

$$x = -\frac{13}{2} \quad \text{or} \quad x = -\frac{11}{6}. \qquad \square$$

Solving Inequalities

Solving inequalities is similar to solving equations, but with inequalities care must be taken when multiplying or dividing by negative numbers.

Properties of Inequalities

- If $a < b$ and $b < c$, then $a < c$.

- If $a < c$, then $a + b < c + b$.

- If $a < b$, and $c > 0$, then $ac < bc$.

- If $a < b$, and $c < 0$, then $ac > bc$.

EXAMPLE 22

Solve the inequality.
(a) $2x - 3 > 2$ (b) $-5x + 2 < 3$ (c) $-1 < 2x - 3 \leq 7$

Solution (a) The second inequality property states that adding the same value to each side does not change the inequality relation, so

$$2x - 3 + 3 > 2 + 3$$

and

$$2x > 5.$$

Then use the third inequality property and divide both sides by 2, so

$$\frac{2x}{2} > \frac{5}{2} \text{ and } x > \frac{5}{2}.$$

In interval notation the solution is $\left(\frac{5}{2}, \infty \right)$.

(b) To isolate x, subtract 2 from each side so that

$$-5x + 2 < 3$$
$$-5x + 2 - 2 < 3 - 2$$
$$-5x < 1.$$

Next divide each side by -5. Be careful to reverse the inequality sign when dividing by a negative number (dividing by -5 is the same as multiplying by $-1/5$). This gives

$$\frac{-5x}{-5} > \frac{1}{-5}$$
$$x > -\frac{1}{5}.$$

In interval notation the solution is $\left(-\frac{1}{5}, \infty\right)$.

(c) The inequality $-1 < 2x - 3 \le 7$ implies the two inequalities

$$-1 < 2x - 3 \quad \text{and} \quad 2x - 3 \le 7.$$

Solving this pair of inequalities simultaneously gives

$$
\begin{array}{lll}
-1 < 2x - 3 & \text{and} & 2x - 3 \le 7 \\
2 < 2x & \text{and} & 2x \le 10 \\
1 < x & \text{and} & x \le 5 \\
1 < x \le 5. & &
\end{array}
$$

In interval notation the solution is $(1, 5]$. □

EXAMPLE 23

Solve the inequality.

(a) $x^2 - x - 2 > 0$ (b) $-x^2 - 2x + 3 \ge 0$ (c) $\dfrac{x^2 + 3x + 2}{x^2 + 1} < 0$

Solution To solve a quadratic inequality, first try to factor the quadratic expression. In the case of a fractional expression, as in part (c), try to factor numerator and denominator.

(a) Factoring gives

$$x^2 - x - 2 = (x - 2)(x + 1) > 0.$$

In order for the product of two linear factors to be positive, either both factors must be positive or both must be negative. For example:

- If $x = 3$, then $(3 - 2)(3 + 1) = (1)(4) = 4 > 0$.
- If $x = -2$, then $(-2 - 2)(-2 + 1) = (-4)(-1) = 4 > 0$.
- But if $x = 1.5$, then $(1.5 - 2)(1.5 + 1) = (-0.5)(2.5) = -1.25 < 0$.

The values that make the linear factors 0, in this case at $x = 2$ and $x = -1$, separate the real line into three intervals, $(-\infty, -1)$, $(-1, 2)$, and $(2, \infty)$. To solve the inequality, select test values from each interval, substitute these values into the equation, and the inequality will have the same sign for all values in the interval. For example, if $x = 1$, $(0 - 2)(0 + 1) = -2 < 0$ and the inequality is negative on the entire interval $(-1, 2)$.

From the chart, the solution to the inequality is all $x < -1$ or all $x > 2$, that is, $(-\infty, -1) \cup (2, \infty)$.

(b) Factoring the quadratic gives

$$-x^2 - 2x + 3 = -(x^2 + 2x - 3)$$

$$-(x+3)(x-1) \geq 0$$
$$(x+3)(x-1) \leq 0$$

and using the sign chart the solution implies that the interval is $[-3, 1]$.

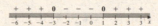

(c) Since the denominator can not be factored, we have

$$\frac{x^2 + 3x + 2}{x^2 + 1} = \frac{(x+2)(x+1)}{x^2 + 1}.$$

We could proceed as we did in parts (a) and (b), but in examples like this it is easier to first observe that the denominator is always greater than 0, so

$$\frac{x^2 + 3x + 2}{x^2 + 1} < 0 \text{ precisely when } (x+2)(x+1) < 0.$$

The sign chart shows that the solution is the interval $(-2, -1)$.   □

Inequalities Involving Absolute Values

To solve inequalities involving absolute values use the two facts:

- $|x| < a$ means $-a < x < a$.

- $|x| > a$ means $x < -a$ or $x > a$.

Often in problems like this, an expression involving x is inside the absolute value. The first statement, for example, could then be written as

$$|\square| < a \text{ means } -a < \square < a,$$

where $\square$ is the expression. A similar interpretation is given to the second statement.

EXAMPLE 24
Solve the inequality.

(a) $|2x - 1| < 2$ (b) $|3x + 2| \geq 1$ (c) $\left|\dfrac{2}{x+1}\right| < 3$

Solution (a) Rewrite the absolute value inequality as the two inequalities

$$-2 < 2x - 1 < 2$$
$$-2 + 1 < 2x < 2 + 1$$
$$-1 < 2x < 3$$
$$-\frac{1}{2} < x < \frac{3}{2}.$$

The solution set is the interval $\left(-\frac{1}{2}, \frac{3}{2}\right)$.

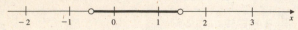

(b) The inequality is equivalent to

$$|3x + 2| \geq 1$$
$$3x + 2 \leq -1 \text{ or } 3x + 2 \geq 1$$
$$3x \leq -3 \text{ or } 3x \geq -1$$
$$x \leq -1 \text{ or } x \geq -\frac{1}{3}.$$

In interval notation the solution set is $(-\infty, -1] \cup \left[-\frac{1}{3}, \infty\right)$.

(c) We have

$$\left|\frac{2}{x+1}\right| < 3$$
$$2 < 3|x+1|$$
$$|x+1| > \frac{2}{3}$$
$$x + 1 < -\frac{2}{3} \quad \text{or} \quad x + 1 > \frac{2}{3}$$
$$x < -\frac{5}{3} \quad \text{or} \quad x > -\frac{1}{3}.$$

The solution set is $(-\infty, -\frac{5}{3}) \cup (-\frac{1}{3}, \infty)$.   □

Exercise Set 1.2 (Page 12)

1   $-1 \leq x \leq 5$

3   $-\sqrt{3} < x \leq \sqrt{2}$

5   $x < 3$

7   $x \geq \sqrt{2}$

9   $[-2, 3]$

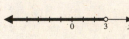

11 $[2, 5)$

13 $(-\infty, 3)$

15 $[3, \infty)$

17. **a.** $|3 - 7| = |-4| = 4$
    b. $\frac{1}{2}(3 + 7) = 5$

19. **a.** $|-3 - 5| = |-8| = 8$
    b. $\frac{1}{2}(-3 + 5) = 1$

21. $x^2 + 3x + 2 = (x + 1)(x + 2)$

23. $x^2 + 5x + 6 = (x + 2)(x + 3)$

25. $x^2 + 4x - 12 = (x - 2)(x + 6)$

27. **a.** $(0, 3]$ **b.** $[-1, 4)$

29. **a.** $(-2, 0)$ **b.** $(-\infty, 3]$

31. The inequality $x + 3 < 5$ implies that $x < 2$, so in interval notation $(-\infty, 2)$.

33. The inequality $2x - 2 \geq 8$ implies that $2x \geq 10$, so $x \geq 5$, and in interval notation $[5, \infty)$.

35. The inequality $-3x + 4 < 5$ implies that $-3x < 1$, so $x > -\frac{1}{3}$, and in interval notation $\left(-\frac{1}{3}, \infty\right)$.

37. The inequality $2x + 9 \leq 5 + x$ implies that $x \leq -4$, so in interval notation $(-\infty, -4]$.

39. The inequality $-1 < 3x - 3 < 6$ implies that $2 < 3x < 9$, so $\frac{2}{3} < x < 3$, and in interval notation $\left(\frac{2}{3}, 3\right)$.

41. The inequality

$$(x + 1)(x - 2) \geq 0$$

is satisfied for x in $(-\infty, -1] \cup [2, \infty)$.

43. The inequality

$$x^2 - 4x + 3 = (x - 1)(x - 3) \leq 0$$

is satisfied for x in $[1, 3]$.

45. The inequality

$$(x - 1)(x - 2)(x + 1) \leq 0$$

is satisfied for x in $(-\infty, -1] \cup [1, 2]$.

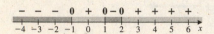

47. The inequality

$$x^3 - 3x^2 + 2x = x(x - 1)(x - 2) \geq 0$$

is satisfied for x in $[0, 1] \cup [2, \infty)$.

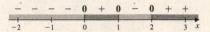

49. The inequality

$$x^3 - 2x^2 = x^2(x - 2) < 0$$

implies that

$$(x - 2) < 0 \text{ and } x \neq 0,$$

which is satisfied for x in $(-\infty, 0) \cup (0, 2)$.

```
–  –  –  –  0  –  0  +  +  +  +
 ┼──┼──┼──┼──┼──┼──┼──┼──┼──┼──┼──
-4 -3 -2 -1  0  1  2  3  4  5  6  x
```

51. The inequality

$$\frac{x + 3}{x - 1} \geq 0$$

is satisfied for x in $(-\infty, -3] \cup (1, \infty)$.

```
+  0  –  –  –  ?  +  +  +  +  +
 ┼──┼──┼──┼──┼──┼──┼──┼──┼──┼──┼──
-4 -3 -2 -1  0  1  2  3  4  5  6  x
```

53. The inequality

$$\frac{x(x + 2)}{x - 2} \leq 0$$

is satisfied for x in $(-\infty, -2] \cup [0, 2)$.

```
–  –  0  +  0  –  ?  +  +  +  +
 ┼──┼──┼──┼──┼──┼──┼──┼──┼──┼──┼──
-4 -3 -2 -1  0  1  2  3  4  5  6  x
```

55. The inequality

$$\frac{(1 - x)(x + 2)}{x(x + 1)} > 0$$

is satisfied for x in $(-2, -1) \cup (0, 1)$.

```
–  –  0  +  ?  –  ?  +  0  –  –  –  –  –
 ┼──┼──┼──┼──┼──┼──┼──┼──┼──┼──┼──┼──
-4 -3 -2 -1  0  1  2  3  4  5  6  x
```

57. The inequality

$$\frac{1}{x} \le 5 \quad \text{implies that} \quad \frac{1}{x} - 5 \le 0,$$

which can be written as

$$\frac{1 - 5x}{x} \le 0,$$

so in interval notation $(-\infty, 0) \cup \left[\frac{1}{5}, \infty\right)$.

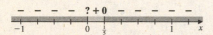

59. The inequality

$$\frac{2}{x - 1} \ge \frac{3}{x + 2}$$

implies that

$$\frac{2}{x - 1} - \frac{3}{x + 2} \ge 0,$$

and

$$\frac{2(x + 2) - 3(x - 1)}{(x - 1)(x + 2)} \ge 0,$$

which can be written as

$$\frac{7 - x}{(x - 1)(x + 2)} \ge 0,$$

so in interval notation $(-\infty, -2) \cup (1, 7]$.

61. We have $|5x - 3| = 2$ when $5x - 3 = 2$ or $5x - 3 = -2$, so $x = 1$ or $x = \frac{1}{5}$.

63. We have $\left|\frac{x-1}{2x+3}\right| = 2$ when $\frac{x-1}{2x+3} = 2$ or $\frac{x-1}{2x+3} = -2$, which implies that $x - 1 = 4x + 6$ or $x - 1 = -4x - 6$ which implies $-7 = 3x$ or $5x = -5$, so $x = -\frac{7}{3}$ or $x = -1$.

65. We have $|x - 4| \le 1$, which implies $-1 \le x - 4 \le 1$, so $3 \le x \le 5$, and in interval notation we have $[3, 5]$.

67. We have $|3 - x| \ge 2$, which implies $3 - x \ge 2$ or $3 - x \le -2$, so $-x \ge -1$ or $-x \le -5$, which implies $x \le 1$ or $x \ge 5$, and in interval notation we have $(-\infty, 1] \cup [5, \infty)$.

69. We have $\frac{1}{|x+5|} > 2$, which implies $|x + 5| < \frac{1}{2}$, so $-\frac{1}{2} < x + 5 < \frac{1}{2}$, which implies $-\frac{11}{2} < x < -\frac{9}{2}$, and in interval notation we have $\left(-\frac{11}{2}, -5\right) \cup \left(-5, -\frac{9}{2}\right)$. We omit $x = -5$ since this is where $|x + 5| = 0$.

71. We have $|x^2 - 4| > 0$, which implies $x^2 - 4 \ne 0$, so $x \ne \pm 2$, and in interval notation we have $(-\infty, -2) \cup (-2, 2) \cup (2, \infty)$.

73. Since $0 < a < b$ we have both $0 < a^2 < ab$ and $0 < ab < b^2$. Hence, $0 < a^2 < ab < b^2$.

75. a. Since $20 \leq F \leq 50$ implies that $-12 \leq F - 32 \leq 18$ we have $-12 \left(\frac{5}{9}\right) \leq \frac{5}{9}(F - 32) \leq 18 \left(\frac{5}{9}\right)$, so $-\frac{20}{3} \leq C \leq 10$.

b. Since $20 \leq C \leq 50$ we have $20 \leq \frac{5}{9}(F - 32) \leq 50$, which implies $\frac{9}{5}(20) \leq F - 32 \leq \frac{9}{5}(50)$ so $68 \leq F \leq 122$.

77. $18,750

. . . .

1.3   The Cartesian Plane

The xy-coordinate system is called the coordinate plane or Cartesian plane. Points are specified using an ordered pair (x, y), with the horizontal or x-coordinate specified first and the vertical or y-coordinate specified second.

Describing Regions in the Plane

EXAMPLE 1
Describe the set of points in the xy-plane satisfying the inequalities $-2 \leq x \leq 1$.

Solution Since the y-coordinate can be any real number, and the x-coordinate is restricted to lie between -2 and 1, the inequality describes all points between the vertical lines determined by $\{(x, y) \mid x = -2\}$ and $\{(x, y) \mid x = 1\}$. For example, $(-2, 0)$, $(-2, 2)$, $(-2, -2)$, $(0, 1)$, $(-1, 1)$ and $(1/2, 3)$ are all in the region, whereas $(-3, 1)$ and $(3, -1)$ are not in the region.

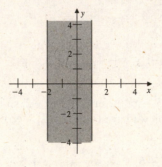

□

EXAMPLE 2
Describe the set of points in the xy-plane with $|x - 1| < 2$ and $|y + 1| < 3$.

Solution Rewriting the inequalities gives

$$|x - 1| < 2$$
$$-2 < x - 1 < 2$$
$$-1 < x < 3$$

and

$$|y + 1| < 3$$
$$-3 < y + 1 < 3$$
$$-4 < y < 2.$$

This means that all the points in the solution set have x-coordinates between -1 and 3, not including the values -1 and 3, and y-coordinates between -4 and 2, not including the values -4 and 2. So the points lie between, but not on, the vertical lines $x = -1$ and $x = 3$ and between, but not on, the horizontal lines $y = -4$ and $y = 2$, describing a rectangular region as shown in the figure.

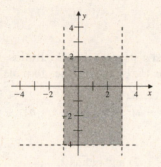

□

The Distance Formula

The **distance** between two points $P(x_1, y_1)$ and $Q(x_2, y_2)$ is given by the formula

$$d(P, Q) = \sqrt{(x_2 - x_1)^2 + (y_2 - y_1)^2}.$$

EXAMPLE 3

Find the distances between the points $(1, 1)$, $(4, 4)$, and $(0, 8)$, and show that they are vertices of a right triangle. At which vertex is the right angle?

Solution Let $A = (1, 1)$, $B = (4, 4)$, and $C = (0, 8)$. The sketch of the points in the figure indicates the right angle is likely at B.

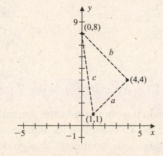

To verify that the triangle is a right triangle, we determine the lengths a, b, and c and verify that $a^2 + b^2 = c^2$. The lengths of the sides are

$$a = d(A, B) = \sqrt{(4-1)^2 + (4-1)^2} = \sqrt{18}$$

$$b = d(B, C) = \sqrt{(4-0)^2 + (4-8)^2} = \sqrt{32}$$

$$c = d(A, C) = \sqrt{(0-1)^2 + (8-1)^2} = \sqrt{50},$$

so

$$a^2 + b^2 = 18 + 32 = 50 = c^2.$$

$\square$

The Equation of a Circle

A **circle** is a set of all points that lie at a fixed distance, called the **radius**, from a fixed point, called the **center**. If the radius is r, and the fixed point in the plane is (h, k), the distance formula implies a point (x, y) will lie on the circle provided

$$\sqrt{(x-h)^2 + (y-k)^2} = r \quad \text{or} \quad (x-h)^2 + (y-k)^2 = r^2.$$

EXAMPLE 4

Find the equation of the circle with center $(-1, 2)$ and radius 2.

Solution Substituting into the formula, the equation of the circle is

$$(x - (-1))^2 + (y - 2)^2 = 4$$

or

$$(x + 1)^2 + (y - 2)^2 = 4.$$

$\square$

EXAMPLE 5

Find the center and radius of the circle $x^2 - 2x + y^2 + 2y + 1 = 0$.

Solution To find the center and radius, rewrite the equation in the general form. To do this, complete the square on the x-terms and on the y-terms of the equation. To complete the square on $x^2 - 2x$, take half the coefficient of the x-term, square it, and then add and subtract this value, so the net change is 0. This gives

$$x^2 - 2x = x^2 - 2x + \left(\frac{2}{2}\right)^2 - \left(\frac{2}{2}\right)^2$$
$$= x^2 - 2x + 1 - 1$$
$$= (x - 1)^2 - 1.$$

Doing the same with the y-terms gives

$$x^2 - 2x + y^2 + 2y + 1 = 0$$
$$\left(x^2 - 2x + 1\right) - 1 + \left(y^2 + 2y + 1\right) - 1 + 1 = 0$$
$$(x - 1)^2 + (y + 1)^2 = 1$$
$$(x - 1)^2 + (y - (-1))^2 = 1.$$

The circle has center $(1, -1)$ and radius $\sqrt{1} = 1$.

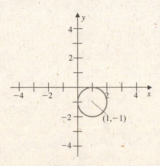

□

EXAMPLE 6

Find an equation of the circle with center $(-1, 2)$ that passes through $(2, 3)$.

Solution　If $(2, 3)$ is on the circle, then the line segment from the center point $(-1, 2)$ to $(2, 3)$ is a radius, and the length of the radius is

$$d((-1, 2), (2, 3)) = \sqrt{(2 + 1)^2 + (3 - 2)^2} = \sqrt{10}.$$

The equation of the circle is given by

$$(x + 1)^2 + (y - 2)^2 = 10.$$

□

EXAMPLE 7

Find the area of the region that lies outside the circle $x^2 + y^2 = 2$ and inside the circle $x^2 + y^2 = 5$.

Solution　This *annular* region is shown in the figure.

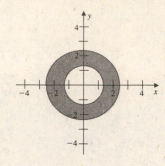

The shaded region consists of all points that simultaneously satisfy $x^2 + y^2 \geq 2$ and $x^2 + y^2 \leq 5$. Since the area of a circle of radius r is πr^2, and the radius of the inner circle is $\sqrt{2}$ and outer circle is $\sqrt{5}$, the area of the annular region is

$$\pi(\sqrt{5})^2 - \pi(\sqrt{2})^2 = 5\pi - 2\pi = 3\pi.$$

$\square$

Exercise Set 1.3 (Page 20)

1. The points are on the graph.

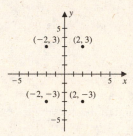

3. The points are on the graph.

5. a. The distance is $\sqrt{(2 - (-1))^2 + (4 - 3)^2} = \sqrt{10}$.

   b. The midpoint is $\left(\frac{2-1}{2}, \frac{4+3}{2}\right) = \left(\frac{1}{2}, \frac{7}{2}\right)$.

7. a. The distance is $\sqrt{(\pi - (-1))^2 + (0 - 2)^2} = \sqrt{(\pi + 1)^2 + 4} = \sqrt{\pi^2 + 2\pi + 5}$.

   b. The midpoint is $\left(\frac{\pi-1}{2}, \frac{0+2}{2}\right) = \left(\frac{\pi-1}{2}, 1\right)$.

9. The values when $x = 5$ are on the graph.

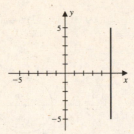

11. The values when $x > 1$ are on the graph.

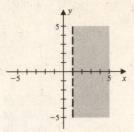

13. The values when $x \geq 1$ and $y \geq 2$ are on the graph.

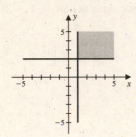

15. The values when $-3 < y \leq 1$ are on the graph.

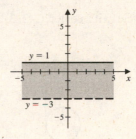

17. The values when $2 \leq |x|$, that is, when $x \geq 2$ or $x \leq -2$, are on the graph.

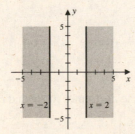

19. The values when $-1 \leq x \leq 2$ and $2 < y < 3$ are on the graph.

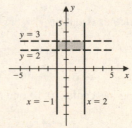

21. The values when $|x - 1| < 3$ and $|y + 1| < 2$, that is, when $-2 < x < 4$ and $-3 < y < 1$ are on the graph.

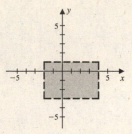

23. The circle with center $(2, 0)$ and radius 3 has equation

$$9 = (x - 2)^2 + y^2.$$

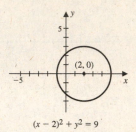

25. The circle with center $(-2, 3)$ and radius 2 has equation

$$4 = (x + 2)^2 + (y - 3)^2.$$

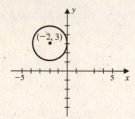

27. The circle with center $(-1, -2)$ and radius 2 has equation

$$4 = (x + 1)^2 + (y + 2)^2.$$

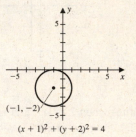

29. The circle has equation $4 = (x - 2)^2 + (y - 2)^2.$

31. The circle has center $(0, 0)$ and radius $\sqrt{9} = 3.$

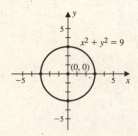

33. The circle has center $(0, 1)$ and radius $\sqrt{1} = 1.$

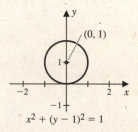

35. The circle has center $(2, -1)$ and radius $\sqrt{9} = 3$.

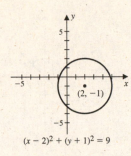

$$(x - 2)^2 + (y + 1)^2 = 9$$

37. The circle has center $(1, 0)$ and radius 2 since

$$3 = x^2 - 2x + y^2 = (x^2 - 2x + 1) - 1 + y^2,$$

so $(x - 1)^2 + y^2 = 4$.

39. The circle has center $(-1, 2)$ and radius 1 since

$$-4 = x^2 + 2x + y^2 - 4y$$
$$= (x^2 + 2x + 1) - 1 + (y^2 - 4y + 4) - 4,$$

so $(x + 1)^2 + (y - 2)^2 = 1$.

41. The circle has center $(2, 1)$ and radius 3 since

$$0 = x^2 - 4x + y^2 - 2y - 4$$
$$= (x^2 - 4x + 4) - 4$$
$$\quad + (y^2 - 2y + 1) - 1 - 4,$$

so $(x - 2)^2 + (y - 1)^2 = 9$.

43. The values when $x^2 + y^2 \le 1$ are on the graph.

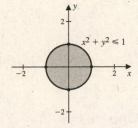

45. The values when $1 < x^2 + y^2 < 4$ are on the graph.

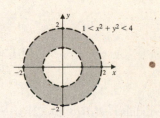

47. The values when $x^2 + y^2 \leq 4$ and $y \geq x$ are on the graph.

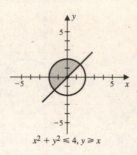

$$x^2 + y^2 \leq 4, \, y \geq x$$

49. The point $(6, 3)$ is closer to the origin, since $d((6, 3), (0, 0)) = \sqrt{45}$ and $d((-7, 2), (0, 0)) = \sqrt{53}$.

51. a. Let $a = d((-3, -4), (2, -1)) = \sqrt{34}$, $\quad b = d((-1, 4), (2, -1)) = \sqrt{34}$, and $c = d((-1, 4), (-3, -4)) = \sqrt{68}$.

Since $a^2 + b^2 = c^2$, the points are the vertices of a right triangle.

b. The right angle is located at vertex $(2, -1)$.

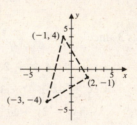

53. The point $(-3, -4)$ is 5 units left and 3 units down from the point $(2, -1)$. The unique point producing a square is the same distance from $(-1, 4)$, that is, $(-1 - 5, 4 - 3) = (-6, 1)$.

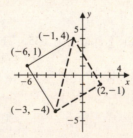

55. The radius $r = d((0, 0), (2, 3)) = \sqrt{13}$, so the equation of the circle is $x^2 + y^2 = 13$.

57. The radius $r = d((3, 7), (0, 7)) = \sqrt{9} = 3$, so the equation of the circle is $(x - 3)^2 + (y - 7)^2 = 9$.

59. A circle with radius 3, tangent to both the x- and y-axes, and center in the second quadrant, has center $(-3, 3)$. The equation is $(x + 3)^2 + (y - 3)^2 = 9$.

61. The area of the shaded region in the figure is $\pi(3)^2 - \pi(1)^2 = 8\pi$.

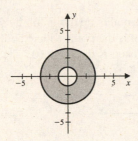

63. The values when $|x| + |y| \leq 4$ are on the graph.

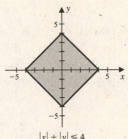

$|x| + |y| \leq 4$

65. The area of the shaded region is the area of the circle minus the area of the square, that is, $\pi(1)^2 - \sqrt{2}\sqrt{2} = \pi - 2$.

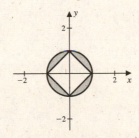

67. The estimated cost is \$771,110.03.

69. The direct route would cost $3(200, 000) + \sqrt{20}(150, 000)$ dollars or approximately 127,082,039 dollars. Going from A to C and then from C to B would cost $9(200, 000) = 1, 800, 000$ dollars.

. . . .

1.4   The Graph of an Equation

The graph of an equation is the set of points (x, y) that satisfy the equation. The variable x is called the *independent* variable since its values can vary freely over a collection of real numbers. The variable y is called the *dependent* variable since its value depends on the particular value of x selected.

   One way to obtain the graph of an equation is to plot several representative points that lie on the graph. The points where the graph crosses the x-axis and the y-axis , called the x- and y-intercepts, are particularly useful. This simple point-plotting approach will work in some cases but usually is not sufficient for our purposes in PreCalculus and Calculus.

Plotting Points

EXAMPLE 1

Sketch the graph of the equation.

(a) $y = x - 1$ (b) $y = x^2 + 1$

Solution (a) To sketch the graph, we first make a table of representative points on the graph by selecting particular values of x and then determining the associated y values. The x-intercept is the point with y-coordinate zero. From the table this is $(1, 0)$. The y-intercept is obtained by setting x equal to 0 and is the point $(0, -1)$. The figure shows the graph of the equation is a straight line.

x	-3	-2	-1	0	1	2	3
$y = x - 1$	-4	-3	-2	-1	0	1	2

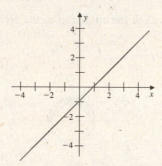

(b) Since $x^2 + 1 \geq 1$, the graph does not cross the x-axis and has no x-intercepts. The table shows that the y-intercept is $(0, 1)$. The graph is the parabola shown in the figure.

x	-2	-1	0	1	2
$y = x^2 + 1$	5	2	1	2	5

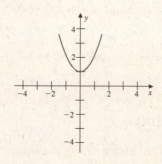

$\square$

Symmetry

The left portion of the graph in part (b) of the preceding example is the reflection through the y-axis of the right portion of the graph. *Symmetries* of this type are often useful in sketching graphs.

EXAMPLE 2

Test the curve for symmetry, find any x- and y-intercepts and sketch the graph of the equation.
(a) $y = x^2 - 3$ (b) $y = x^3 + 2$

Solution (a)

- y-axis symmetry: Yes, since if we replace x with $-x$,

$$(-x)^2 - 3 = x^2 - 3,$$

and the equation remains unchanged.

- x-axis symmetry: No, since if y is replaced with $-y$ the equation is changed. For example, $(2, 1)$ is on the curve but $(2, -1)$ is not on the curve.
- origin symmetry: No, since, for example, $(2, 1)$ is on the curve but $(-2, -1)$ is not on the curve.
- x-intercepts: Solve the equation

$$x^2 - 3 = 0$$
$$x^2 = 3$$
$$x = \pm\sqrt{3}.$$

The graph crosses the x-axis at $(-\sqrt{3}, 0)$ and $(\sqrt{3}, 0)$.

- y-intercepts: Set $x = 0$, then $y = -3$ and the y-intercept is the point $(0, -3)$.

Since the graph is symmetric with respect to the y-axis, we can plot the graph for positive values of x and then reflect the graph through the y-axis as shown in the figure.

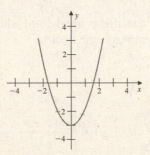

(b) If x is replaced with $-x$, the equation becomes

$$y = (-x)^3 + 2 = -x^3 + 2,$$

which has altered the original equation, so the graph is not symmetric with respect to the y-axis. Replacing y with $-y$ gives

$$-y = x^3 + 2$$

$$y = -x^3 - 2,$$

so the graph is not symmetric with respect to the y-axis. Replacing x with $-x$ and y with $-y$ gives

$$-y = (-x)^3 + 2 = -x^3 + 2$$
$$y = x^3 - 2,$$

and the curve is also not symmetric with respect to the origin. The curve is sketched in the figure. Notice that although there is no axis or origin symmetry, this particular curve is symmetric with respect to the point $(0,2)$.

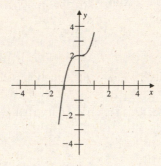

☐

EXAMPLE 3

Sketch the graph of $y = \dfrac{x^2 - x - 6}{x - 3}$.

Solution First factor the numerator to obtain

$$y = \frac{x^2 - x - 6}{x - 3} = \frac{(x - 3)(x + 2)}{(x - 3)} = x + 2, \quad \text{for} \quad x \neq 3.$$

It is important to recognize that the original equation is not defined when x is 3. This is illustrated by removing the point $(3, 5)$ from the graph of $y = x + 2$, as shown in the figure.

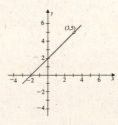

☐

EXAMPLE 4

Find the distance between the points of intersection of the graphs $y = x^2 - 2x + 2$ and $y = x + 2$.

Solution The points of intersection of the two curves are given by those x values that yield the same y value when substituted in both equations. To find the x values of the points of intersection, solve for x in the following equation.

$$x^2 - 2x + 2 = x + 2$$
$$x^2 - 3x = 0$$
$$x(x - 3) = 0$$
$$x = 0, x = 3$$

The y-coordinates of the points of intersection are found by substituting these x-values into either of the equations for y. This gives

$$(0, 2) \quad \text{and} \quad (3, 5)$$

as the points of intersection.
The distance between the two points is

$$d((0, 2), (3, 5)) = \sqrt{(3 - 0)^2 + (5 - 2)^2} = \sqrt{9 + 9} = \sqrt{18} = 3\sqrt{2}.$$

□

The two curves are shown in the figure.

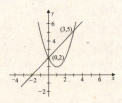

Exercise Set 1.4 (Page 28)

1. Symmetry with respect to the x-axis.

3. Symmetry with respect to the x-axis, y-axis, and origin.

5. Symmetry with respect to the origin.

7. Symmetry with respect to the x-axis.

9. For $y = x + 3$:
Intercepts: $(-3, 0)$, $(0, 3)$;
Symmetry: No axis or origin symmetry.

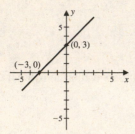

11. For $x + y = 1$, or $y = 1 - x$:
Intercepts: $(1, 0)$, $(0, 1)$;
Symmetry: No axis or origin symmetry.

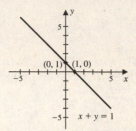

13. For $y = x^2 - 3$:
Intercepts: $(-\sqrt{3}, 0)$, $(\sqrt{3}, 0)$, $(0, -3)$;
Symmetry: y-axis.

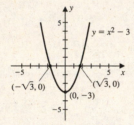

15. For $y = 1 - x^2$:
Intercepts: $(-1, 0)$, $(1, 0)$, $(0, 1)$;
Symmetry: y-axis.

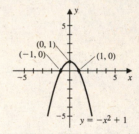

17. For $y = -2x^2$:
Intercepts: $(0, 0)$;
Symmetry: y-axis.

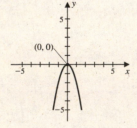

19. For $x = y^2 - 1$:
Intercepts: $(-1, 0)$, $(0, 1)$, $(0, -1)$;
Symmetry: x-axis.

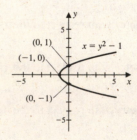

21. For $y = x^3 + 8$:
Intercepts: $(-2, 0)$, $(0, 8)$;
Symmetry: No axis or origin symmetry.

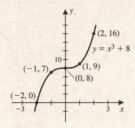

23. For $y = -x^3$:
Intercepts: $(0, 0)$;
Symmetry: origin.

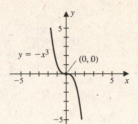

25. For $y = \frac{(x+3)(x-3)}{x-3} = x + 3$, when $x \neq 3$:
Intercepts: $(-3, 0)$, $(0, 3)$;
Symmetry: No axis or origin symmetry.

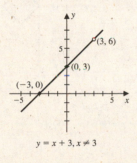

27. For $y = \frac{x^2 - x - 6}{x+2} = \frac{(x-3)(x+2)}{x+2} = x - 3$, when $x \neq -2$:
Intercepts: $(3, 0)$, $(0, -3)$;
Symmetry: No axis or origin symmetry.

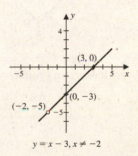

29. For $y = \sqrt{x} + 2$:
Intercepts: $(0, 2)$;
Symmetry: No axis or origin symmetry.

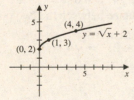

31. For $x^2 + y^2 = 4$:
Intercepts: $(-2, 0)$, $(2, 0)$, $(0, -2)$, $(0, 2)$;
Symmetry: x-axis, y-axis, and origin.

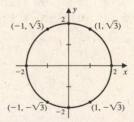

33. For $y = \sqrt{9 - x^2}$:
Intercepts: $(-3, 0), (3, 0), (0, 3)$;
Symmetry: y-axis.

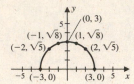

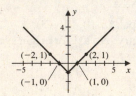

35. For $y = |x|$:
Intercepts: $(0, 0)$;
Symmetry: y-axis.

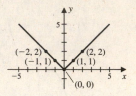

37. For $y = |x| - 1$:
Intercepts: $(-1, 0), (1, 0), (0, -1)$;
Symmetry: y-axis.

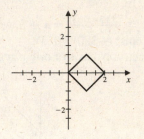

39. This graph has x-axis symmetry.

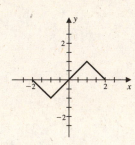

41. This graph has origin symmetry.

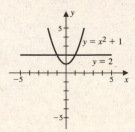

43. From the graph we can see the points of intersection of $y = x^2 + 1$ and $y = 2$ are $(1, 2)$ and $(-1, 2)$. Algebraically we have $x^2 + 1 = 2$ so $x^2 = 1$ and $x = \pm 1$. The distance between the two points of intersection is

$$d((1, 2), (-1, 2)) = \sqrt{(1 - (-1))^2 + (2 - 2)^2} = \sqrt{4} = 2.$$

45. If the three consecutive numbers are x, $x + 1$, and $x + 2$, then $x + x + 1 + x + 2 = 156$ which implies $3x + 3 = 156$ so $x = 51$. The three numbers are 51, 52, and 53.

47. Let x denote the length of the rectangle and y denote the width. Then $xy = 12$ and $2x + 2y = 14$, which implies $x + y = 7$. So $y = 7 - x$ which implies $x(7 - x) = 12$ giving $x^2 - 7x + 12 = 0$. Factoring the quadratic we have $(x - 4)(x - 3) = 0$, so $x = 4$ or $x = 3$. The rectangle has dimension 3 by 4.

49. Let x denote the number of quarts of 10% solution that are required. Then $0.1x + (10 - x) = 0.3(10)$, which implies $x = \frac{7}{0.9} \approx 7.8$ quarts. Approximately $10 - 7.8 = 2.2$ quarts must be drained and replaced with pure antifreeze.

51. Assume the graph has symmetry with respect to both the x- and y-axes. To show the graph is symmetric with respect to the origin, we need to show that if (x, y) is on the graph, then $(-x, -y)$ is also on the graph. If (x, y) is on the graph and the graph is symmetric with respect to the x-axis, then $(x, -y)$ is also on the graph. But if $(x, -y)$ is on the graph and the graph is symmetric with respect to the y-axis, then $(-x, -y)$ is also on the graph. Hence the graph is symmetric with respect to the origin.

It is also true that if the graph has symmetry with respect to the origin and to one of the axes, then it has symmetry with respect to the other axis. Consider, for example, the situation of symmetry with respect to the origin and the x-axis. To show the graph is symmetric with respect to the y-axis, we need to show that if (x, y) is on the graph then $(-x, y)$ is also on the graph. If (x, y) is on the graph and the graph is symmetric with respect to the x-axis, then $(x, -y)$ is also on the graph. But if $(x, -y)$ is on the graph and the graph is symmetric with respect to the origin, then $(-x, -(-y)) = (-x, y)$ is also on the graph. Hence the graph is symmetric with respect to the y-axis.

1.5   Using Technology to Graph Equations

Graphing devices sketch curves by plotting many points which are connected by very small line segments. The *viewing rectangle* or *viewing window* of a graphing device is the rectangular portion of the plane in which the plot is displayed. Selecting the appropriate viewing rectangle when using a graphing device is very important. A viewing rectangle specified as $[a, b] \times [c, d]$, defines the rectangular region in the plane with (x, y) restricted by $a \leq x \leq b$ and $c \leq y \leq d$.

EXAMPLE 1
Use a graphing device to sketch the graph of $y = x^3 - 12x + 20$ with the following viewing rectangles, and determine which gives the best representation for the graph of the equation.
  (a) $[-1, 1] \times [-1, 1]$ (b) $[-5, 5] \times [-5, 5]$
  (c) $[-10, 10] \times [-50, 50]$ (d) $[-100, 100] \times [-200, 200]$

Solution The viewing rectangle in (a) is too small to show any portion of the graph. We expect the graph to be a smooth curve, so we also reject the graph in part (b). The viewing rectangle in part (c) gives a nice smooth curve which we accept with confidence as a good representative of the graph. The viewing rectangle used in part (d) is too large in both the x and y directions and clearly has distorted the graph. It does illustrate, however, that the curve in part (c) shows the important features of the graph.

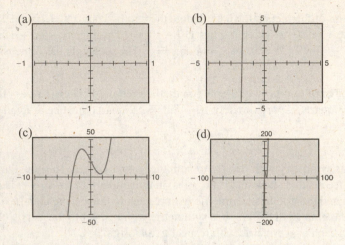

EXAMPLE 2

Determine an appropriate viewing rectangle for the graph of the equation $y = \dfrac{x+6}{x^2-1}$.

Solution   We start by selecting a viewing rectangle of $[-5, 5] \times [-5, 5]$. The graph in part (a) appears to be a reasonable representation of the graph. The two vertical lines are not part of the graph but show how the curve approaches these lines.

We need to be careful when using a graphing device to be certain we have selected a viewing rectangle that shows all the important features of the graph. In this case, if we set $x = 0$, then the y-value of the point on the curve is -6, which we do not see on the graph in part (a). In part (b) a viewing rectangle of $[-5, 5] \times [-10, 5]$ shows a great deal of the curve is missing in part (a).

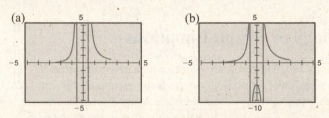

EXAMPLE 3

Graphically approximate the solutions to the inequality $x^2 + 3x - 2 < 0$.

Solution   In (a), the equation $y = x^2 + 3x - 2$ is graphed in a viewing rectangle $[-5, 5] \times [-5, 5]$. The inequality is negative, which corresponds to the portion of the graph that lies below the x-axis. Clicking on the x-intercepts of the graph gives values of $x \approx -3.56$ and $x \approx 0.56$. So $x^2 + 3x - 2 < 0$ for approximately $-3.56 < x < 0.56$.

In (b), the viewing rectangle is $[0, 1] \times [-1, 1]$, giving 0.56, which is accurate to within 2 decimal places. For more accuracy, *zoom* closer in to the x-intercept by shrinking the viewing rectangle.

(a) 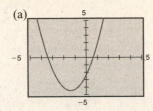 (b)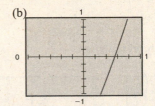

EXAMPLE 4

The number of birds in a sanctuary at time t is given by

$$n = 5000 \left(\frac{4t^2 + 1}{2t^2 + 1} \right).$$

As the time t increases, does the size of the population become stable? If so, what is the stabilizing level?

Solution The equation along with the horizontal line at 10000 are plotted using a viewing rectangle of $[0, 25] \times [0, 12000]$. The figure indicates that the population appears to level off to a stabilizing value of 10000.

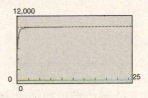

Exercise Set 1.5 (Page 35)

1. Part (d) gives the best representation.

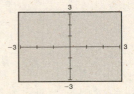 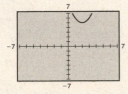 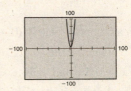 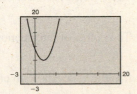

3. Part (c) gives the best representation.

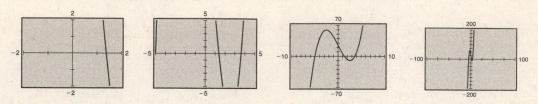

5. $[-10, 10] \times [-10, 10]$ **7.** $[0, 20] \times [0, 10]$

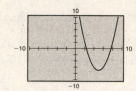

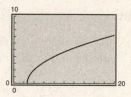

9. $[-10, 10] \times [-10, 10]$

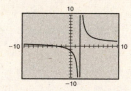

11. The graph of $y = x^3 + x^2 + 3x - 4$ crosses the x-axis at $x \approx 0.86$.

13. The graph of $y = x^4 - 3x^3 + x^2 - 4$ crosses the x-axis at $x \approx -0.93$ and $x \approx 2.82$.

15. The points of intersection are approximately $(1, 6)$ and $(-1, 0)$.

17. The points of intersection are approximately $(0, -1)$ and $(0.46, -0.79)$.

19. **a.** We have $x \le -3.56$ or $x \ge 0.56$, and   **b.** $x < -2.3$ or $1.3 < x < 3$.

21. For x very large the graphs are almost identical. That is, as x grows without bound $x^4 - 4x^3 + 3x^2$ is approximately x^4.

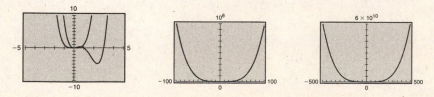

23. The graph appears to approach the horizontal line $y = \frac{a}{b}$ and the vertical line $x = \frac{1}{b}$.

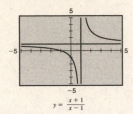
$y = \frac{x+1}{x-1}$

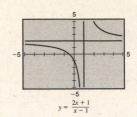
$y = \frac{2x+1}{x-1}$

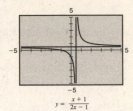
$y = \frac{x+1}{2x-1}$

25. If $a = b = 0$, the graph is the standard parabola $y = x^2$. The constants a and b shift $y = x^2$ either horizontally or vertically. If $a > 0$, the shift is to the right a units. If $a < 0$, the shift is to the left $|a|$ units. If $b > 0$, the shift is upward b units, and if $b < 0$, the shift is downward $|b|$ units.

1.6 Functions

A *function* is a process that for each admissible input returns another *unique* real number. If a function is f and x is an input value, then the output is denoted as $f(x)$, read "f of x." Functions play a central role in all quantitative study, and you should become comfortable with the key ideas and notation that is used.

Evaluation of Functions

EXAMPLE 1
Find each value of the function f defined by $f(x) = x^2 - 2x + 1$.
(a) $f(3)$ (b) $f(-2)$ (c) $f\left(\frac{1}{2}\right)$

Solution   (a) $f(3) = (3)^2 - 2(3) + 1 = 4$
(b) $f(-2) = (-2)^2 - 2(-2) + 1 = 9$
(c)
$$f\left(\frac{1}{2}\right) = \left(\frac{1}{2}\right)^2 - 2\left(\frac{1}{2}\right) + 1 = \left(\frac{1}{4}\right) - 1 + 1 = \frac{1}{4}$$

                                                           □

The input to a function can be any expression that represents a real number. The variable used in the definition of the function is simply a place holder, and evaluating a function at a specific value means replacing the variable everywhere it appears with the input value. If

$$f(x) = \frac{x^3 - x + 1}{2x - 1},$$

then the function can be thought of as

$$f(\square) = \frac{\square^3 - \square + 1}{2\square - 1}.$$

Provided that what is in the

                                         □

is not

$$\frac{1}{2}$$

, which could make the quotient undefined. For example,

$$f\left(\boxed{2x-1}\right) = \frac{\boxed{2x-1}^{3} - \boxed{2x-1} + 1}{2\boxed{2x-1} - 1}$$

provided that

$$2x - 1 \neq \frac{1}{2}$$

, that is,

$$x \neq \frac{3}{4}$$

. Then, if necessary, the expression can be simplified.

EXAMPLE 2
If $f(x) = |x - 4|$, find the value of the function.
(a) $f(5)$ (b) $f(1)$ (c) $f(0)$ (d) $f(x+3)$ (e) $f(4 - x^2)$ (f) $f(-x)$

Solution
(a) $f(5) = |5 - 4| = 1$
(b) $f(1) = |1 - 4| = |-3| = 3$
(c) $f(0) = |0 - 4| = |-4| = 4$
(d) $f(x+3) = |(x+3) - 4| = |x - 1|$
(d) $f(4 - x^2) = |(4 - x^2) - 4| = |-x^2| = x^2$
(e) $f(-x) = |-x - 4| = |-(x+4)| = |x + 4|$ □

Domain and Range

The *domain* of a function is the collection of all real numbers that can be input to the function. The *range* is the collection of all outputs from the function. If the domain of a function is not explicitly stated, we assume it is the largest set of real numbers for which the function is defined.

EXAMPLE 3
Determine the domain and range of the function f given in the figure.

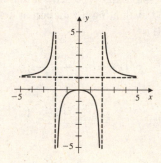

Solution The vertical lines $x = -2$ and $x = 2$ indicate the function is not defined at the values -2 and 2 so the domain is $(-\infty, -2) \cup (-2, 2) \cup (2, \infty)$.

Horizontal lines between $y = 0$, not including 0, and $y = 1$, including 1, do not cross the curve, so the range is $(-\infty, 0] \cup (1, \infty)$. $\square$

EXAMPLE 4

Find the domain of the function.

(a) $f(x) = \dfrac{3}{x+1}$ (b) $f(x) = \sqrt{x+3}$

Solution (a) The only values of x for which the function is not defined are when $x + 1 = 0$. So the domain is all real numbers except $x = -1$, that is, $(-\infty, -1) \cup (-1, \infty)$.

(b) For the function to be defined, the expression under the radical must be greater than or equal to 0. That is, the domain is all x with $x + 3 \geq 0$, that is, $x \geq -3$. In interval notation this is $[-3, \infty)$. $\square$

EXAMPLE 5

Find the domain of the function.

(a) $f(x) = \sqrt{x(x+2)}$ (b) $f(x) = \sqrt{\dfrac{x}{x+2}}$

Solution (a) The function is defined only when the expression under the radical is greater than or equal to 0. To solve the inequality

$$x(x+2) \geq 0$$

make a sign chart as shown in the figure.

The chart indicates that the inequality is positive on $(-\infty, -2) \cup (0, \infty)$. Since $\sqrt{0} = 0$, the values -2 and 0 are also in the domain, so the domain is $(-\infty, -2] \cup [0, \infty)$.

(b) For x to be in the domain of f, the quotient under the radical must satisfy

$$\frac{x}{x+2} \geq 0.$$

The sign chart

shows that $\frac{x}{x+2} > 0$, if $x < -2$ or $x > 0$. In addition, $\frac{x}{x+2} = 0$ when $x = 0$, which is permissible, but $\frac{x}{x+2}$ does not exist, indicated by the ?, when $x = -2$. Hence the domain of f is $(-\infty, -2) \cup [0, \infty)$. $\square$

Vertical and Horizontal Line Tests

If every vertical line crosses the graph of an equation at most once, then the equation defines a function. If a vertical line crosses the graph of a function, then the x-intercept is in the domain. If a horizontal line crosses the graph of a function, then the y-intercept is in the range of the function.

EXAMPLE 6

Determine if the graph defines a function and if so find the domain and the range.

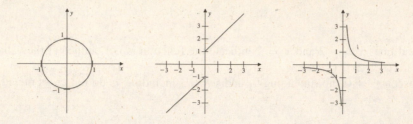

Solution (a) Any vertical line drawn between $x = -1$ and $x = 1$ crosses the curve in exactly two places, so the graph does not define a function. In this example, the curve is the circle with center the origin and radius 1, which has the equation $x^2 + y^2 = 1$. Solving for y gives

$$y = \pm\sqrt{1 - x^2}$$

so

$$y = \sqrt{1 - x^2} \quad \text{or} \quad y = -\sqrt{1 - x^2}.$$

Each equation defines a function of y in terms of x. Choosing the positive square root defines a function that describes the upper semi-circle. Choosing the negative square root gives us a function that describes the lower semi-circle.
(b) Every vertical line crosses the curve in at most only one place, so the graph does define a function. Since every vertical line crosses the curve, the domain is the set of all real numbers, $(-\infty, \infty)$. The horizontal lines that do *not* cross the curve are those between -1 and 1 including the line $y = -1$. The range is consequently $(-\infty, -1) \cup [1, \infty)$.
(c) The curve defines a function with domain all real numbers except $x = 0$. The range is also all real numbers except 0. So the domain is $(-\infty, 0) \cup (0, \infty)$ and the range is $(-\infty, 0) \cup (0, \infty)$.   □

Difference Quotients

In calculus we frequently encounter an important quantity called the *difference quotient* of a function. The difference quotient for the function f at a number x in its domain is

$$\frac{f(x + h) - f(x)}{h},$$

where h represents an arbitrary small positive number.

EXAMPLE 7
Let $f(x) = 3x^2 + 4x + 1$. Find

$$f(x + h) \quad \text{and} \quad \frac{f(x + h) - f(x)}{h},$$

assuming that $h \neq 0$.

Solution Replacing every instance of x with $x + h$ gives

$$\begin{aligned}
f(x + h) &= 3(x + h)^2 + 4(x + h) + 1 \\
&= 3(x^2 + 2hx + h^2) + 4x + 4h + 1 \\
&= 3x^2 + 6hx + 3h^2 + 4x + 4h + 1,
\end{aligned}$$

and

$$\begin{aligned}
\frac{f(x + h) - f(x)}{h} &= \frac{3(x + h)^2 + 4(x + h) + 1 - (3x^2 + 4x + 1)}{h} \\
&= \frac{3(x^2 + 2hx + h^2) + 4x + 4h + 1 - 3x^2 - 4x - 1}{h}
\end{aligned}$$

$$= \frac{3x^2 - 3x^2 + 4x - 4x + 1 - 1 + 6hx + 3h^2 + 4h}{h}$$

$$= \frac{6hx + 3h^2 + 4h}{h} = \frac{h(6x + 3h + 4)}{h}$$

$$= 6x + 3h + 4.$$

Be careful to distribute the minus sign when subtracting $f(x)$ from $f(x+h)$. Note in the example that we have $-(3x^2 + 4x + 1) = -3x^2 - 4x - 1$. □

EXAMPLE 8

Let $f(x) = \sqrt{2x+1}$. Find

$$f(x+h) \quad \text{and} \quad \frac{f(x+h) - f(x)}{h},$$

assuming that $h \neq 0$.

Solution Since

$$f(x+h) = \sqrt{2(x+h) + 1} = \sqrt{2x + 2h + 1},$$

we have

$$\frac{f(x+h) - f(x)}{h} = \frac{\sqrt{2x + 2h + 1} - \sqrt{2x + 1}}{h}.$$

To simplify this expression, we again use the algebraic relation $(a - b)(a + b) = a^2 - b^2$, which will permit us to eliminate the radicals in the numerator. Multiplying the numerator and denominator by $\sqrt{2x + 2h + 1} + \sqrt{2x + 1}$ gives

$$\frac{f(x+h) - f(x)}{h} = \frac{\sqrt{2x + 2h + 1} - \sqrt{2x + 1}}{h} \cdot \frac{\sqrt{2x + 2h + 1} + \sqrt{2x + 1}}{\sqrt{2x + 2h + 1} + \sqrt{2x + 1}}$$

$$= \frac{\left(\sqrt{2x + 2h + 1}\right)^2 - \left(\sqrt{2x + 1}\right)^2}{h\left(\sqrt{2x + 2h + 1} + \sqrt{2x + 1}\right)}$$

$$= \frac{2x + 2h + 1 - 2x - 1}{h\left(\sqrt{2x + 2h + 1} + \sqrt{2x + 1}\right)}$$

$$= \frac{2h}{h\left(\sqrt{2x + 2h + 1} + \sqrt{2x + 1}\right)}$$

$$= \frac{2}{\sqrt{2x + 2h + 1} + \sqrt{2x + 1}}.$$

You might argue that this final result is not really simpler than the original form. In a sense this is true, but we want to determine the value of the difference quotient as h approaches zero, which we can now do. The difference quotient approaches

$$\frac{2}{\sqrt{2x + 0 + 1} + \sqrt{2x + 1}} = \frac{2}{2\sqrt{2x + 1}} = \frac{1}{\sqrt{2x + 1}}.$$

Notice that if we substitute $h = 0$ in the original expression for the difference quotient, the result is a quantity of the form $\frac{0}{0}$, which is undefined. In the original form we can not determine what happens to the difference quotient as h approaches 0. □

Odd and Even Functions

A function f is *odd* provided that for all x in its domain $f(-x) = -f(x)$, and a function is *even* provided $f(-x) = f(x)$. Geometrically, this means

- an odd function is symmetric with respect to the origin

- an even function is symmetric with respect to the y-axis.

EXAMPLE 9
Determine whether the function is even, odd, or neither even nor odd.
(a) $f(x) = x^2 + 3$ (b) $f(x) = x^3 + 1$
(c) $f(x) = x^3 + 2x$ (d) $f(x) = \sqrt[3]{x}$

Solution (a)

- f even: $f(-x) = (-x)^2 + 3 = x^2 + 3 = f(x)$
- f not odd: $-f(x) = -x^2 - 3 \neq f(-x)$

(b)

- f not even: $f(-x) = (-x)^3 + 1 = -x^3 + 1 \neq f(x)$
- f not odd: $-f(x) = -x^3 - 1 \neq f(-x)$

(c)

- f not even: $f(-x) = (-x)^3 + 2(-x) = -x^3 - 2x \neq f(x)$
- f odd: $-f(x) = -x^3 - 2x = f(-x)$

(d)

- f not even: $f(-x) = \sqrt[3]{-x} = -\sqrt[3]{x} \neq f(x)$
- f odd: $-f(x) = -\sqrt[3]{x} = f(-x)$

$\square$

EXAMPLE 10
The graph of a function f is given for $x \geq 0$. Extend the graph for $x < 0$ if
(a) f is even   (b) f is odd.

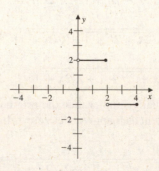

Solution (a) The graph is symmetric with respect to the y-axis, so reflect the curve over the y-axis, which gives the following graph.

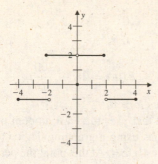

(b) The graph must be symmetric with respect to the origin, which gives the following graph. Notice that the point $(2, 2)$ is on the original graph, so $(-2, -2)$ is also on the graph of the odd function. Similarly, reflecting the circle at $(2, -1)$ through the origin gives the circle at $(-2, 1)$ on the final graph.

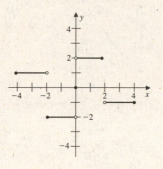

$\square$

Applications

EXAMPLE 11

A rectangular plot of ground containing 100 feet^2 is to be fenced within a large plot.
(a) Express the perimeter of the plot as a function of the width. What is the domain of the function?
(b) Use a graphing device to approximate the dimensions of the plot that requires the least amount of fence.

$$l \quad\boxed{\quad A = 100 \text{ ft}^2 \quad}\quad w$$

Solution (a) The perimeter of the plot is

$$P = 2l + 2w.$$

Since we need to express the perimeter in terms of w only, we use the information for the area of the plot to eliminate the variable l from the equation. That is,

$$A = lw = 100, \quad \text{so} \quad l = \frac{100}{w}.$$

Substituting into the equation for P gives

$$P(w) = 2w + \frac{200}{w}.$$

The domain of the function is all real numbers $w > 0$.

(b) To find the exact dimensions of the plot that uses the least amount of fencing requires concepts from calculus. However, we can approximate the dimensions using a graphing device to plot the perimeter P with respect to the width w. The figure shows a low point at $w \approx 10$ feet. At this point the perimeter is smallest, and hence the amount of fencing required is minimized. Then $\ell = w \approx 10$ feet, and the plot that minimizes the amount of fencing is a square.

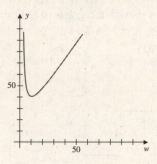

$\square$

EXAMPLE 12

A manufacturer estimates the profit on producing x units of their product at $P(x) = 200x - x^2$.

(a) What is the average rate of change in the profit as the number of units changes from x to $x + h$?

(b) Use the result in part (a) to find the average rate of change in the profit as the number of units produced changes from 40 to 60.

(c) Sketch the graph of $y = P(x)$ and the line that passes through the points $(40, P(40))$ and $(60, P(60))$.

Solution (a) The average rate of change is the change in the values of the function as the independent variable varies over some interval. That is, the average rate of change is the difference quotient,

$$\frac{P(x+h) - P(x)}{h} = \frac{200(x+h) - (x+h)^2 - (200x - x^2)}{h}$$

$$= \frac{200x + 200h - (x^2 + 2hx + h^2) - 200x + x^2}{h}$$

$$= \frac{200h - 2hx - h^2}{h} = \frac{h(200 - 2x - h)}{h}$$

$$= 200 - 2x - h.$$

(b) When $x = 40$ and $h = 20$, we have $x + h = 60$, so the average rate of change can be computed using the final formula from part (a) as

$$200 - 2x - h = 200 - 2(40) - 20 = 100.$$

When the number of units produced increases from 40 to 60, the profit increases at an average rate of 100 per unit increase in production.

(c)

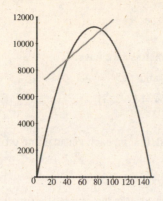

□

Exercise Set 1.6 (Page 50)

1. For $f(x) = 2x^2 + 3$ we have:

   a. $f(2) = 2(2)^2 + 3 = 11$

   b. $f\left(\sqrt{3}\right) = 2\left(\sqrt{3}\right)^2 + 3 = 9$

   c. $f\left(2 + \sqrt{3}\right) = 2\left(2 + \sqrt{3}\right)^2 + 3 = 2\left(7 + 4\sqrt{3}\right) + 3 = 17 + 8\sqrt{3}$

   d. $f(2) + f\left(\sqrt{3}\right) = 11 + 9 = 20$

   e. $f(2x) = 2(2x)^2 + 3 = 2(4x^2) + 3 = 8x^2 + 3$

   f. $f(1 - x) = 2(1 - x)^2 + 3 = 2(1 - 2x + x^2) + 3 = 2x^2 - 4x + 5$

   g. $f(x + h) = 2(x + h)^2 + 3 = 2(x^2 + 2hx + h^2) + 3 = 2x^2 + 4hx + 2h^2 + 3$

   h. $f(x + h) - f(x) = 2x^2 + 4hx + 2h^2 + 3 - (2x^2 + 3) = 4hx + 2h^2$

3. For $f(t) = |t - 2|$ we have:

   a. $f(4) = |4 - 2| = |2| = 2$

   b. $f(1) = |1 - 2| = |-1| = 1$

   c. $f(0) = |0 - 2| = |-2| = 2$

   d. $f(t + 2) = |t + 2 - 2| = |t|$

   e. $f(2 - t^2) = |2 - t^2 - 2| = |-t^2| = |t^2| = t^2$

   f. $f(-t) = |-t - 2| = |-(t + 2)| = |t + 2|$

5. Reading from the graph given in the exercise we have:

    a. $f(-1) = 0$, $f(0) = -\frac{1}{2}$, $f(1) = 0$, and $f(3) = -2$

    b. Domain$(f) = [-2, 3]$ and Range$(f) = [-2, 3]$

7. The graph satisfies the vertical line test, that is, each vertical line crosses the curve in at most one place, so it is the graph of a function.

9. The graph satisfies the vertical line test, so it is the graph of a function.

11. **a.** Since the only vertical line that does not intersect the curve is $x = 0$, the domain is $(-\infty, 0) \cup (0, \infty)$.

    b. Since the only horizontal line that does not intersect the curve in at least one place is $y = 0$, the range is $(-\infty, 0) \cup (0, \infty)$.

13. For this function we have

    a. Domain$(f) = (-\infty, \infty)$

    b. Range$(f) = (1, \infty) \cup \{-1\}$

15. For $f(x) = x^2 - 1$ we have:

    a. Domain$(f) = (-\infty, \infty)$.

    b. Range$(f) = [-1, \infty)$.

17. For $f(x) = \sqrt{x} + 4$ we have:
    a. Domain$(f) = [0, \infty)$.
    b. Range$(f) = [4, \infty)$.

19. For $f(x) = x^2 - 2x + 1 = (x-1)^2$ we have:
    a. Domain$(f) = (-\infty, \infty)$.
    b. Range$(f) = [0, \infty)$.

21. For $f(x) = \begin{cases} 1, & \text{if } x \geq 0 \\ -1, & \text{if } x < 0 \end{cases}$ we have:
    a. Domain$(f) = (-\infty, \infty)$.
    b. Range$(f) = \{-1, 1\}$.

23. For $f(x) = x^2 - 1$, $b = 0$, and $f(a) = a^2 - 1$

$$f(a) = 0 \quad \text{implies} \quad a^2 - 1 = 0$$

so $a^2 = 1$ and $a = \pm 1$.

25. For $f(x) = \sqrt{x - 1}$, $b = \frac{1}{2}$, and $f(a) = \sqrt{a - 1}$

$$f(a) = \frac{1}{2} \quad \text{implies} \quad \sqrt{a - 1} = \frac{1}{2}$$

so $a - 1 = \frac{1}{4}$ and $a = \frac{5}{4}$.

27. a. $\text{Domain}(f) = (-\infty, \infty)$

    b. $\text{Domain}(f) = \{x : 2 - x \neq 0\} = \{x : x \neq 2\} = (-\infty, 2) \cup (2, \infty)$

    c. $\text{Domain}(f) = \{x : 2 - x \geq 0\} = \{x : x \leq 2\} = (-\infty, 2]$

    d. $\text{Domain}(f) = \{x : 2 - x > 0\} = (-\infty, 2)$

29. a. $\text{Domain}(f) = \{x : x^2 - 2 \neq 0\} = \{x : x \neq \pm\sqrt{2}\} = (-\infty, -\sqrt{2}) \cup (-\sqrt{2}, \sqrt{2}) \cup (\sqrt{2}, \infty)$

    b. For $f(x) = \frac{x-2}{x^2-2}$, the domain is still $(-\infty, -\sqrt{2}) \cup (-\sqrt{2}, \sqrt{2}) \cup (\sqrt{2}, \infty)$

    c. We have

$$\text{Domain}(f) = \left\{ x : \frac{2x^2}{x^2 - 2} \geq 0 \right\} = \{0\} \cup \left\{ x : \frac{2x^2}{x^2 - 2} > 0 \right\}$$

$$= \{0\} \cup \{x : x^2 - 2 > 0\}$$

$$= \{0\} \cup (-\infty, -\sqrt{2}) \cup (\sqrt{2}, \infty).$$

31. a. For $f(x) = \sqrt{x(x-2)}$, $\text{Domain}(f) = \{x : x(x-2) \geq 0\} = (-\infty, 0] \cup [2, \infty)$.

    b. For $f(x) = \sqrt{x(2-x)}$, $\text{Domain}(f) = \{x : x(2-x) \geq 0\} = \{x : x(x-2) \leq 0\} = [0, 2]$.

    c. For $f(x) = \frac{x^2}{x^2-2x} = \frac{x^2}{x(x-2)}$,
$\text{Domain}(f) = \{x : x(x-2) \neq 0\} = \{x : x \neq 0, x \neq 2\} = (-\infty, 0) \cup (0, 2) \cup (2, \infty)$.

    d. For $f(x) = \sqrt{\frac{x^2}{x^2-2x}} = \sqrt{\frac{x^2}{x(x-2)}}$, $\text{Domain}(f) = \{x : x(x-2) > 0\} = (-\infty, 0) \cup (2, \infty)$.

33. For $f(x) = x^2 + 2$ we have
$f(-x) = (-x)^2 + 2 = x^2 + 2$;
$-f(x) = -(x^2 + 2) = -x^2 - 2$;
$f\left(\frac{1}{x}\right) = \left(\frac{1}{x}\right)^2 + 2 = \frac{1}{x^2} + 2$;
$\frac{1}{f(x)} = \frac{1}{x^2+2}$;
$f(\sqrt{x}) = (\sqrt{x})^2 + 2 = x + 2$;
$\sqrt{f(x)} = \sqrt{x^2 + 2}$.

35. For $f(x) = \frac{1}{x}$ we have
$f(-x) = \frac{1}{-x} = -\frac{1}{x}$;
$-f(x) = -\frac{1}{x}$;
$f\left(\frac{1}{x}\right) = \frac{1}{\frac{1}{x}} = x$;
$\frac{1}{f(x)} = \frac{1}{\frac{1}{x}} = x$;
$f(\sqrt{x}) = \frac{1}{\sqrt{x}} = \frac{1}{\sqrt{x}} \cdot \frac{\sqrt{x}}{\sqrt{x}} = \frac{\sqrt{x}}{x}$;
$\sqrt{f(x)} = \sqrt{\frac{1}{x}} = \frac{1}{\sqrt{x}} = \frac{\sqrt{x}}{x}$.

37. For $f(x) = 3x - 2$:

    a. $f(x + h) = 3(x + h) - 2 = 3x + 3h - 2$

    b. $f(x + h) - f(x) = 3x + 3h - 2 - 3x + 2 = 3h$

    c. The difference quotient is
$$\frac{f(x+h) - f(x)}{h} = \frac{3h}{h} = 3.$$

    d. As h approaches 0, $\frac{f(x+h)-f(x)}{h}$ approaches 3. In fact, it is always 3 if $h \neq 0$.

39. For $f(x) = x^2$:

a. $f(x+h) = (x+h)^2 = x^2 + 2hx + h^2$

b. $f(x+h) - f(x) = x^2 + 2xh + h^2 - x^2 = 2xh + h^2$

c. The difference quotient is

$$\frac{f(x+h) - f(x)}{h} = \frac{2hx + h^2}{h} = \frac{h(2x+h)}{h} = 2x + h.$$

d. As h approaches 0, $\frac{f(x+h)-f(x)}{h}$ approaches $2x$.

41. For $f(x) = 2 - x - x^2$:

a. $f(x+h) = 2 - (x+h) - (x+h)^2 = 2 - x - h - x^2 - 2hx - h^2$

b. $f(x+h) - f(x) = 2 - x - h - x^2 - 2hx - h^2 - 2 + x + x^2 = -h - 2hx - h^2$

c. The difference quotient is

$$\frac{f(x+h) - f(x)}{h} = \frac{-h - 2hx - h^2}{h} = \frac{h(-1 - 2x - h)}{h} = -1 - 2x - h.$$

d. As h approaches 0, $\frac{f(x+h)-f(x)}{h}$ approaches $-1 - 2x$.

43. For $f(x) = \frac{1}{x}$:

a. $f(x+h) = \frac{1}{x+h}$

b. $f(x+h) - f(x) = \frac{1}{x+h} - \frac{1}{x} = \frac{x - (x+h)}{x(x+h)} = -\frac{h}{x(x+h)}$

c. The difference quotient is

$$\frac{f(x+h) - f(x)}{h} = \frac{-\frac{h}{x(x+h)}}{h} = \frac{-h}{hx(x+h)} = -\frac{1}{x(x+h)}.$$

d. As h approaches 0, $\frac{f(x+h)-f(x)}{h}$ approaches $-\frac{1}{x^2}$.

45. For $f(x) = \frac{x}{x-3}$:

a. $f(x+h) = \frac{x+h}{x+h-3}$

b.

$$f(x+h) - f(x) = \frac{x+h}{x+h-3} - \frac{x}{x-3} = \frac{x^2 - 3x + hx - 3h - x^2 - xh + 3x}{(x+h-3)(x-3)}$$

$$= -\frac{3h}{(x+h-3)(x-3)}$$

c. The difference quotient is

$$\frac{f(x+h) - f(x)}{h} = \frac{-\frac{3h}{(x+h-3)(x-3)}}{h} = \frac{-3h}{h(x+h-3)(x-3)} = -\frac{3}{(x+h-3)(x-3)}.$$

d. As h approaches 0, $\frac{f(x+h)-f(x)}{h}$ approaches $-\frac{3}{(x-3)^2}$.

47. For $f(x) = x^3$:

a. $f(x+h) = (x+h)^3 = x^3 + 3x^2h + 3xh^2 + h^3$

b. $f(x+h) - f(x) = x^3 + 3x^2h + 3xh^2 + h^3 - x^3 = 3x^2h + 3xh^2 + h^3$

c. The difference quotient is

$$\frac{f(x+h) - f(x)}{h} = \frac{3x^2h + 3xh^2 + h^3}{h} = \frac{h(3x^2 + 3xh + h^2)}{h}$$
$$= 3x^2 + 3xh + h^2.$$

d. As h approaches 0, $\frac{f(x+h)-f(x)}{h}$ approaches $3x^2$.

49. a. For $f(x) = \sqrt{x}$, we have $f(x+h) = \sqrt{x+h}$ and

$$\frac{f(x+h) - f(x)}{h} = \frac{\sqrt{x+h} - \sqrt{x}}{h}$$
$$= \frac{\sqrt{x+h} - \sqrt{x}}{h} \cdot \frac{\sqrt{x+h} + \sqrt{x}}{\sqrt{x+h} + \sqrt{x}} = \frac{(x+h) - x}{h(\sqrt{x+h} + \sqrt{x})}$$
$$= \frac{h}{h(\sqrt{x+h} + \sqrt{x})} = \frac{1}{\sqrt{x+h} + \sqrt{x}}.$$

b. As h approaches 0, the value of the difference quotient $\frac{f(x+h)-f(x)}{h}$ approaches

$$\frac{1}{\sqrt{x} + \sqrt{x}} = \frac{1}{2\sqrt{x}} = \frac{\sqrt{x}}{2x}.$$

51. The function is even, since the graph is symmetric with respect to the y-axis.

53. The function is neither even nor odd, since the graph is not symmetric with respect to either the x-axis or the origin.

55. The left figure is even; the right figure is odd.

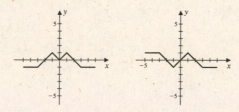

57. The transformations are shown in the figures.

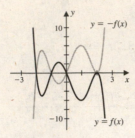 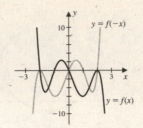 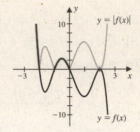

59. The distance between the ships is

$$d = \sqrt{(10(3+t))^2 + (15t)^2}$$
$$= \sqrt{325t^2 + 600t + 900}.$$

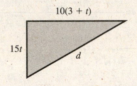

61. A rectangle with sides of length s and r has area $A = rs$ and perimeter $P = 2s + 2r$. Since $A = 64$, $r = 64/s$ and $P(s) = 2s + 128/s$.

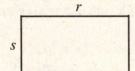

63. Let the length of the side of the cube be x. Since there are six faces of a cube, and each face of the cube has area x^2, the total surface area is $S = 6x^2$. The cube has volume $V = x^3$, so $x = \sqrt[3]{V}$. Hence $S(V) = 6(\sqrt[3]{V})^2 = 6V^{\frac{2}{3}}$.

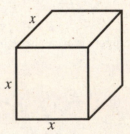

65. Let r be the radius of the circle and x and y be the lengths of the sides of the rectangle. Then $x^2 + y^2 = (2r)^2 = 4r^2$, so $y = \sqrt{4r^2 - x^2}$. Hence $A = xy = x\sqrt{4r^2 - x^2}$.

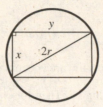

67. For $P(x) = 300x - 2x^2$:

a. Average rate of change:

$$\frac{P(x + h) - P(x)}{h} = \frac{300(x + h) - 2(x + h)^2 - (300x - 2x^2)}{h}$$

$$= \frac{300x + 300h - 2(x^2 + 2hx + h^2) - 300x + 2x^2}{h}$$

$$= \frac{300h - 4hx - 2h^2}{h} = \frac{h(300 - 4x - 2h)}{h} = 300 - 4x - 2h$$

b. If $h = 25$ and $x = 25$, then $x + h = 50$, and the average rate of change in the profit as the number of units changes from 25 to 50 is

$$300 - 4(25) - 2(25) = 150.$$

c. As h approaches 0, $\frac{P(x+h)-P(x)}{h}$ approaches $300 - 4x$, which is the instantaneous rate of change. And when $x = 25$, the instantaneous rate of change is $300 - 4(25) = 200$.

d. The graph shows $P(x)$ and the line joining $(25, P(25))$ and $(50, P(50))$.

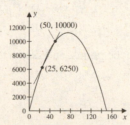

69. a. Let r and h be the radius and height of the cylinder, respectively. Since the volume is 900 and $V = \pi r^2 h$, we have $900 = \pi r^2 h$ and $h = 900/(\pi r^2)$. The amount of material needed to construct the can is $M = 2(\pi r^2) + 2\pi r h$, so

$$M(r) = 2\pi r^2 + 2\pi r(900/\pi r^2) = 2\pi r^2 + 1800/r.$$

b. A graphing device shows that the minimum material occurs when $r \approx 5.2$, $h \approx 10.6$, and in this case $M \approx 516.1$.

71. a. The combined area of the semi-circle regions is πr^2, and the area of the rectangle making up the remainder is $2rl$, so the total area is $A = \pi r^2 + 2rl$. Since the perimeter is 1 mile and is also $2\pi r + 2l$, we have $1 = 2\pi r + 2l$ and $l = (1 - 2\pi r)/2$. Hence

$$A(r) = \pi r^2 + 2r(1 - 2\pi r)/2 = \pi r^2 + r - 2\pi r^2 = r - \pi r^2.$$

b. A graphing device shows that the maximum area occurs when $r \approx 0.16$ and $A(0.16) \approx 0.08$.

73. a. Let x be the number of tickets exceeding 10. Then for $0 \leq x$, the revenue $R(x)$ is

$$R(x) = (10 + x)(80 - 2x) = 800 + 60x - 2x^2.$$

b. The maximum revenue occurs when $x = 15$, and the maximum revenue is $R(15) = 25 \cdot 50 = \$1250$.

. . . .

1.7   Linear Functions

The Slope of a Line

The *slope* of a line describes the inclination of the line and is a number that can be determined from any two points on the line. If two points on the line are (x_1, y_1) and (x_2, y_2), then the slope is

$$m = \frac{y_2 - y_1}{x_2 - x_1}.$$

In the slope formula, it does not matter which point is considered first and which is second provided we are consistent in the order of subtraction in the numerator and the denominator, since

$$\frac{y_1 - y_2}{x_1 - x_2} = \frac{-(y_2 - y_1)}{-(x_2 - x_1)} = \frac{y_2 - y_1}{x_2 - x_1}.$$

EXAMPLE 1

Find the slope of the line that passes through the points, and sketch the graph of the line.
(a) $(5, 2)$, $(-3, 1)$ (b) $(3, 4)$, $(6, 1)$

Solution (a) $m = \dfrac{1 - 2}{-3 - 5} = \dfrac{-1}{-8} = \dfrac{1}{8}$

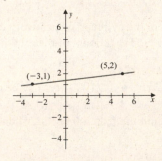

(b) $m = \dfrac{4 - 1}{3 - 6} = \dfrac{3}{-3} = -1$

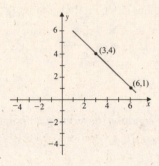

EXAMPLE 2

Find, if possible, the slope of the line that passes through the points, and sketch the line.

(a) $(-2, -1), (7, -1)$ (b) $(-1, -4), (-1, 8)$

Solution (a) $m = \dfrac{-1 - (-1)}{-2 - 7} = \dfrac{-1 + 1}{-9} = 0$

Since the slope of a line is 0, the line is horizontal. It has y-intercept $(0, -1)$.

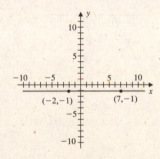

(b) The denominator of the slope formula is $-1 - (-1) = 0$, so the quotient for the slope is undefined. The line is vertical, with x-intercept $(-1, 0)$.

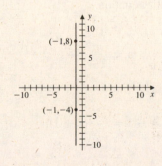

Point-Slope Equation of a Line

If a line passes through the point (x_0, y_0) and has slope m, then a point (x, y) will lie on the line precisely when

$$m = \frac{y - y_0}{x - x_0},$$

that is, when

$$y - y_0 = m(x - x_0).$$

This gives an equation called the *point-slope equation*, describing all points that are on the line.

EXAMPLE 3

Find an equation of the line that passes through the point $(-1, 1)$ and has slope 2.

Solution Substituting directly into the point-slope form gives

$$y - 1 = 2(x - (-1)) \quad \text{or} \quad y = 2x + 3.$$

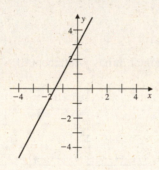

EXAMPLE 4

Find the equation of the line passing through the two points $(-1, 1)$ and $(2, -3)$.

Solution First use the two points to find the slope. Either of the given points along with the slope can then be substituted into the point-slope form to find the equation.

$$m = \frac{-3 - 1}{2 - (-1)} = -\frac{4}{3}$$

$$y - 1 = -\frac{4}{3}(x - (-1))$$

$$y - 1 = -\frac{4}{3}(x + 1)$$

$$y = -\frac{4}{3}x - \frac{1}{3}$$

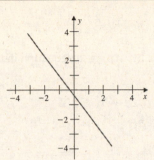

Slope-Intercept Equation of a Line

The *slope-intercept* equation of a line is a special case of the point-slope equation of the line where the *y*-intercept $(0, b)$ is the given point on the line. The slope-intercept equation then has the form

$$y = mx + b.$$

This was the form given in the final reduction in Examples 3 and 4. When the equation is in this form, the slope is m, the coefficient of x, and the line crosses the *y*-axis at b, the constant term.

EXAMPLE 5
Find an equation of the line that has slope -2 and *y*-intercept 1.

Solution Substituting directly into the slope-intercept form of a line gives

$$y = -2x + 1.$$

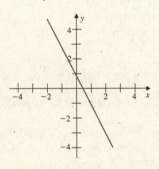

Parallel and Perpendicular Lines

Suppose two non-vertical lines, l_1 and l_2, have equations

$$l_1 : y = m_1 x + b_1 \quad \text{and} \quad l_2 : y = m_2 x + b.$$

- l_1 and l_2 are *parallel* precisely when $m_1 = m_2$.
- l_1 and l_2 are *perpendicular* precisely when $m_1 = -\frac{1}{m_2}$.

EXAMPLE 6
Find the line that passes through $(1, 1)$ and is parallel to the line $2x + 3y = -2$. Sketch the two lines.

Solution   First find the slope of the given line by writing the equation in the form $y = mx + b$.

$$2x + 3y = -2$$
$$3y = -2x - 2$$
$$y = -\frac{2}{3}x - \frac{2}{3}$$

The slope of the line is $-\frac{2}{3}$, as is the slope of any line that is parallel to it. Since the line we want passes through the point $(1, 1)$, the equation is

$$y - 1 = -\frac{2}{3}(x - 1)$$

so

$$y = -\frac{2}{3}x + \frac{5}{3}.$$

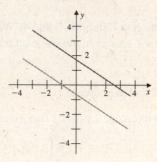

$\square$

EXAMPLE 7

Find the line that passes through $(-1, -1)$ and is perpendicular to the line $-3x + 5y = 3$. Sketch the two lines.

Solution   Rewriting the equation of the line in slope-intercept form gives

$$-3x + 5y = 3$$
$$5y = 3x + 3$$
$$y = \frac{3}{5}x + \frac{3}{5}.$$

The two lines are perpendicular, so the slope of the line we want is the negative reciprocal of $\frac{3}{5}$, that is, $-\frac{5}{3}$. Since the line we want also passes through the point $(-1, -1)$ the equation is

$$y + 1 = -\frac{5}{3}(x + 1) \quad \text{so} \quad y = -\frac{5}{3}x - \frac{8}{3}.$$

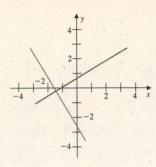

□

EXAMPLE 8

Find an equation of the line that is tangent to the circle $x^2 + y^2 = 9$ at the point $(2\sqrt{2}, 1)$. At what other point on the circle will the tangent line be parallel to this line?

Solution The circle $x^2 + y^2 = 9$ has center $(0, 0)$ and radius 3. The slope of the radius from $(0, 0)$ to $(2\sqrt{2}, 1)$ is $m = \frac{1}{2\sqrt{2}} = \frac{\sqrt{2}}{4}$. The tangent line, which is perpendicular to the radius, has slope $-\frac{4}{\sqrt{2}} = -2\sqrt{2}$. The equation of the tangent line is

$$y - 1 = -2\sqrt{2}(x - 2\sqrt{2}),$$

so

$$y = -2\sqrt{2}x + 8 + 1 = -2\sqrt{2}x + 9.$$

The other point on the circle where the tangent line is parallel is at the point diametrically opposite $(2\sqrt{2}, 1)$, which is the point $(-2\sqrt{2}, -1)$.

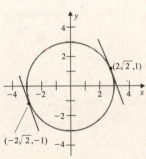

□

Application

EXAMPLE 9

A new Corvette costs $54,000.00. At the end of 5 years, it will be worth an estimated $34,000.00. Assume that the value of the car depreciates linearly.
(a) Find a linear equation that expresses the value V of the car as a function of time t, where $0 \le t \le 5$.
(b) How much will the car be worth after 3 years?
(c) What is the average rate of change in the value of the car from 2 to 4 years?

Solution (a) At time $t = 0$ the car is worth 54000 dollars, and at time $t = 5$ it is worth 34000 dollars. If a linear model is used to represent value against time, the two points $(0, 54000)$ and $(5, 34000)$ lie on the line. The slope is then

$$m = \frac{54000 - 34000}{0 - 5} = -\frac{20000}{5} = -4000,$$

and the equation of the line is

$$y = -4000t + 54000.$$

The value, $V(t)$, at time t is

$$V(t) = -4000t + 54000.$$

(b) Substitute $t = 3$ in the equation obtained in part (a) to get the value of the Corvette after 3 years to be

$$V(3) = -4000(3) + 54000 = 42000 \text{ dollars.}$$

(c) The average rate of change is the difference quotient

$$\frac{V(4) - V(2)}{4 - 2} = \frac{38000 - 46000}{2} = -4000.$$

Exercise Set 1.7 (Page 62)

1. Slope: $m = \frac{4-1}{2-1} = 3$;
Equation: $y = 3x - 2$

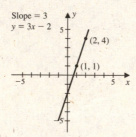

3. Slope: $m = \frac{-4-1}{2-(-2)} = -\frac{5}{4}$;
Equation: $y - 1 = -\frac{5}{4}(x + 2)$ or $y = -\frac{5}{4}x - \frac{3}{2}$

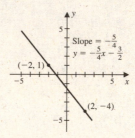

5. Slope: $m = \frac{3-3}{-1-2} = 0$;
Equation: $y = 3$

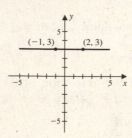

7. Slope: $m = \frac{6-(-3)}{-1-(-1)}$ which is undefined;
Equation: $x = -1$

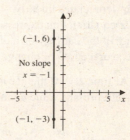

9. For $y = x - 3$ the slope is $m = 1$.

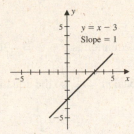

11. For $3x - y = 2$, or $y = 3x - 2$, the slope is $m = 3$.

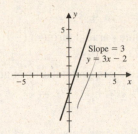

13. For $-3x - 4y = 2$, or $-4y = 3x + 2$ so $y = -\frac{3}{4}x - \frac{1}{2}$, the slope is $m = -\frac{3}{4}$.

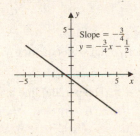

15. For $y = 2$ the slope is 0.

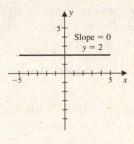

17. a. $y - 4 = x - 1$ implies $y = x + 3$

b. $y - 4 = -(x - 1)$ implies $y = -x + 5$

c. $y - 4 = 3(x - 1)$ implies $y = 3x + 1$

d. $y - 4 = \frac{1}{3}(x - 1)$ implies $y = \frac{1}{3}x + \frac{11}{3}$

(a) 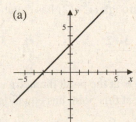 (b) 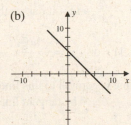 (c) 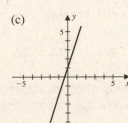 (d)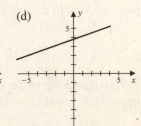

19. The slopes of the lines are:
(a) 0 (b) 1 (c) −1 (d) 1 (e) 2 (f) 0 (g) −1 (h) 2 (i) 1 (j) 1
Parallel Lines: (a) and (f); (c) and (g); (b), (d), (i), and (j); (e) and (h).

21. For $y = 2x + 1$:

a. The line parallel through $(0, 0)$ has same slope $m = 2$, so the equation is $y = 2x$.

b. The line perpendicular through $(0, 0)$ has slope negative reciprocal $m = -\frac{1}{2}$, so the equation is $y = -\frac{1}{2}x$.

23. For $y = -2x + 3$:

   a. The line parallel through $(-1, 2)$ has same slope $m = -2$, so the equation is $y - 2 = -2(x + 1)$ or $y = -2x$.

   b. The line perpendicular through $(-1, 2)$ has slope negative reciprocal $m = \frac{1}{2}$, so the equation is $y - 2 = \frac{1}{2}(x + 1)$ or $y = \frac{1}{2}x + \frac{5}{2}$.

25. Since the y-intercept is 2, and the slope is -1, the equation is $y = -x + 2$.

27. We have $y - (-2) = 3(x - 1)$ or $y = 3x - 5$.

29. The line passes through $(1, 0)$ and $(0, 3)$, so the slope is $m = \frac{3-0}{0-1} = -3$, and the equation is $y - 0 = -3(x - 1)$ or $y = -3x + 3$.

31. Since the line is parallel to the x-axis, its slope is 0, so the line is horizontal with equation $y = -1$.

33. We have $2x - 3y = 2$, which implies $-3y = -2x + 2$, so $y = \frac{2}{3}x - \frac{2}{3}$. The parallel line also has slope $\frac{2}{3}$ and equation $y - 3 = \frac{2}{3}(x - 4)$ or $y = \frac{2}{3}x + \frac{1}{3}$.

35. A line perpendicular to $y = 2 - x$ has slope 1, and if it passes through $(1, 1)$, has equation $y - 1 = x - 1$ or $y = x$.

37. **a.** The line tangent to the circle $x^2 + y^2 = 3$ at the point $(1, \sqrt{2})$ is perpendicular to the radius line passing through $(0, 0)$ and $(1, \sqrt{2})$. The slope of the radius line is $m = \frac{\sqrt{2}-0}{1-0} = \sqrt{2}$. A line tangent to the circle is perpendicular to this radius, so it has slope $-\frac{1}{\sqrt{2}} = -\frac{\sqrt{2}}{2}$, and if it passes through $(1, \sqrt{2})$ has equation

$$y - \sqrt{2} = -\frac{\sqrt{2}}{2}(x - 1) \quad \text{or} \quad y = -\frac{\sqrt{2}}{2}x + \frac{3\sqrt{2}}{2}.$$

   b. The other point on the circle where the tangent line is parallel to this line is at the point diametrically opposite the point $(1, \sqrt{2})$, which is the point $(-1, -\sqrt{2})$. The tangent line at this point has equation

$$y + \sqrt{2} = -\frac{\sqrt{2}}{2}(x + 1) \quad \text{or} \quad y = -\frac{\sqrt{2}}{2}x - \frac{3\sqrt{2}}{2}.$$

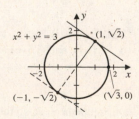

39. Since $v(t) = -32t$, $v(7) = -224$, and the rock is traveling downward at a speed of 224 feet/second.

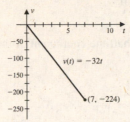

41. a. For $W(n) = 500 - 0.5n$ we have the following graph.

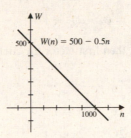

b. The total fish weight is

$$T(n) = n \cdot W(n) = n(500 - 0.5n) = 500n - 0.5n^2.$$

c. At $n = 1000$ the total weight is 0, so the fish population has disappeared. This is the predicted limiting value of the number of fish in the pond. For $n > 1000$, the total weight given by the equation is negative, which is physically unreasonable.

43. a. $V(t) = 28,000 - 4000t$ dollars.

b. $V(5) \approx \$8000$.

c. In 7 years.

1.8 Quadratic Functions

A *quadratic function* has the form $f(x) = ax^2 + bx + c$, where a, b, and c are real numbers and $a \neq 0$. Every quadratic function can be rewritten in the *standard form* $f(x) = a(x - h)^2 + k$. The graph of a quadratic function is called a *parabola*.

Completing the Square

The process of writing a quadratic function in standard form is called *completing the square*.

EXAMPLE 1

Write the quadratic function in standard form.

(a) $f(x) = x^2 - 4x + 5$ (b) $f(x) = 2x^2 - 4x - 1$

Solution (a) To complete the square, take half the coefficient of the x term, square it and both add and subtract the value, so the net result is to add 0.

$$f(x) = x^2 - 4x + 5$$
$$= x^2 - 4x + \left(\frac{4}{2}\right)^2 - \left(\frac{4}{2}\right)^2 + 5$$
$$= x^2 - 4x + 4 + (-4 + 5)$$
$$= (x - 2)^2 + 1$$

(b) If the coefficient of the x^2 term is not 1, then first factor the coefficient from both the x^2 term and the x term and proceed as in part (a).

$$f(x) = 2x^2 - 4x - 1$$
$$= 2(x^2 - 2x) - 1$$
$$= 2\left(x^2 - 2x + \left(\frac{2}{2}\right)^2 - \left(\frac{2}{2}\right)^2\right) - 1$$
$$= 2\left(x^2 - 2x + 1 - 1\right) - 1$$
$$= 2(x - 1)^2 - 2 - 1$$
$$= 2(x - 1)^2 - 3$$

$\square$

Horizontal and Vertical Shifts

Let h and $k > 0$.

- The graph of $y = f(x - h)$ is the graph of $y = f(x)$ shifted to the right h units.

- The graph of $y = f(x + h)$ is the graph of $y = f(x)$ shifted to the left h units.

- The graph of $y = f(x) + k$ is the graph of $y = f(x)$ shifted upward k units.

- The graph of $y = f(x) - k$ is the graph of $y = f(x)$ shifted downward k units.

EXAMPLE 2

Use the graph of $y = f(x) = x^2$ given in the figure to sketch the graph.

(a) $y = f(x - 1)$ (b) $y = f(x + 2)$ (c) $y = f(x) + 1$
(d) $y = f(x) - 2$ (e) $y = f(x - 2) + 3$

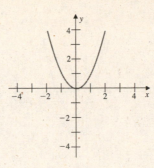

Solution Part (a) is a horizontal shift of $y = x^2$ to the right 1 unit, (b) to the left 2 units, (c) upward 1 unit, (d) downward 2 units. Part (e) is a combination of shifts to the right 2 units and upward 3 units.

    (a) $y = f(x - 1)$ (b) $y = f(x + 2)$ (c) $y = f(x) + 1$

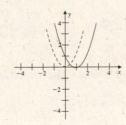 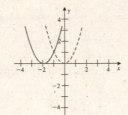 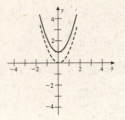

       (d) $y = f(x) - 2$ (e) $y = f(x - 2) + 3$.

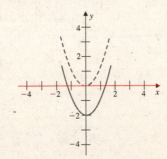 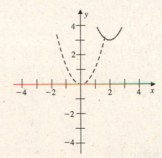

□

EXAMPLE 3
Sketch the graph of the quadratic function.
(a) $f(x) = x^2 - 2x - 1$ (b) $f(x) = x^2 + 4x + 5$

Solution To use the basic graph $y = x^2$ and the shifting properties, the quadratic function is first put in standard form by completing the square.
(a)

$$\begin{aligned} f(x) &= x^2 - 2x - 1 \\ &= (x^2 - 2x + 1 - 1) - 1 \\ &= (x - 1)^2 - 2 \end{aligned}$$

To obtain the graph, first shift $y = x^2$, to the right 1 unit and then downward 2 units. The vertex of the parabola is at $(1, -2)$.

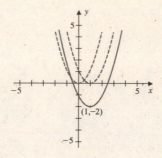

(b)

$$f(x) = x^2 + 4x + 5$$
$$= (x^2 + 4x + 4 - 4) + 5$$
$$= (x + 2)^2 + 1$$

To obtain the graph first shift $y = x^2$, to the left 2 units and then upward 1 unit. The vertex of the parabola is at $(-2, 1)$.

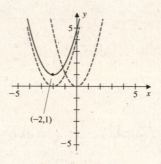

□

Vertical Scaling and Reflection

Let $a > 1$.

- The graph of $y = af(x)$, is a vertical stretching, by a factor of a, of the graph of $y = f(x)$.

- The graph of $y = \dfrac{1}{a}f(x)$, is a vertical shrinking, by a factor of a, of the graph of $y = f(x)$.

- The graph of $y = -f(x)$ is the reflection through the x-axis of the graph of $y = f(x)$.

For example, if $y = x^2$, then the point $(1, 1)$ on the curve lies

- directly below the point $(1, 2)$ on $y = 2x^2$,

- directly above the point $\left(1, \frac{1}{2}\right)$ on $y = \frac{1}{2}x^2$,

- and when $x = 1$, the point $(1, -1)$ on $y = -x^2$ is one unit below the x-axis.

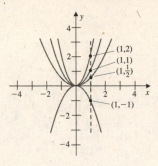

EXAMPLE 4

Sketch the graph of the function and find the range.

(a) $f(x) = 2x^2 - 8x + 10$ (b) $f(x) = -4x^2 - 4x + 3$

Solution (a) Completing the square gives

$$\begin{aligned}
f(x) &= 2x^2 - 8x + 10 \\
&= 2(x^2 - 4x) + 10 \\
&= 2(x^2 - 4x + 4 - 4) + 10 \\
&= 2(x - 2)^2 - 8 + 10 \\
&= 2(x - 2)^2 + 2.
\end{aligned}$$

To obtain the graph, start with $y = x^2$, stretch it by a factor of 2, and shift the result to the right 2 units and upward 2 units. The range of the function is then $[2, \infty)$, with a minimum point at $(2, 2)$.

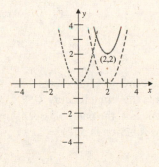

(b) Completing the square gives

$$\begin{aligned}
f(x) &= -4x^2 - 4x + 3 \\
&= -4(x^2 + x) + 3 \\
&= -4\left(x^2 + x + \frac{1}{4} - \frac{1}{4}\right) + 3 \\
&= -4\left(x + \frac{1}{2}\right)^2 + 4.
\end{aligned}$$

To obtain the graph, start with $y = x^2$, stretch it by a factor of 4, reflect the result about the x-axis, and shift the new result to the left $\frac{1}{2}$ unit and upward 4 units. The range of the function is $(-\infty, 4]$, with a maximum point at $\left(-\frac{1}{2}, 4\right)$.

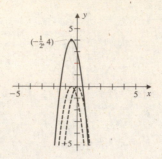

EXAMPLE 5

Let the quadratic function $f(x) = x^2 - 4x + 5$.
(a) Express the quadratic in standard form.
(b) Find any x- and y-intercepts.
(c) Find any maximum or minimum value of the function.

Solution (a) Completing the square gives

$$
\begin{aligned}
f(x) &= x^2 - 4x + 5 \\
&= x^2 - 4x + 4 - 4 + 5 \\
&= (x - 2)^2 + 1.
\end{aligned}
$$

(b) For the x-intercepts solve $f(x) = 0$.

$$
\begin{aligned}
(x - 2)^2 + 1 &= 0 \\
(x - 2)^2 &= -1
\end{aligned}
$$

Since the left hand side is always positive, there are no x-intercepts. (See the figure.)
For the y-intercepts: Set $x = 0$, then $y = 5$ and the y-intercept is $(0, 5)$.
(c) Since the leading coefficient of the quadratic is 1, the parabola opens upward, and the vertex $(2, 1)$ is the minimum point on the curve. The minimum value of the function is $f(2) = 1$.

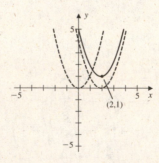

EXAMPLE 6

Use the quadratic formula to find any x-intercepts of the parabola $y = 2x^2 - 4x + 1$.

Solution The x-intercepts are those values of x that make $y = 0$, so solve $2x^2 - 4x + 1 = 0$. Notice that the quadratic can not be factored directly, so we will use the quadratic formula with $a = 2$, $b = -4$, and $c = 1$. This gives

$$2x^2 - 4x + 1 = 0,$$

$$x = \frac{-(-4) \pm \sqrt{(-4)^2 - 4(2)(1)}}{2(2)}$$

$$= \frac{4 \pm \sqrt{8}}{4} = \frac{4 \pm 2\sqrt{2}}{4}$$

$$= \frac{2 \pm \sqrt{2}}{2} = 1 \pm \frac{\sqrt{2}}{2}.$$

Equivalently the x-intercepts can also be found by writing the quadratic in standard form,

$$y = 2x^2 - 4x + 1$$

$$= 2(x^2 - 2x) + 1$$

$$= 2(x^2 - 2x + 1 - 1) + 1$$

$$= 2(x - 1)^2 - 1.$$

Then

$$2(x - 1)^2 - 1 = 0$$

$$2(x - 1)^2 = 1$$

$$(x - 1)^2 = \frac{1}{2}$$

$$x - 1 = \pm\sqrt{\frac{1}{2}} = \pm\frac{\sqrt{2}}{2}$$

$$x = 1 \pm \frac{\sqrt{2}}{2}.$$

$\square$

EXAMPLE 7

Match the curve with the equation in the figure.

(a) $y = x^2 + 1$ (b) $y = x^2 - 2x + 1$ (c) $y = (x + 1)^2$

(d) $y = (x + 1)^2 - 1$ (e) $y = x^2 - 2x - 1$ (f) $y = -x^2 + 1$

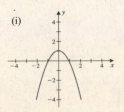 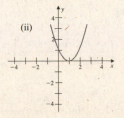 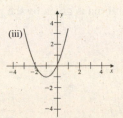

(i)   (ii)   (iii)

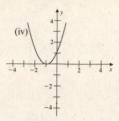
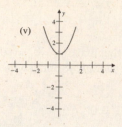
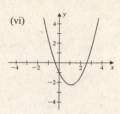

Solution The curves are all transformations of $y = x^2$, as follows:

(a) An upward shift by 1 unit, so it is (v).

(b) Since

$$y = x^2 - 2x + 1 = (x - 1)^2,$$

the curve is a shift right by 1 unit, so it is (ii).

(c) A shift left by 1 unit, so it is (iv).

(d) A shift left by 1 unit, followed by a shift downward of 1 unit, so it is (iii).

(e) Note that

$$
\begin{aligned}
y &= x^2 - 2x - 1 \\
&= \left(x^2 - 2x + 1 - 1\right) - 1 \\
&= (x - 1)^2 - 2.
\end{aligned}
$$

The transformation is to the right 1 unit and downward 2 units, so it is (vi).

(f) A reflection about the x-axis followed by a shift upward by 1 unit, so it is (i). Note also (f) is the only equation with a negative leading coefficient, so it is the only equation with a curve that opens downward.

$\square$

EXAMPLE 8

What can you say about a, b, and c in $f(x) = ax^2 + bx + c$ if

(a) $(1, 0)$ is on the graph?

(b) the y-intercept is 3?

(c) $(1, 0)$ is the vertex?

(d) conditions (a), (b), and (c) are all satisfied?

Solution (a) If $(1, 0)$ is on the graph, then

$$0 = f(1) \quad \text{so} \quad 0 = a(1)^2 + b(1) + c \quad \text{and} \quad a + b + c = 0.$$

(b) If $(0, 3)$ is on the graph, then

$$3 = f(0) \quad \text{so} \quad 3 = a(0)^2 + b(0) + c \quad \text{and} \quad c = 3.$$

(c) The vertex of a parabola is given by the point

$$\left(-\frac{b}{2a}, \frac{4ac - b^2}{4a}\right)$$

so

$$-\frac{b}{2a} = 1, \quad \text{and} \quad \frac{4ac - b^2}{4a} = 0.$$

Combining these equations permits us to conclude that since $b = -2a$,

$$0 = \frac{4ac - b^2}{4a} = \frac{4ac - 4a^2}{4a} = \frac{4a}{4a}(a - c) = a - c$$

Hence $a = c$.

(d) From (c) we know that $a = c$, and from (b) we know that $c = 3$. Hence $a = c = 3$. From (a) we know that

$$0 = a + b + c = 3 + b + 3 = b + 6.$$

Hence $b = -c$ and $f(x) = 3x^2 - 6x + 3$.

$\square$

Exercise Set 1.8 (Page 73)

1. $y = x^2 + 1$

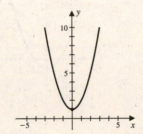

3. $y = -x^2 + 1$

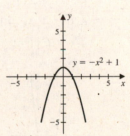

5. $y = (x - 3)^2$

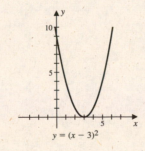

7. $y = (x + 1)^2 - 1$

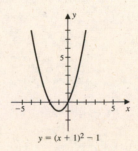

9. Completing the square gives

$$y = x^2 - 4x + 3 = x^2 - 4x + 4 - 4 + 3$$
$$= (x - 2)^2 - 1.$$

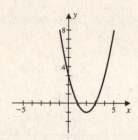

11. Completing the square gives

$$y = -x^2 - 2x$$
$$= -(x^2 + 2x)$$
$$= -(x^2 + 2x + 1 - 1)$$
$$= -(x + 1)^2 + 1.$$

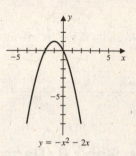

$$y = -x^2 - 2x$$

13. Completing the square gives

$$y = 3x^2 + 6x$$
$$= 3(x^2 + 2x)$$
$$= 3(x^2 + 2x + 1 - 1)$$
$$= 3(x + 1)^2 - 3.$$

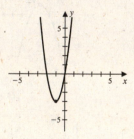

15. $y = \frac{1}{2}x^2 - 1$

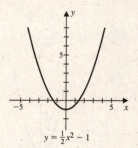

$$y = \frac{1}{2}x^2 - 1$$

17. Completing the square gives

$$
\begin{aligned}
y &= \frac{1}{2}x^2 - x + 3 \\
&= \frac{1}{2}\left(x^2 - 2x\right) + 3 \\
&= \frac{1}{2}(x^2 - 2x + 1 - 1) + 3 \\
&= \frac{1}{2}(x - 1)^2 + \frac{5}{2}.
\end{aligned}
$$

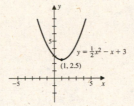

19. a. $f(x) = x^2 - 6x + 7 = (x^2 - 6x + 9 - 9) + 7 = (x - 3)^2 - 2$

b. To find the x-intercepts solve

$$(x - 3)^2 - 2 = 0, \quad \text{that is} \quad (x - 3)^2 = 2, \quad \text{so} \quad x = \pm\sqrt{2} + 3,$$

and the x-intercepts are $(3 + \sqrt{2}, 0)$ and $(3 - \sqrt{2}, 0)$. The y-intercept is $(0, 7)$.

c. Since the parabola opens upward, the vertex $(3, -2)$ is a minimum point. So the minimum value of the function is $f(3) = -2$.

21. a. We have $f(x) = -x^2 + 4x + 6 = -(x^2 - 4x + 4 - 4) + 6 = -(x - 2)^2 + 10$.

b. To find the x-intercepts solve

$$-(x - 2)^2 + 10 = 0, \quad \text{that is} \quad (x - 2)^2 = 10 \quad \text{so} \quad x = \pm\sqrt{10} + 2.$$

and the x-intercepts are $(\sqrt{10} + 2, 0)$ and $(-\sqrt{10} + 2, 0)$. The y-intercept is $(0, 6)$.

c. Since the parabola opens downward the vertex $(2, 10)$ is a maximum point. So the maximum value of the function is $f(2) = 10$.

23. For $6x^2 - 5x + 1 = 0$,

$$x = \frac{-(-5) \pm \sqrt{(-5)^2 - 4(6)(1)}}{2(6)}$$

$$= \frac{5 \pm \sqrt{25 - 24}}{12} = \frac{5 \pm 1}{12}$$

so the x-intercepts are

$$x = \frac{1}{2} \quad \text{and} \quad x = \frac{1}{3}.$$

25. For $2x^2 - 4x + 1 = 0$,

$$x = \frac{-(-4) \pm \sqrt{(-4)^2 - 4(2)(1)}}{2(2)}$$

$$= \frac{4 \pm \sqrt{16 - 8}}{4}$$

$$= \frac{4 \pm \sqrt{8}}{4} = \frac{4 \pm 2\sqrt{2}}{4}$$

so the x-intercepts are

$$x = 1 + \frac{\sqrt{2}}{2} \quad \text{and} \quad x = 1 - \frac{\sqrt{2}}{2}.$$

27. For $2x^2 + 4x + 3 = 0$,

$$x = \frac{-4 \pm \sqrt{(4)^2 - 4(2)(3)}}{2(2)}$$

$$= \frac{-4 \pm \sqrt{16 - 24}}{4}$$

$$= \frac{-4 \pm \sqrt{-8}}{4}.$$

Since the square root of a negative number is not a real number, the quadratic has no solutions and there are no x-intercepts.

29. The various graphs are shown.

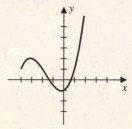
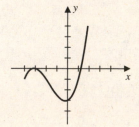
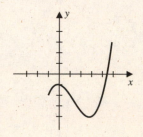

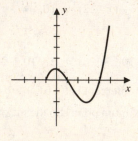
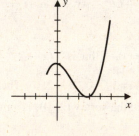
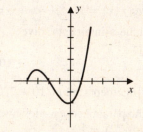

31. The function will have the form $f(x) = a(x-1)^2 + 3$ and if the graph passes through the point $(-2, 5)$ we have

$$5 = a(-2-1)^2 + 3, \quad \text{that is} \quad 5 = 9a + 3 \quad \text{so} \quad a = \frac{2}{9},$$

and $f(x) = \frac{2}{9}(x-1)^2 + 3$.

33. a. If x is in the domain of $f(x) = \sqrt{x^2 - 3}$, then

$$x^2 - 3 = \left(x - \sqrt{3}\right)\left(x + \sqrt{3}\right) \geq 0 \quad \text{so} \quad x \leq -\sqrt{3} \quad \text{or} \quad x \geq \sqrt{3},$$

which can be written $\left(-\infty, -\sqrt{3}\right] \cup \left[\sqrt{3}, \infty\right)$.

b. The domain of $f(x) = \sqrt{x^2 - \frac{1}{2}x} = \sqrt{x\left(x - \frac{1}{2}\right)}$ is all x with $x\left(x - \frac{1}{2}\right) \geq 0$ so $x \leq 0$ or $x \geq \frac{1}{2}$, which can be written $(-\infty, 0] \cup \left[\frac{1}{2}, \infty\right)$.

35. Let $f(x) = ax^2 + bx + c$.

a. Since $(1, 1)$ is on the graph, we have $1 = f(1) = a + b + c$. So $a + b + c = 1$.

b. The y-intercept is 6, which implies $6 = f(0) = c$. So $c = 6$.

c. The vertex of a parabola is $\left(-\frac{b}{2a}, \frac{4ac - b^2}{4a}\right)$. The vertex is $(1, 1)$, which implies $-\frac{b}{2a} = 1$ and $\frac{4ac - b^2}{4a} = 1$, so $b = -2a$ and

$$1 = \frac{4ac - b^2}{4a} = \frac{4ac - (-2a)^2}{4a} = \frac{4ac - 4a^2}{4a} = \frac{4a(c-a)}{4a} = c - a, \quad \text{when } a \neq 0.$$

Since $c - a = 1$, we have $c = a + 1$.

d. $a + b + c = 1$ and $c = 6$ so $a + b = -5$. But $c = a + 1$ so $a = 5$ and $b = -10$. So $a = 5, b = -10, c = 6$ and $f(x) = 5x^2 - 10x + 6$.

37. a. Let $v(t) = 144 - 32t$.

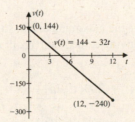

b. From Exercise 36(b), the rock strikes the ground at $t = 12$, so the velocity of the rock is $v(12) = 144 - 32(12) = -240$ feet/second. The minus sign indicates the rock is falling to the earth.

c. Domain$(v) = [0, 12]$; Range$(v) = [-240, 144]$

39. Let $P(x) = -0.1x^2 + 160x - 20000$.

a. The parabola opens downward, so the vertex is the maximum point on the curve. To find the number of terminals to produce in order to maximize the profit, complete the square and find the vertex.

$$-0.1x^2 + 160x - 20000 = -0.1(x^2 - 1600x) - 20000$$
$$= -0.1(x^2 - 1600x + 640000 - 640000) - 20000$$
$$= -0.1(x - 800)^2 + 64000 - 20000$$
$$= -0.1(x - 800)^2 + 44000$$

The vertex is at (800, 44000), so the company should produce 800 terminals to maximize the profit.

b. The maximum profit is $44,000.00.

41. **a.** The area of the rectangle, for $x > 0$, is $A(x) = 2x(4 - x^2) = 8x - 2x^3$.

b. A graphing device indicates that the maximum area occurs when the width, w, and the height, h, satisfy $w \approx 2.3$, $h \approx 2.7$, and $A(w/2) \approx 6.2$.

43. $R = kP(M - P)$, where k is the constant of proportionality.

44. **a.** Let $P(x) = 200x - x^2$. Completing the square gives

$$P(x) = -x^2 + 200x = -(x^2 - 200x)$$
$$= -(x^2 - 200x + 10000 - 10000) = -(x - 100)^2 + 10000.$$

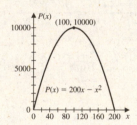

b. The vertex of the parabola will give the maximum profit, so from part (a) the vertex is at (100, 10000) and hence 100 units should be produced giving a maximum profit is $10,000.00.

c. We have

$$\frac{P(x + h) - P(x)}{h} = \frac{200(x + h) - (x + h)^2 - (200x - x^2)}{h}$$
$$= \frac{200x + 200h - (x^2 + 2hx + h^2) - 200x + x^2}{h}$$
$$= \frac{200h - 2hx - h^2}{h} = \frac{h(200 - 2x - h)}{h}$$
$$= 200 - 2x - h.$$

As h approaches 0, the difference quotient approaches $200 - 2x$.

45. a. Let $v(t) = -4t^2 - 4t + 80$. Completing the square gives

$$-4t^2 - 4t + 80 = -4(t^2 + t) + 80 = -4\left(t^2 + t + \frac{1}{4} - \frac{1}{4}\right) + 80 = -4\left(t + \frac{1}{2}\right)^2 + 81.$$

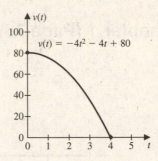

b. The car is at rest when

$$v(t) = -4t^2 - 4t + 80 = -4(t^2 + t - 20) = -4(t + 5)(t - 4) = 0 \quad \text{so} \quad t = -5, t = 4.$$

Since t is time, $t = -5$ is not a solution, and the car comes to rest after $t = 4$ seconds.

c.

t	0	$\frac{1}{2}$	1	$\frac{3}{2}$	2	$\frac{5}{2}$	3	$\frac{7}{2}$	4
$v(t)$	80	77	72	65	56	45	32	17	0

d. slowest: 77 feet/second; fastest: 80 feet/second

e.

	slowest	fastest
second 1/2 sec	72	77
third 1/2 sec	65	72

f.

	minimum distance	maximum distance
first 1/2 sec	$77 \times 1/2 = 38.5$ ft	$80 \times 1/2 = 40$ ft
second 1/2 sec	$72 \times 1/2 = 36$ ft	$77 \times 1/2 = 38.5$ ft

g. Lower bound:

$$\frac{1}{2}[77 + 72 + 65 + 56 + 45 + 32 + 17] = \frac{1}{2} \cdot 364 = 182$$

Upper bound:

$$\frac{1}{2}[80 + 77 + 72 + 65 + 56 + 45 + 32 + 17] = \frac{1}{2} \cdot 444 = 222$$

Review Exercises For Chapter 1 (Page 76)

1. **a.** $-1 \le x \le 7$
   b.

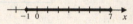

3. **a.** $x < 7$
   b.

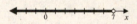

5. **a.** $(-4, \infty)$
   b.

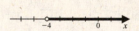

7. **a.** $[2, 10)$
   b.

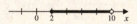

9. $2x + 3 \ge 4$ implies $2x \ge 1$ so $x \ge \frac{1}{2}$

11. $x^2 + 2x + 1 \ge 1$ implies $x^2 + 2x \ge 0$, so $x(x + 2) \ge 0$. The solution set is $(-\infty, -2] \cup [0, \infty)$.

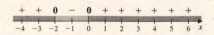

13. $(x - 1)(x + 2)(x - 2) \ge 0$
    The solution set is $[-2, 1] \cup [2, \infty)$.

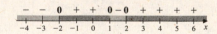

15. $x^2 + 3x > 0$ implies $x(x+3) > 0$.
The solution set is $(-\infty, -3) \cup (0, \infty)$.

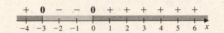

17. $\frac{2x-1}{x+1} \leq -2$ implies $\frac{2x-1}{x+1} + 2 \leq 0$ so
$\frac{2x-1+2(x+1)}{x+1} \leq 0$ and $\frac{4x+1}{x+1} \leq 0$. The solution set is
$\left(-1, -\frac{1}{4}\right]$.

19. a. $|2x - 3| < 5$ implies $-5 < 2x - 3 < 5$ so
$-2 < 2x < 8$ and $-1 < x < 4$

b.

21. a. $|3 - x| \leq 4$ implies $-4 \leq 3 - x \leq 4$ so $-x \leq 1$
and $-x \geq -7$ or $x \geq -1$ and $x \leq 7$. That is,
$[-1, 7]$.

b.

23. $2 < y \leq 3$

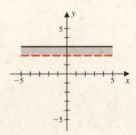

25. $|x - 1| < 2$ implies $-2 < x - 1 < 2$ so
$-1 < x < 3$

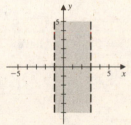

27. $|x| + |y| = 1$

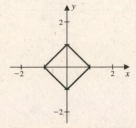

29. $|x| \leq |y|$

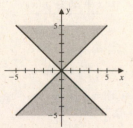

31. a. For $f(x) = x^2 - 3$ the domain is the set of all real numbers.

b. Since the graph is obtained from the basic graph of $y = x^2$, shifted 3 units downward, the vertex is
$(0, -3)$, so the range is $[-3, \infty)$.

33. a. For $f(x) = \sqrt{x-2} + 2$ the domain is the set of all real numbers for which the expression under the radical is nonnegative, so the domain is $\{x : x - 2 \geq 0\} = [2, \infty)$.

b. The range is $[2, \infty)$.

35. a. For

$$f(x) = \frac{1}{x^2 - 6x + 8}$$

the domain is

$$\{x \mid x^2 - 6x + 8 \neq 0\} = \{x : (x-4)(x-2) \neq 0\} = (-\infty, 2) \cup (2, 4) \cup (4, \infty).$$

b. For

$$f(x) = \frac{x-2}{x^2 - 6x + 8} = \frac{x-2}{(x-4)(x-2)} = \frac{1}{x-4}$$

the domain is still $(-\infty, 2) \cup (2, 4) \cup (4, \infty)$, since $x = 2$ can not be substituted into the original equation.

c. For

$$f(x) = \sqrt{\frac{x^2}{x^2 - 6x + 8}}$$

the domain is

$$\{x \mid (x-4)(x-2) > 0 \text{ or } x = 0\} = (-\infty, 2) \cup (4, \infty).$$

37. For $f(x) = 5x + 3$ we have:

a. $f(x + h) = 5(x + h) + 3 = 5x + 5h + 3$

b.
$$\frac{f(x+h) - f(x)}{h} = \frac{5x + 5h + 3 - (5x + 3)}{h} = \frac{5h}{h} = 5.$$

39. For $f(x) = x^2 - 1$ we have:

a. $f(x + h) = (x + h)^2 - 1 = x^2 + 2hx + h^2 - 1$

b.
$$\frac{f(x+h) - f(x)}{h} = \frac{x^2 + 2hx + h^2 - 1 - (x^2 - 1)}{h} = \frac{2hx + h^2}{h} = \frac{h(2x + h)}{h} = 2x + h.$$

41. For $f(x) = \frac{1}{x-1}$ we have:

a. $f(x + h) = \frac{1}{x+h-1}$

b.
$$\frac{f(x+h) - f(x)}{h} = \frac{\frac{1}{x+h-1} - \frac{1}{x-1}}{h} = \frac{\frac{x-1-(x+h-1)}{(x+h-1)(x-1)}}{h}$$

$$= \frac{-h}{h(x+h-1)(x-1)} = \frac{-1}{(x+h-1)(x-1)}.$$

43. a. (ii)   **b.** (v)   **c.** (vi)   **d.** (iv)   **e.** (i)   **f.** (iii)

45. a., d.

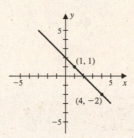

b. $d = \sqrt{(4-1)^2 + (-2-1)^2} = \sqrt{18} = 3\sqrt{2}$ **c.** $\left(\frac{4+1}{2}, \frac{-2+1}{2}\right) = \left(\frac{5}{2}, -\frac{1}{2}\right)$

e. $m = \frac{-2-1}{4-1} = -\frac{3}{3} = -1$; $y - 1 = -(x-1)$ implies $y = -x + 2$

47. a., d.

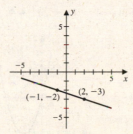

b. $d = \sqrt{(-1-2)^2 + (-2-(-3))^2} = \sqrt{10}$ **c.** $\left(\frac{-1+2}{2}, \frac{-2-3}{2}\right) = \left(\frac{1}{2}, -\frac{5}{2}\right)$

e. $m = \frac{-3-(-2)}{2-(-1)} = -\frac{1}{3}$; $y - (-2) = -\frac{1}{3}(x - (-1))$ implies $y = -\frac{1}{3}x - \frac{7}{3}$

49. a. $m = 4$; $y - 0 = 4(x - 0)$ implies $y = 4x$ **b.** $m = -\frac{1}{4}$; $y = -\frac{1}{4}x$

51. $-7x - 5y = -1$ implies $y = -\frac{7}{5}x + \frac{1}{5}$

    a. $m = -\frac{7}{5}$; $y - (-3) = -\frac{7}{5}(x - (-1))$ implies $y = -\frac{7}{5}x - \frac{22}{5}$

    b. $m = \frac{5}{7}$; $y + 3 = \frac{5}{7}(x + 1)$ implies $y = \frac{5}{7}x - \frac{16}{7}$

53. a. For $y = (x - 1)^2 - 2$, we have the following graph.

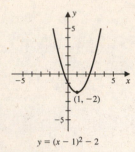

$y = (x - 1)^2 - 2$

b. The range is $[-2, \infty)$.
c. The x-intercepts occur when $0 = (x - 1)^2 - 2$ so $x = 1 \pm \sqrt{2}$. The y-intercept is $(0, -1)$.
d. The minimum value of y is -2 when $x = 1$.

55. a. For $y = -(x + 1)^2 - 1$, we have the following graph.

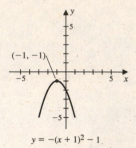

$(-1, -1)$

$y = -(x + 1)^2 - 1$

b. The range is $(-\infty, -1]$.
c. There are no x-intercepts. The y-intercept is $(0, -2)$.
d. The maximum value of y is -1 when $x = -1$.

57. a. We have $y = x^2 - 4x = x^2 - 4x + 4 - 4 = (x - 2)^2 - 4$.

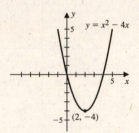

$y = x^2 - 4x$

$(2, -4)$

b. The range is $[-4, \infty)$.

c. x-intercepts: $(0, 0)$, $(4, 0)$ since $0 = x^2 - 4x = x(x - 4)$ implies $x = 0$, $x = 4$;

y-intercept: $(0, 0)$;

d. A minimum occurs at the point $(2, -4)$.

59. a. We have $y = 2x^2 - 12x + 18 = 2(x^2 - 6x) + 18 = 2(x^2 - 6x + 9 - 9) + 18 = 2(x - 3)^2$.

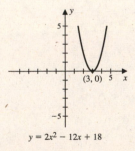

$(3, 0)$

$y = 2x^2 - 12x + 18$

b. The range is $[0, \infty)$.

c. x-intercepts: $(3, 0)$; y-intercept: $(0, 18)$;

d. A minimum occurs at the point $(3, 0)$.

61. a. We have

$$y = -\frac{1}{2}x^2 + 3x - 3 = -\frac{1}{2}(x^2 - 6x) - 3 = -\frac{1}{2}\left(x^2 - 6x + 9 - 9\right) - 3 = -\frac{1}{2}(x-3)^2 + \frac{3}{2}.$$

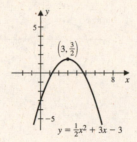

b. The range is $\left(-\infty, \frac{3}{2}\right]$.

c. x-intercepts: $\left(3 + \sqrt{3}, 0\right)$, $\left(3 - \sqrt{3}, 0\right)$, since $-\frac{1}{2}(x-3)^2 = -\frac{3}{2}$ implies $(x-3)^2 = 3$ and $x = 3 \pm \sqrt{3}$. y-intercept: $(0, -3)$;

d. A maximum occurs at the point $\left(3, \frac{3}{2}\right)$.

63.   center: $(0, 0)$; radius: 4   **65.**   center: $(-2, 1)$; radius: 3

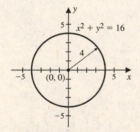 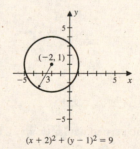

67. Completing the square on the x- and y- terms gives

$$4 = x^2 + 4x + 4 - 4 + y^2 + 2y + 1 - 1, \text{ which implies } (x + 2)^2 + (y + 1)^2 = 9.$$

center: $(-2, -1)$; radius: 3

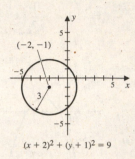

$(x + 2)^2 + (y + 1)^2 = 9$

69. To show that the points $(-3, 3)$, $(3, -5)$, and $(7, -2)$ are vertices of a rectangle, show that the line passing through $(-3, 3)$ and $(3, -5)$ is perpendicular to the line passing through $(3, -5)$ and $(7, -2)$. Computing the slopes of the lines gives

$$m_1 = \frac{3 - (-5)}{-3 - 3} = -\frac{4}{3} \quad \text{and} \quad m_2 = \frac{-5 - (-2)}{3 - 7} = \frac{3}{4}.$$

Since the slopes are negative reciprocals of one another, the lines are perpendicular. To find the fourth vertex, first note that $(7, -2)$ is 4 to the right and 3 units upward from $(3, -5)$. Moving this distance from $(-3, 3)$ places the other vertex of the rectangle at $(-3 + 4, 3 + 3) = (1, 6)$.

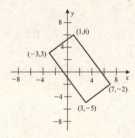

71. The line perpendicular to $y = 2x - 2$ and passing through the point $(-2, -1)$ has slope $m = -\frac{1}{2}$ and equation $y + 1 = -\frac{1}{2}(x + 2)$, or $y = -\frac{1}{2}x - 2$. The point of intersection of $y = 2x - 2$ and $y = -\frac{1}{2}x - 2$ is found from the equation

$$2x - 2 = -\frac{1}{2}x - 2, \text{ so } \frac{5}{2}x = 0 \text{ and } x = 0, y = -2.$$

Then $d((-2, -1), (0, -2)) = \sqrt{(-2 - 0)^2 + (-1 - (-2))^2} = \sqrt{5}$.

73. If the circle has center $(4, 2)$ and is tangent to the x-axis, then the circle also passes through $(2, 0)$. So the radius is 2 and the equation is $(x - 4)^2 + (y - 2)^2 = 4$.

75. We need to find $(x, 0)$ so that

$$\sqrt{(x - (-2))^2 + (0 - (-3))^2} = \sqrt{(x - 3)^2 + (0 - 5)^2}$$

so $\sqrt{(x + 2)^2 + 9} = \sqrt{(x - 3)^2 + 25}$ and $x^2 + 4x + 13 = x^2 - 6x + 34$.

This reduces to $10x = 21$, so $x = \frac{21}{10}$, and the point is $\left(\frac{21}{10}, 0\right)$.

77. a. The data points appear to lie on a parabola with vertex $(3, 10)$ that opens downward and has been vertically elongated by a factor of 4. This implies that the equation of the parabola is $y = -4(x - 3)^2 + 10$.

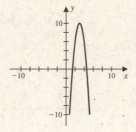

b. For each single unit change in x, the change in y is -2, so the data points represent a line with slope -2. The line passes through the point $(0, 1)$, so the equation is $y - 1 = -2(x - 0)$, that is, $y = -2x + 1$.

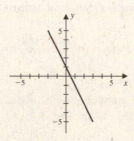

79. Let x denote the price over $36.00 for a complete dinner and let $P(x)$ denote the amount taken in by the restaurant, so that

$$P(x) = (36 + x)(200 - 4x) = 7200 + 56x - 4x^2.$$

From the graph of $P(x) = 7200 + 56x - 4x^2$ we see that the maximum return occurs at approximately $x = 7$. So the owner should charge $36 + 7 = \$43.00$ per dinner. By completing the square on the quadratic we have

$$-4x^2 + 56x + 7200 = -4(x^2 - 14x) + 7200$$

$$= -4(x^2 - 14x + 49 - 49) + 7200 = -4(x - 7)^2 + 7396,$$

so the exact maximum return is when $x = 7$.

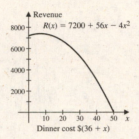

81. a. The amount of fence required is the perimeter of the plot plus the length of fence down the middle. If ℓ and w are the length and width of the plot, then the amount of fence, F, is $F = 3\ell + 2w$, where the assumption is the length of the middle section of fence is l. Since the area of the plot is 432, we have $432 = \ell w$. This implies that $w = \frac{432}{\ell}$ and that

$$F(\ell) = 3\ell + 2 \cdot \frac{432}{\ell} = 3\ell + \frac{864}{\ell}.$$

b. The domain of the function is $(0, \infty)$.

83. a. If the relationship between cost and number of units produced per day is linear, then the points $(100, 3200)$ and $(500, 9600)$ are on the line. The slope of the line is

$$m = \frac{9600 - 3200}{500 - 100} = \frac{6400}{400} = 16$$

and the equation of the line is $y - 3200 = 16(x - 100) = 16x - 1600$.
Hence $y = 16x + 1600$.

b. The slope of line indicates that for each one unit increase in the number of units produced, results in an increase of $16.00 in the cost.

c. The y-intercept is 1600 and represents the fixed costs of production.

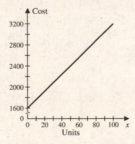

85. Part (d) gives the best representation.

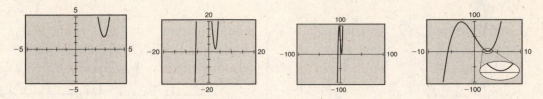

87. a. Between A and B and to the right of E.   **b.** Between C and D.   **c.** Between B and C.
   d. Between D and E.

Chapter 1 Exercises for Calculus (Page 79)

1. For each increase in x of 4 units, the y-coordinate $y = g(x)$ always increases by 12 units, so g is linear and f is quadratic. The slope of the line $y = g(x)$ is $m = \frac{12}{4} = 3$ and the line passes through $(-2, 2)$, so

$$y - 2 = 3(x + 2) \text{ implies } y = g(x) = 3x + 8.$$

The quadratic $y = f(x) = ax^2$ passes through $(-2, 2)$, so $2 = f(-2) = 4a \Rightarrow a = \frac{1}{2}$ and $f(x) = \frac{1}{2}x^2$.

3. a. The table shows that the solution is $14 \leq n \leq 16$.

n	12	13	14	15	16	17
$\dfrac{n(n+1)}{2}$	78	91	105	120	136	153

b. The table shows that the solution is $8 \leq n \leq 9$.

n	5	6	7	8	9	10
$\dfrac{n(n+1)(2n+1)}{6}$	55	91	140	204	285	385

5. The graph of the temperature might be as follows.

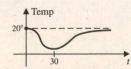

7. Graphs for the airplane are shown below.

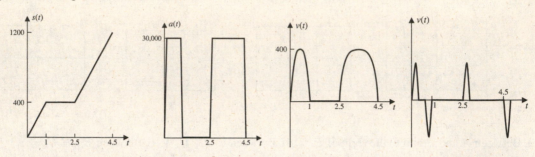

9. a. Let x be the number purchased. Then the price per item is

$$P(x) = \begin{cases} 300, & \text{if } 0 \le x \le 100, \\ 300 - (x - 100) = 400 - x, & \text{if } 100 < x \le 150, \\ 225, & \text{if } 150 < x. \end{cases}$$

b. The graph of the price per item is given below.

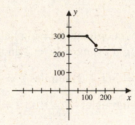

11. a. The length of the rod at a temperature t above $0°$ C, is $L(t) = \ell + a\ell t$, where ℓ is the original length of the rod and $a = 11 \times 10^{-6}$ is the coefficient of linear expansion. Since the length of the rod is 2 meters at $0°$C, the length of the rod is $L(t) = 2 + (22 \times 10^{-6})t$.

b. $L(1000) = 2 + (22 \times 10^{-6})(1000) = 2 + 22 \times 10^{-3}$.

Chapter 1 Chapter Test (Page 81)

1. False. Since $3x - 2 = 4$ implies $3x = 6$, the solution is $x = 2$.

2. False. Since $x^2 - 3x + 2 = (x - 1)(x - 2)$, the solutions are $x = 2$ and $x = 1$.

3. True.

4. True.

5. True.

6. True.

7. False. Another solution is $x = 2$.

8. True.

9. True.

10. True.

11. False. Since $x - 3 \geq 0$ implies $x \geq 3$, the domain is $[3, \infty)$.

12. False. Since the denominator cannot be 0, $x = 3$ must be excluded, resulting in the domain $(3, \infty)$.

13. True.

14. True.

15. False. The center of the circle $(x-2)^2 + y^2 = 4$ is $(2, 0)$, but the radius $\sqrt{4} = 2$.

16. False. Completing the square on x and on y gives

$$x^2 + 2x + 1 + y^2 - 4y + 4 = 4 + 1 + 4$$

or

$$(x + 1)^2 + (y - 2)^2 = 9.$$

This is the equation of a circle with radius 3, but the center is at $(-1, 2)$.

17. False. Since $y = \frac{2}{3}x - \frac{4}{3}$, the slope of the line is 2/3.

18. False. The lines $x + y = 2$ and $3x - 2y = 1$ intersect at the point $(1, 1)$.

19. False. Since $-3x + 2y = 5$ has slope 3/2 and $4y = 6x + 7$ has slope 3/2, the slopes are the same and the lines are parallel.

20. False. Since $x - 3y = 3$ has slope 1/3 and $4x - 6y = 5$ has slope 2/3, the slopes differ and the lines are not parallel.

21. False. Since $2x + y = 2$ has slope -2 and $2y + x = -1$ has slope $-1/2$, the products of the slopes $(-2)(-1/2) = 1 \neq -1$, so the lines are not perpendicular.

22. False. Since $x + 2y = 1$ has slope $-1/2$ and $-2x + y = 3$ has slope 2, the slopes differ and the lines are not parallel.

23. False. The equation of the line that has slope -3 and passes through the point $(0, 1)$ is $y - 1 = -3(x - 0)$, or $y = -3x + 1$.

24. False. The equation of the line that passes through the two points $(2, 1)$ and $(-3, 2)$ has slope $(2 - 1)/(-3 - 2) = -1/5$ and equation

$$-\frac{1}{5} = \frac{y - 1}{x - 2}$$

which simplifies to $5y + x = 7$.

25. False. Since $y = 1$ is not included, the shaded region is described by $y < 1$.

26. False. Since $y = 1$ is included, the region outside the shaded region is described by $y \geq 1$.

27. True.

28. True.

29. True.

30. True.

31. True.

32. False. The parabola $y = -(x + 1)^2 - 2$ has a maximum point at $(-1, -2)$.

33. True.

34. True.

35. True.

36. True.

37. False. The graph of an even function is symmetric with respect to the y-axis and the graph of an odd function is symmetric with respect to the origin.

38. False. The graph of $y = -f(x)$ is the reflection of the graph of $y = f(x)$ about the x-axis.

39. False. A curve will describe the graph of a function provided every vertical line crosses the curve at most one time.

40. False. The graph of $y = (x - 1)^2 + 2$ is obtained by shifting the graph of $y = x^2$ to the right 1 unit and upward 2 units.

41. True.

42. True.

43. True.

44. True.

45. True.

CHAPTER 2: New Functions from Old

2.1 Introduction

Now that we are equipped with a few basic functions, like the linear and quadratic functions we reviewed in the first chapter, more complicated functions can be constructed. There are a variety of methods for constructing new functions from old. The standard algebraic operations can be used to add, subtract, multiply and divide functions, but there are other methods, some we have already seen, to combine functions. Translating and scaling functions, as was done for quadratic functions, can be used to create new functions from any given function. Often, complicated functions can be decomposed into a sequence of translations and or scalings starting from a basic and familiar function. In this way a complicated problem can be reduced to solving a sequence of much simpler problems.

In this chapter we review the algebraic combinations of functions, translations and scalings of arbitrary functions, the method of combining functions called composition, and when possible, finding inverses of functions.

2.2 Other Common Functions

The Absolute Value Function

The absolute value function is defined as

$$f(x) = |x| = \begin{cases} x, & \text{if } x \geq 0 \\ -x, & \text{if } x < 0. \end{cases}$$

For $x \geq 0$, the graph of the absolute value function is the same as the straight line $y = x$. For $x < 0$, the graph is the same as $y = -x$, as shown in the figure.

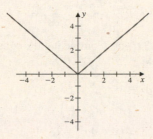

EXAMPLE 1

Use the graph of $y = |x|$ to sketch the graph of the function.
(a) $f(x) = |x - 2| + 2$ (b) $f(x) = |2x - 5|$

93

Solution (a) The shifting properties from the previous chapter applied to quadratic functions work equally well for any function. Changing x to $x - 1$ in $y = |x|$, shifts the graph to the right 1 unit. Adding 2 shifts the resulting graph upward 2 units.

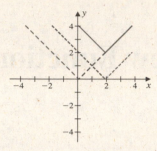

(b) First note that

$$f(x) = |2x - 5| = 2 \left| x - \frac{5}{2} \right|.$$

To sketch the graph, we reduce the problem to three easier problems.
- Sketch $y = |x|$.
- Sketch $y = 2|x|$.
- Sketch $y = 2 \left| x - \frac{5}{2} \right|$.

The graph is a vertical stretching, by a factor of 2, of the graph of $y = |x|$, followed by a shift to the right of $\frac{5}{2}$ units.

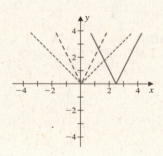

$\square$

EXAMPLE 2
Sketch the graph of $f(x) = |x^2 - 2x - 1|$.

Solution First sketch the graph of $y = x^2 - 2x - 1$. Completing the square to find the vertex of the parabola gives

$$\begin{aligned} y &= x^2 - 2x - 1 \\ &= x^2 - 2x + 1 - 1 - 1 \\ &= (x - 1)^2 - 2. \end{aligned}$$

Taking the absolute value of the y-coordinates will leave any portions of the graph that lie above the x-axis unchanged and will reflect about the x-axis any portions of the graph that lie below the x-axis. For example, if $y = x^2 - 2x - 1$ and $x = 1$, then $y = -2$ and if $y = |x^2 - 2x - 1|$ and $x = 1$, then $y = 2$. So, $(1, -2)$ is on the graph of $y = x^2 - 2x - 1$, and $(1, 2)$ is on the graph of $y = |x^2 - 2x - 1|$.

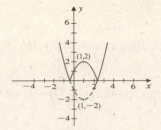

The Square Root Function

EXAMPLE 3

(a) Use the graph of $y = \sqrt{x}$ to sketch the graph of the function $f(x) = 1 - \sqrt{x + 1}$.

(b) Find the domain and range of the function.

Solution (a) The graph of $y = -\sqrt{x}$ is the reflection of the graph of $y = \sqrt{x}$ about the x-axis. Then shift the graph of $y = -\sqrt{x}$ to the left 1 unit and upward 1 unit.

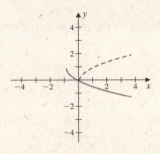

(b) The domain of the function is all real numbers satisfying

$$x + 1 \geq 0$$
$$x \geq -1.$$

In interval notation, the domain is $[-1, \infty)$. From the graph, we see that the range is the interval $(-\infty, 1]$.

The Greatest Integer Function

The greatest integer function is defined by

$$\lfloor x \rfloor = \text{the largest integer less than or equal to } x.$$

EXAMPLE 4

Evaluate the greatest integer.

(a) $\lfloor 2 \rfloor$

(b) $\lfloor 1.6 \rfloor$

(c) $\lfloor -4 \rfloor$

(d) $\lfloor -3.5 \rfloor$

Solution (a) The greatest integer less than or equal to an integer is the integer itself, so $\lfloor 2 \rfloor = 2$.
(b) Since $1 \leq 1.6 < 2$, we have $\lfloor 1.6 \rfloor = 1$.
(c) Since -4 is an integer, we have $\lfloor -4 \rfloor = -4$.
(d) Since $-4 \leq -3.5 < -3$, the smaller integer less than -3.5 is -4, we have $\lfloor -3.5 \rfloor = -4$. □

EXAMPLE 5
Sketch the graph of $f(x) = x + \lfloor x \rfloor$.

Solution First write the definition for $f(x)$ on a selection of intervals. The meaning of $f(x)$ depends on the values of $\lfloor x \rfloor$.

$$\lfloor x \rfloor = \begin{cases} 0, & \text{if } 0 \leq x < 1, \\ 1, & \text{if } 1 \leq x < 2, \\ 2, & \text{if } 2 \leq x < 3, \\ 3, & \text{if } 3 \leq x < 4, \\ -1, & \text{if } -1 \leq x < 0, \\ -2, & \text{if } -2 \leq x < -1 \end{cases} \qquad f(x) = x + \lfloor x \rfloor = \begin{cases} x, & \text{if } 0 \leq x < 1, \\ x+1, & \text{if } 1 \leq x < 2, \\ x+2, & \text{if } 2 \leq x < 3, \\ x+3, & \text{if } 3 \leq x < 4, \\ x-1, & \text{if } -1 \leq x < 0, \\ x-2, & \text{if } -2 \leq x < -1 \end{cases}$$

Adding the graphs of $y = x$ and $y = \lfloor x \rfloor$ shown on the left gives the graphs of $y = f(x) = x + \lfloor x \rfloor$ shown on the right.

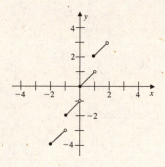

□

Exercise Set 2.2 (Page 91)

1. $f(x) = |x - 4|$

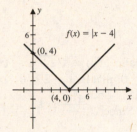

3. $f(x) = |x + 2| - 2$

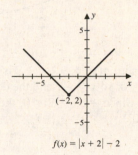

5. $f(x) = -2|x|$

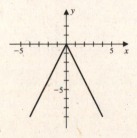

7. $f(x) = -3|x - 1| + 1$

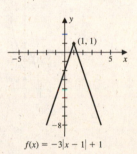

9. a. $g(x) = \sqrt{x} + 3$

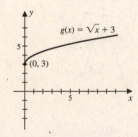

b. Domain: $g(x) = \sqrt{x} + 3$ is defined when $x \geq 0$, so the domain is $[0, \infty)$.
Range: Since $\sqrt{x} + 3 \geq 3$, the range is $[3, \infty)$.

11. a. $g(x) = \sqrt{x + 2} - 2$

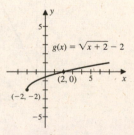

b. Domain: $g(x) = \sqrt{x + 2} - 2$ is defined when $x + 2 \geq 0$, so $x \geq -2$ and the domain is $[-2, \infty)$.
Range: Since $\sqrt{x + 2} - 2 \geq -2$, the range is $[-2, \infty)$.

13. a. $g(x) = -\sqrt{x + 2}$

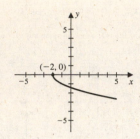

b. Domain: $g(x) = -\sqrt{x + 2}$ is defined when $x + 2 \geq 0$, so $x \geq -2$ and the domain is $[-2, \infty)$. Range: Since $-\sqrt{x + 2} \leq 0$, the range is $(-\infty, 0]$.

15. a. $g(x) = \sqrt{2 - x} - 1 = \sqrt{-(x - 2)} - 1$

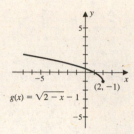

b. Domain: $g(x) = \sqrt{2 - x} - 1$ is defined when $2 - x \geq 0$, that is, $x - 2 \leq 0$, so $x \leq 2$ and the domain is $(-\infty, 2]$. Range: Since $\sqrt{2 - x} - 1 \geq -1$, the range is $[-1, \infty)$.

17. a. $g(x) = x^3 + 2$ **b.** $g(x) = x^3 - 2$ **c.** $g(x) = (x + 2)^3$ **d.** $g(x) = (x - 2)^3$
e. $g(x) = 3x^3$ **f.** $g(x) = -3x^3$

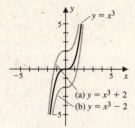 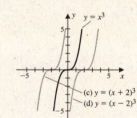 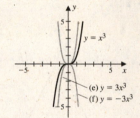

19. $f(x) = 2x - 3$

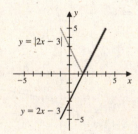

21. $f(x) = -x^2 - 4$

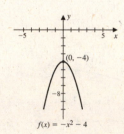 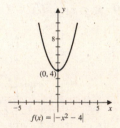

23. $f(x) = -(x-1)^2 + 1$

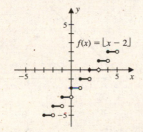

$f(x) = -(x-1)^2 + 1$

$f(x) = |-(x-1)^2 + 1|$

25.

$$f(x) = x^2 - 6x + 7$$
$$= x^2 - 6x + 9 - 9 + 7$$
$$= (x-3)^2 - 2$$

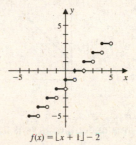

$f(x) = x^2 - 6x + 7$

$f(x) = |x^2 - 6x + 7|$

27. $f(x) = \lfloor x - 2 \rfloor$

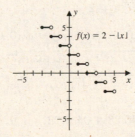

$f(x) = \lfloor x - 2 \rfloor$

29. $f(x) = \lfloor x + 1 \rfloor - 2$

$f(x) = \lfloor x + 1 \rfloor - 2$

31. $f(x) = 2 - \lfloor x \rfloor$

$f(x) = 2 - \lfloor x \rfloor$

33. a. $y = f(x) + 1$ **b.** $y = f(x - 2)$ **c.** $y = 2f(x)$ **d.** $y = f(2x)$
   e. $y = -f(x)$ **f.** $y = f(-x)$ **g.** $y = |f(x)|$ **h.** $y = f(|x|)$

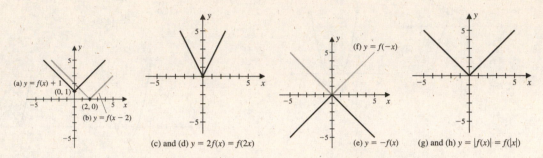

35. a. The graphs of parts (a), (b), and (c) are shown in left, center, and right, respectively.

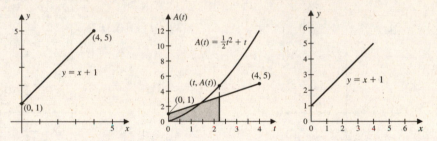

b. The area under the curve is the area of a trapezoid, which implies

$$A(t) = \frac{1}{2}(1 + (t + 1)) \cdot t = t + \frac{1}{2}t^2.$$

c.
$$d(x) = d((0, 0), (x, f(x))) = d((0, 0), (x, x + 1))$$
$$= \sqrt{(x - 0)^2 + (x + 1 - 0)^2} = \sqrt{x^2 + x^2 + 2x + 1}$$
$$= \sqrt{2x^2 + 2x + 1}$$

37. Domain$(f) = (0, 13]$; Range$(f) = \{0.88, 1.05, 1.22, .., 2.92\} = \{0.88 + 0.17n, n = 0, 1, ..., 13\}$

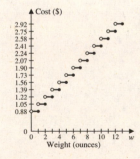

39. a. For approximating $\sqrt{2}$ start with $a = 1.4$. The table gives four iterations.

a	$b = \frac{1}{2}\left(a + \frac{2}{a}\right)$
1.4	1.414285714
1.414285714	1.414213564
1.414213564	1.414213562
1.414213562	1.414213562

So $\sqrt{2} \approx 1.414313562 \approx 1.414$.

b. For approximating $\sqrt{13}$ start with $a = 3.5$. The table gives four iterations.

a	$b = \frac{1}{2}\left(a + \frac{13}{a}\right)$
3.5	3.607142857
3.607142857	3.605551627
3.605551627	3.605551276
3.605551276	3.605551276

So $\sqrt{3} \approx 3.605551276 \approx 3.606$.

Notice that the number of decimal places of accuracy approximately doubles with each iteration of the technique. This provides an extremely efficient technique for determining square roots.

41. a. Let $f_0(x) = \lfloor x - +0.5 \rfloor$ and n be an integer.

If $n \leq x < n + 0.5$, then $n + 0.5 \leq x + 0.5 < n + 1$, so $f_0(x) = n$.

If $n + 0.5 \leq x < n + 1$, then $n + 1 \leq x + 0.5 < n + 1.5$, so $f_0(x) = n + 1$.

b. Let $f_1(x) = (0.1)\lfloor 10x + 0.5 \rfloor$ and n and m be an integers, with $0 \leq m \leq 9$.

If

$$n + m/10 \leq x < n + m/10 + 0.05, \quad \text{then} \quad 10n + m \leq 10x < 10n + m + 0.5,$$

so

$$10n + m \leq 10x + 0.5 < 10n + m + 1 \quad \text{and} \quad \lfloor 10x + 0.5 \rfloor = 10n + m.$$

If

$$n + m/10 + 0.05 \leq x < n + (m + 1)/10. \quad \text{then} \quad 10n + m + 0.5 \leq 10x < 10n + m + 1,$$

so

$$10n + m + 1 \leq 10x + 0.5 < 10n + m + 1.5 \quad \text{and} \quad \lfloor 10x + 0.5 \rfloor = 10n + m + 1.$$

Hence, in either case,

$$f_1(x) = \frac{1}{10}\lfloor 10x + 0.5 \rfloor$$

rounds to the nearest decimal point.

c. Let $f_n(x) = (0.1)^n \lfloor 10^n x + 0.5 \rfloor$. By an argument similar to that in (b), this will round an real number to n decimal places.

43. a. $7.97, rounded to nearest cent.

b. $9.11, rounded to nearest cent.

c. $12.26, rounded to nearest cent.

....

2.3   Arithmetic Combinations of Functions

Suppose that f and g are functions and x is in the domain of both. Then $f(x)$ and $g(x)$ are both real numbers, and the quantities $f(x) + g(x)$, $f(x) - g(x)$, $f(x) \cdot g(x)$, and if $g(x) \neq 0$, $\frac{f(x)}{g(x)}$, are also real numbers. We can define the *arithmetic combinations* of two functions by these formulas.

- Addition: $(f + g)(x) = f(x) + g(x)$

- Subtraction: $(f - g)(x) = f(x) - g(x)$

- Multiplication: $(fg)(x) = f(x) \cdot g(x)$

- Division: $\left(\frac{f}{g}\right)(x) = \frac{f(x)}{g(x)}$

As you might expect, the symbols $f + g$, $f - g$, fg, and f/g are the names used for the new functions, whereas the expressions on the right of the equation indicate how to evaluate these functions at specific inputs.

For x to be in the domain of $f + g$, $f - g$, or fg, the value x must be in both f and g. So, if the domain of the function f is A and the domain of g is B, the domain of $f + g$, $f - g$, and fg consists of all real numbers common to both domains which is the intersection $A \cap B$. The domain of f/g consists of those values in $A \cap B$, excluding any values of x for which $g(x) = 0$.

EXAMPLE 1
Let $f(x) = x^2$ and $g(x) = x + 1$. Find $f + g$, $f - g$, fg, and f/g, and give the domains of each new function.

Solution

$$(f + g)(x) = f(x) + g(x) = x^2 + x + 1$$
$$(f - g)(x) = f(x) - g(x) = x^2 - (x + 1) = x^2 - x - 1$$
$$(fg)(x) = f(x) \cdot g(x) = x^2(x + 1) = x^3 + x^2$$
$$\left(\frac{f}{g}\right)(x) = \frac{f(x)}{g(x)} = \frac{x^2}{x + 1}$$

Since the domain of f is the set of all real numbers, and the domain of g is the set of all real numbers, the domain of $f + g$, $f - g$ and fg is the set of all real numbers. The domain of f/g is the set of all real numbers satisfying $g(x) = x + 1 \neq 0$, that is, $x \neq -1$. Hence the domain is $(-\infty, -1) \cup (-1, \infty)$. $\quad\square$

EXAMPLE 2

Let $f(x) = 1/x$ and $g(x) = \sqrt{x+2}$. Find $f+g$, $f-g$, fg, and f/g and give the domains of each new function.

Solution

$$(f+g)(x) = f(x) + g(x) = \frac{1}{x} + \sqrt{x+2}$$

$$(f-g)(x) = f(x) - g(x) = \frac{1}{x} - \sqrt{x+2}$$

$$(fg)(x) = f(x) \cdot g(x) = \frac{1}{x} \cdot \sqrt{x+2} = \frac{\sqrt{x+2}}{x}$$

$$\left(\frac{f}{g}\right)(x) = \frac{f(x)}{g(x)} = \frac{\frac{1}{x}}{\sqrt{x+2}} = \frac{1}{x\sqrt{x+2}}$$

- Domain of f : All real numbers except $x = 0$, that is, $(-\infty, 0) \cup (0, \infty)$.
- Domain of g : All x for which $\sqrt{x+2}$ is defined, so

$$x + 2 \geq 0, \quad x \geq -2, \quad \text{and we have } [-2, \infty).$$

- Domain of $f+g$, $f-g$, fg : The intersection of the domains of f and g. Since the interval $[-2, \infty)$ contains 0, the domain is $[-2, 0) \cup (0, \infty)$.
- Domain of f/g : All x in $[-2, 0) \cup (0, \infty)$, except those values that make the denominator of f/g equal to 0. Since

$$x\sqrt{x+2} = 0 \quad \text{implies} \quad x = 0 \quad \text{or} \quad x = -2,$$

and 0 has already been removed, we remove -2 and the domain of f/g is $(-2, 0) \cup (0, \infty)$.

$\square$

EXAMPLE 3

Let $f(x) = x^2 + x - 2$ and $g(x) = x - 1$. Find $f+g$, $f-g$, fg, and f/g and give the domains of each new function.

Solution

$$(f+g)(x) = f(x) + g(x) = x^2 + x - 2 + x - 1 = x^2 + 2x - 3$$

$$(f-g)(x) = f(x) - g(x) = x^2 + x - 2 - (x - 1)$$

$$= x^2 + x - 2 - x + 1 = x^2 - 1$$

$$(fg)(x) = f(x) \cdot g(x) = (x^2 + x - 2)(x - 1)$$

$$= x^3 - x^2 + x^2 - x - 2x + 2 = x^3 - 3x + 2$$

$$\left(\frac{f}{g}\right)(x) = \frac{f(x)}{g(x)} = \frac{x^2 + x - 2}{x - 1}$$

$$= \frac{(x-1)(x+2)}{x-1} = x + 2$$

- Domains of f and g : All real numbers.
- Domain of $f+g$, $f-g$, fg : All real numbers.
- Domain of f/g : It appears from the simplification for f/g, that the domain is all real numbers. This is *not* the case, since for x to be in the domain of f/g it must be in both the domains of f and g and $g(x)$ *must be nonzero*. Since $g(1) = 0$ $x = 1$, is removed from the domain of f/g. The domain of f/g is $(-\infty, 1) \cup (1, \infty)$.

□

EXAMPLE 4
Let
$$f(x) = \begin{cases} x^2 + 1, & \text{if } x \geq 0 \\ x, & \text{if } x < 0 \end{cases} \quad \text{and} \quad g(x) = \begin{cases} x + 1, & \text{if } x \geq 0 \\ -1, & \text{if } x < 0. \end{cases}$$

Find $f + g$, $f - g$, fg, and f/g and give the domains of each new function.

Solution The functions are plotted in the figure.

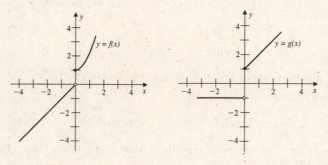

To combine these two functions, consider the definitions in the two separate parts of the domains of f and g, that is, for $x \geq 0$ and for $x < 0$.

$$(f + g)(x) = \begin{cases} x^2 + x + 2, & \text{if } x \geq 0 \\ x - 1, & \text{if } x < 0 \end{cases}$$

$$(f - g)(x) = \begin{cases} x^2 - x, & \text{if } x \geq 0 \\ x + 1, & \text{if } x < 0 \end{cases}$$

$$(fg)(x) = \begin{cases} x^3 + x^2 + x + 1, & \text{if } x \geq 0 \\ -x, & \text{if } x < 0 \end{cases}$$

$$\left(\frac{f}{g}\right)(x) = \begin{cases} \frac{x^2 + 1}{x + 1}, & \text{if } x \geq 0 \\ -x, & \text{if } x < 0 \end{cases}$$

- Domains of f and g : All real numbers, since all real numbers are specified in the two part definitions.
- Domain of $f + g$, $f - g$, fg : All real numbers.
- Domain of f/g : The quotient in the first part of the definition is undefined when $x = -1$. However, this part of the definition is only valid for $x \geq 0$, so this is not a problem. The domain is the set of all real numbers.

□

EXAMPLE 5
Let $f(x) = \sqrt{x^2 - 1}$ and $g(x) = \sqrt{9 - x^2}$. Find $f + g$, $f - g$, fg, and f/g and give the domains of each new function.

Solution

$$(f + g)(x) = f(x) + g(x) = \sqrt{x^2 - 1} + \sqrt{9 - x^2}$$

$$(f - g)(x) = f(x) - g(x) = \sqrt{x^2 - 1} - \sqrt{9 - x^2}$$

$$(fg)(x) = f(x) \cdot g(x) = \sqrt{x^2 - 1} \cdot \sqrt{9 - x^2}$$

$$\left(\frac{f}{g}\right)(x) = \frac{f(x)}{g(x)} = \frac{\sqrt{x^2 - 1}}{\sqrt{9 - x^2}}$$

- Domain of f : The expression under the radical must be greater than or equal to 0. To solve the inequality, factor and set up a chart.

$$x^2 - 1 = (x - 1)(x + 1) \geq 0$$

Since $x^2 - 1 = 0$ for $x = \pm 1$, the domain of f is $(-\infty, -1] \cup [1, \infty)$.

- Domain of g : The expression under the radical must be greater than or equal to 0.

$$9 - x^2 = (3 - x)(3 + x) \geq 0$$

Since $9 - x^2 = 0$ for $x = \pm 3$, the domain of g is $[-3, 3]$.

- Domain of $f + g$, $f - g$, fg : $((-\infty, -1] \cup [1, \infty)) \cap [-3, 3] = [-3, -1] \cup [1, 3]$.
- Domain of f/g : Remove the values that make the denominator 0, that is, $g(x) = 0$. Solve

$$\sqrt{9 - x^2} = 0$$
$$9 - x^2 = 0$$
$$x = \pm 3.$$

The domain of f/g is $(-3, -1] \cup [1, 3)$.

$\square$

EXAMPLE 6

Functions f and g are defined by $f(x) = x^2 - 9$ and $g(x) = x + 3$. Sketch the graph of $y = f(x)/g(x)$, and sketch the graph of $y = x - 3$. How do the graphs differ?

Solution Factoring $f(x)$ gives

$$\frac{f(x)}{g(x)} = \frac{x^2 - 9}{x + 3} = \frac{(x + 3)(x - 3)}{x + 3} = x - 3, \text{ for } x \neq -3.$$

The important observation here is that the quotient function $f(x)/g(x)$ is not defined at $x = -3$ since the denominator is 0 when $x = -3$. The fact that the fraction simplifies does not change the fact that $\frac{x^2 - 9}{x + 3}$ is undefined for $x = -3$. The only difference in the graphs of $y = x - 3$, shown at left, and $y = f(x)/g(x)$, shown at right, is that the point $(-3, -6)$ is not included on the graph of $y = f(x)/g(x)$.

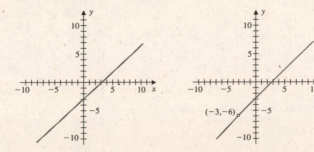

□

The Reciprocal Graphing Technique

Properties of the function $g(x) = \frac{1}{f(x)} = [f(x)]^{-1}$, called the *reciprocal of f,* can be obtained from the graph of the function f. It is most important to determine the values of x that make $f(x) = 0$.

EXAMPLE 7

Let $f(x) = x^2 + 2x - 3$. Use the results about the graph of the reciprocal of a function to sketch the graph of $g(x) = 1/f(x)$.

Solution The points where $f(x) = 0$, which is where the graph of $y = f(x)$ crosses the x-axis, are the points where $g(x)$ is undefined.

$$f(x) = x^2 + 2x - 3 = 0$$
$$(x + 3)(x - 1) = 0$$
$$x = -3, x = 1.$$

The graph of $f(x) = x^2 + 2x - 3 = (x + 1)^2 - 4$ in the figure shows

$$f(x) > 0, \text{ for } x > 1 \text{ or } x < -3,$$

and

$$f(x) < 0, \text{ for } -3 < x < 1.$$

As x gets close to -3 or 1, $f(x)$ gets close to 0, so the magnitude of $g(x) = \frac{1}{f(x)}$ becomes arbitrarily large. This implies that

$$g(x) \to \infty \quad \text{as} \quad x \to -3^-$$
$$g(x) \to -\infty \quad \text{as} \quad x \to -3^+$$
$$g(x) \to -\infty \quad \text{as} \quad x \to 1^-$$
$$g(x) \to \infty \quad \text{as} \quad x \to 1^+.$$

As $f(x)$ gets large in magnitude, $g(x) = \dfrac{1}{f(x)}$ gets very small.

$$f(x) \to \infty \text{ as } x \to \infty \quad \text{so} \quad g(x) \to 0 \text{ as } x \to \infty$$
$$f(x) \to \infty \text{ as } x \to -\infty \quad \text{so} \quad g(x) \to 0 \text{ as } x \to -\infty$$

The vertex of the parabola is $(-1, -4) = (-1, f(-1))$, so the point $\left(-1, \frac{1}{f(-1)}\right) = (-1, -\frac{1}{4})$ is on the graph of $y = g(x)$. This is generally a good additional point to plot.

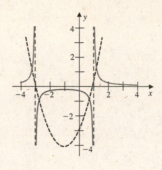

EXAMPLE 8
Sketch the graphs of the functions in the order given and observe the difference in the graph that each successive complication introduces.

(a) $f_1(x) = x^2$ (b) $f_2(x) = (x+1)^2$ (c) $f_3(x) = x^2 + 2x$

(d) $f_4(x) = |x^2 + 2x|$ (e) $f_5(x) = \frac{1}{|x^2+2x|}$ (f) $f_6(x) = \frac{-1}{|x^2+2x|}$

Solution (a) The graph of $y = f_1(x)$ is a straight line passing through $(0, 1)$ and $(-1, 0)$.
(b) The graph of $y = f_2(x)$ is a parabola with vertex at $(-1, 0)$ and opening upward.
(c) Since

$$f_3(x) = x^2 + 2x$$
$$= x^2 + 2x + 1 - 1$$
$$= (x+1)^2 - 1,$$

the graph of $y = f_3(x)$ is obtained by shifting the graph of $y = f_2(x)$ downward 1 unit.
(d) The graph of $y = f_4(x) = |x^2 + 2x|$ is obtained by reflecting about the x-axis any points (x, y) with $y = x^2 + 2x < 0$.
(e) To obtain the graph of

$$y = f_5(x) = \frac{1}{|x^2 + 2x|},$$

use the reciprocal graphing technique.
- Solve $x^2 + 2x = 0$.

$$x^2 + 2x = 0$$
$$x(x+2) = 0$$
$$x = 0, x = -2$$

- Test $f_5(x)$ near 0 and -2.

$$f_5(x) \to \infty \text{ as } x \to 0^-$$
$$f_5(x) \to \infty \text{ as } x \to 0^+$$
$$f_5(x) \to \infty \text{ as } x \to -2^-$$
$$f_5(x) \to \infty \text{ as } x \to -2^+$$

• Plot the point when $x = -1$ to find the vertex of $y = x^2 + x$.

$$f_5(-1) = 1$$

(f) The graph of

$$y = f_6(x) = \frac{-1}{|x^2 + 2x|}$$

is a reflection about the x-axis of the graph of

$$y = f_5(x) = \frac{1}{|x^2 + 2x|}.$$

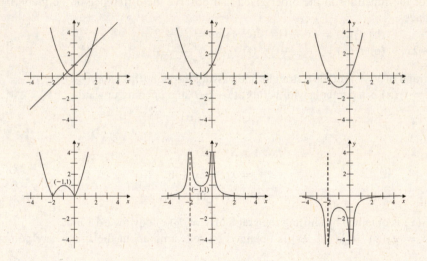

□

Exercise Set 2.3 (Page 100)

1.   For $f(x) = 2x$ and $g(x) = x^2 + 1$ we have

$(f + g)(x) = f(x) + g(x) = 2x + (x^2 + 1) = x^2 + 2x + 1$;

$(f - g)(x) = f(x) - g(x) = 2x - (x^2 + 1) = 2x - x^2 - 1 = -x^2 + 2x - 1$;

$(f \cdot g)(x) = f(x)g(x) = 2x(x^2 + 1) = 2x^3 + 2x$;

$(f/g)(x) = \frac{f(x)}{g(x)} = \frac{2x}{x^2+1}$.

The domain of $f + g$, $f - g$, $f \cdot g$, and f/g is

Domain$(f) \cap$ Domain$(g) = (-\infty, \infty) \cap (-\infty, \infty) = (-\infty, \infty)$.

3. For $f(x) = \frac{1}{x}$ and $g(x) = \frac{x}{x-2}$ we have

$$(f+g)(x) = f(x) + g(x) = \frac{1}{x} + \frac{x}{x-2} = \frac{(x-2)+x^2}{x(x-2)} = \frac{x^2+x-2}{x^2-2x};$$

$$(f-g)(x) = f(x) - g(x) = \frac{1}{x} - \frac{x}{x-2} = \frac{(x-2)-x^2}{x(x-2)} = \frac{-x^2+x-2}{x^2-2x};$$

$$(f \cdot g)(x) = f(x)g(x) = \frac{x}{x(x-2)} = \frac{1}{x-2};$$

$$(f/g)(x) = \frac{f(x)}{g(x)} = \frac{\frac{1}{x}}{\frac{x}{x-2}} = \frac{1}{x} \cdot \frac{x-2}{x} = \frac{x-2}{x^2}.$$

Domain of $f : x \neq 0$; Domain of $g : x - 2 \neq 0$ implies $x \neq 2$

So $\text{Domain}(f) \cap \text{Domain}(g) = \{x \mid x \neq 0 \text{ and } x \neq 2\} = (-\infty, 0) \cup (0, 2) \cup (2, \infty)$.

The domain of $f + g$, $f - g$, and $f \cdot g$ is $\text{Domain}(f) \cap \text{Domain}(g) = (-\infty, 0) \cup (0, 2) \cup (2, \infty)$.

The domain of f/g is $\text{Domain}(f) \cap \text{Domain}(g)$ excluding any values that make the denominator 0, which is $x = 0$, which has already been excluded. So $\text{Domain}(f/g) = (-\infty, 0) \cup (0, 2) \cup (2, \infty)$.

5. For $f(x) = \sqrt{x+1}$ and $g(x) = \sqrt{3-x}$ we have

$$(f+g)(x) = f(x) + g(x) = \sqrt{x+1} + \sqrt{3-x};$$

$$(f-g)(x) = f(x) - g(x) = \sqrt{x+1} - \sqrt{3-x};$$

$$(f \cdot g)(x) = f(x)g(x) = \sqrt{x+1} \cdot \sqrt{3-x} = \sqrt{(x+1)(3-x)} = \sqrt{3 + 2x - x^2};$$

$$(f/g)(x) = \frac{f(x)}{g(x)} = \frac{\sqrt{x+1}}{\sqrt{3-x}}.$$

Domain of $f : x + 1 \geq 0$ implies $x \geq -1$; Domain of $g : 3 - x \geq 0$ implies $x \leq 3$

So $\text{Domain}(f) \cap \text{Domain}(g) = [-1, 3]$.

The domain of $f + g$, $f - g$, and $f \cdot g$ is $\text{Domain}(f) \cap \text{Domain}(g) = [-1, 3]$.

The domain of f/g is $\text{Domain}(f) \cap \text{Domain}(g)$, excluding any values that make the denominator 0, which is $x = 3$, so $\text{Domain}(f/g) = [-1, 3)$.

7. For $f(x) = \begin{cases} -1, & \text{if } x < 0 \\ 1, & \text{if } x \geq 0 \end{cases}$ and $g(x) = \begin{cases} 1, & \text{if } x < 0 \\ 0, & \text{if } x \geq 0 \end{cases}$ we perform the operations separately for $x < 0$ and for $x \geq 0$.

$$(f+g)(x) = f(x) + g(x) = \begin{cases} 0, & \text{if } x < 0 \\ 1, & \text{if } x \geq 0 \end{cases};$$

$$(f-g)(x) = f(x) - g(x) = \begin{cases} -2, & \text{if } x < 0 \\ 1, & \text{if } x \geq 0 \end{cases};$$

$$(f \cdot g)(x) = f(x)g(x) = \begin{cases} -1, & \text{if } x < 0 \\ 0, & \text{if } x \geq 0 \end{cases};$$

The quotient is defined only for $x < 0$, since $g(x) = 0$ for $x \geq 0$. So $(f/g)(x) = -1$, for $x < 0$.

The domain of $f + g$, $f - g$, $f \cdot g$ is $(-\infty, \infty)$ and the domain of f/g is $(-\infty, 0)$.

9. $g(x) = x - 3; h(x) = \frac{1}{x-3}$

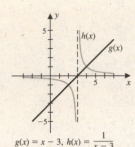

$g(x) = x - 3, h(x) = \frac{1}{x-3}$

11. $g(x) = |x|; h(x) = \frac{1}{|x|}$

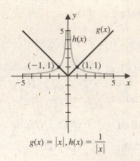

$g(x) = |x|, h(x) = \frac{1}{|x|}$

13. $g(x) = x^2 - 1; h(x) = \frac{1}{x^2-1}$

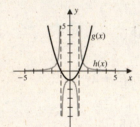

15.

$$g(x) = x^2 - 4x + 3$$
$$= x^2 - 4x + 4 - 4 + 3$$
$$= (x - 2)^2 - 1$$

and

$$h(x) = \frac{1}{(x - 2)^2 - 1}$$

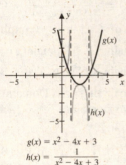

$g(x) = x^2 - 4x + 3$
$h(x) = \frac{1}{x^2 - 4x + 3}$

17. The figure shows the graphs of $f + g$ and $f - g$.

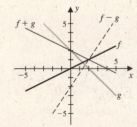

19. The figure shows the graphs of $f + g$ and $f - g$.

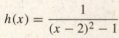

21. We have $f(x) = \frac{x^2-4}{x-2} = \frac{(x-2)(x+2)}{x-2} = x + 2$, for $x \neq 2$. The denominator of the original expression is 0 at

$x = 2$, making the quotient undefined. To graph $y = f(x) = \frac{x^2-4}{x-2}$, simply graph $y = x + 2$ and remove the point with x-coordinate 2. That is, remove the point $(2, 4)$. So the graphs of $f(x) = \frac{x^2-4}{x-2}$ and $g(x) = x + 2$ are identical except for the point $(2, 4)$ removed from the graph of $y = f(x)$.

23. The figure shows the graphs of $y = \sqrt{x - 1}$, $y = \sqrt{x + 2}$, and $y = \sqrt{x - 1} + \sqrt{x + 2}$.

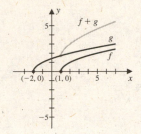

25. The figure shows the graphs of $y = x^3 - 2x^2 - x + 2$, $y = x^3 - 7x + 6$, and $y = 2x^3 - 2x^2 - 8x + 8$.

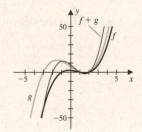

27. Let $R(t)$ denote the rate of spread of the disease at time t, $I(t)$ the number of infected, and $H(t)$ the number healthy. Then $R(t) = kI(t)H(t) = \dfrac{1}{74550}(70)(4930) \approx 4.6$.

. . . .

2.4   Composition of Functions

The *composition* of two functions is a special way of combining the two processes. For example, the function given by $f(x) = (x^2 + 1)^3$ can be interpreted as

$$x \to \boxed{x}^2 + 1 = x^2 + 1 \to \boxed{x^2 + 1}^3 = (x^2 + 1)^3.$$

So x is input to the process that squares the number and adds one, and then this new value is used as input to the process that cubes a number. The final result is the process that does both operations in the prescribed order.

The composition of the two functions f and g is written as

$$(f \circ g)(x) = f(g(x)),$$

with the domain being all x in the domain of g so that $g(x)$ is in the domain of f.

EXAMPLE 1
Let $f(x) = x^2 + 1$ and $g(x) = \sqrt{x}$.
(a) Find $(f \circ g)(2)$ and $(g \circ f)(-3)$.
(b) Write expressions for $f(g(x))$ and $g(f(x))$.
(c) Can $g \circ f$ and $f \circ g$ both be evaluated at -2?

Solution (a)

- $(f \circ g)(2)$: First evaluate g at 2, giving $g(2) = \sqrt{2}$. Then substitute this value into f.

$$(f \circ g)(2) = f(g(2)) = f(\sqrt{2}) = (\sqrt{2})^2 + 1 = 3$$

- $(g \circ f)(-3)$:

$$(g \circ f)(-3) = g(f(-3)) = g(10) = \sqrt{10}$$

(b) Replacing x with $g(x)$ in the definition of f gives

$$f(g(x)) = f(\sqrt{x}) = (\sqrt{x})^2 + 1 = x + 1,$$

and replacing x with $f(x)$ in the definition of g gives

$$g(f(x)) = g(x^2 + 1) = \sqrt{x^2 + 1}.$$

(c) We have

$$(g \circ f)(-2) = g(f(-2)) = g(5) = \sqrt{5}.$$

But $f \circ g$ can not be evaluated at -2 since -2 is not in the domain of g.

EXAMPLE 2
Let $f(x) = \sqrt{x - 6}$ and $g(x) = x^2 + 5x$. Find $f(g(x))$ and specify the domain of $f \circ g$.

Solution

$$f(g(x)) = f(x^2 + 5x) = \sqrt{x^2 + 5x - 6}$$

The domain of $f \circ g$ is the set of all x in the domain of g with $g(x)$ also in the domain of f. The domain of g is all real numbers and the domain of f is the interval $[6, \infty)$. So x is in the domain of $f \circ g$ provided that

$$x^2 + 5x \geq 6$$
$$x^2 + 5x - 6 \geq 0$$
$$(x - 1)(x + 6) \geq 0.$$

The sign chart for the inequality shows the domain of $f \circ g$ is $(-\infty, -6] \cup [1, \infty)$.

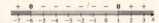

$\square$

EXAMPLE 3
(a) Use the graphs of f and g in the figure to evaluate each expression.
(i) $(f \circ g)(0)$
(ii) $(g \circ f)(0)$
(iii) $(g \circ f)(2)$
(iv) $(f \circ g)(2)$
(b) Determine the values of x with $(f \circ g)(x) = 0$.

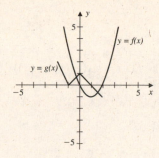

Solution (a) (i) Reading from the graph, we have $g(0) = 1$ and

$$(f \circ g)(0) = f(g(0)) = f(1) = -1.$$

(ii) Similarly,

$$(g \circ f)(0) = g(f(0)) = g(0) = 1.$$

(iii)

$$(g \circ f)(2) = g(f(2)) = g(0) = 1$$

(iv)

$$(f \circ g)(2) = f(g(2)) = f(-1) = 3$$

(b) To solve $(f \circ g)(x) = f(g(x)) = 0$ first observe the values which make f zero. Then find those x that make $g(x)$ one of these values. Note that

$$f(a) = 0 \quad \text{when} \quad a = 0 \quad \text{or} \quad a = 2,$$

and

$$g(x) = 0 \quad \text{when} \quad x = -1 \quad \text{or} \quad x = 1, \quad \text{and} \quad g(x) = 2 \quad \text{when} \quad x = -2.$$

So $(f \circ g)(x) = 0$ when $x = -2$, $x = -1$, or $x = 1$.   □

EXAMPLE 4

Find functions f and g so that $h = f \circ g$.
(a) $h(x) = (x^5 + 3x^3 - 2x + 1)^8$
(b) $h(x) = \sqrt{x^2 - 2x + 1}$

Solution (a) A natural separation of the process h is to first perform the *inside* operation of $x^5 + 3x^3 - 2x + 1$ and then perform the *outside* operation of raising a number to the 8th power. Since we want g to be the inside operation and f to be the outside operation, we can let

$$f(x) = x^8 \quad \text{and} \quad g(x) = x^5 + 3x^3 - 2x + 1.$$

Then

$$f(g(x)) = f(x^5 + 3x^3 - 2x + 1) = (x^5 + 3x^3 - 2x + 1)^8 = h(x).$$

(b) Let $f(x) = \sqrt{x}$ and $g(x) = x^2 - 2x + 1$. Then

$$f(g(x)) = f(x^2 - 2x + 1) = \sqrt{x^2 - 2x + 1} = h(x).$$

□

EXAMPLE 5

(a) Show that $f(x) = \frac{1}{|x^2-2x|}$ can be written as $f(x) = g_4(g_3(g_2(g_1(x))))$, where $g_1(x) = (x-1)^2$, $g_2(x) = x - 1$,

$g_3(x) = |x|$, and $g_4(x) = \frac{1}{x}$.

(b) Sketch the graphs of (i) $y = g_1(x)$,

(ii) $y = g_2(g_1(x))$,

(iii) $y = g_3(g_2(g_1(x)))$,

(iv) $y = f(x)$.

Solution A strategy that can be used to graph $y = f(x)$ is to first recognize that the quadratic in the denominator can be graphed easily by writing it in standard form. Then use the reciprocal graphing technique to graph the quotient.

(a) First notice that

$$x^2 - 2x = x^2 - 2x + 1 - 1 = (x-1)^2 - 1$$

so

$$f(x) = \frac{1}{|(x-1)^2 - 1|}.$$

So

$$g_1(x) = (x-1)^2,$$

$$g_2(g_1(x)) = g_2((x-1)^2) = (x-1)^2 - 1,$$

$$g_3(g_2(g_1(x))) = g_3((x-1)^2 - 1) = |(x-1)^2 - 1|,$$

and

$$g_4(g_3(g_2(g_1(x)))) = \frac{1}{|(x-1)^2 - 1|} = \frac{1}{|x^2 - 2x|} = f(x)$$

(b)

(i) The graph of $y = g_1(x) = (x-1)^2$ is obtained by shifting $y = x^2$ to the right 1 unit.

(ii) The graph of $y = g_2(g_1(x)) = (x-1)^2 - 1$ is obtained by shifting $y = (x-1)^2$ downward 1 unit.

(iii) The graph of $y = g_3(g_2(g_1(x))) = |(x-1)^2 - 1|$ is obtained by reflecting the portions of the graph of $y = g_2(g_1(x))$ that lie below the x-axis above the x-axis.

(iv) Use the reciprocal graphing technique to plot

$$\frac{1}{|(x-1)^2 - 1|}.$$

The important points are those where $y = |(x-1)^2 - 1|$ crosses the x-axis. Solving the quadratic equal to 0 gives

$$(x-1)^2 - 1 = 0$$

$$(x-1)^2 = 1$$

$$x - 1 = \pm 1$$

$$x = 0, x = 2.$$

Near these points,

$$x^2 - 2x \to 0 \text{ as } x \to 0 \quad \text{so} \quad \frac{1}{|x^2 - 2x|} \to \infty \text{ as } x \to 0$$

$$x^2 - 2x \to 0 \text{ as } x \to 2 \quad \text{so} \quad \frac{1}{|x^2 - 2x|} \to \infty \text{ as } x \to 2.$$

Also

$$\frac{1}{|x^2 - 2x|} \to 0 \text{ as } x \to \infty \quad \text{and} \quad \frac{1}{|x^2 - 2x|} \to 0 \text{ as } x \to -\infty.$$

One useful point to plot is the vertex of the parabola $y = x^2 - 2x = (x - 1)^2 - 1$. If $x = 1$, then $|(x - 1)^2 - 1| = |-1| = 1$, so $\frac{1}{|(x^2 - 2x|} = 1$, and the point $(1, 1)$ is on the final curve.

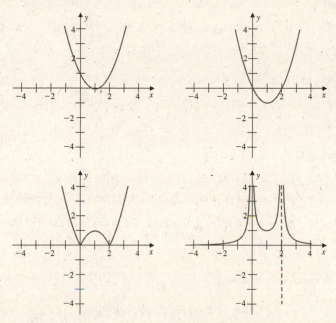

$\square$

EXAMPLE 6

A cube with sides of length 2 feet is being filled with water, and the height of the water at the end of t seconds is $h(t) = \sqrt{t}$ feet, $0 \le t \le 4$. Express the volume of the water in the cube as a function of time.

Solution The volume of the water in the cube as a function of the height of the water is

$$V(h) = l \cdot w \cdot h = 2 \cdot 2 \cdot h = 4h.$$

Since $h(t) = \sqrt{t}$, the volume as a function of time is the composition

$$V(t) = V(h(t)) = 4h(t) = 4\sqrt{t} \text{ feet}^3.$$

$\square$

Exercise Set 2.4 (Page 109)

1. For $f(x) = 2x - 3$ and $g(x) = x^2 + 2$ we have

    $$(f \circ g)(3) = f(g(3)) = f(3^2 + 2) = f(11) = 2(11) - 3 = 19.$$

3. For $f(x) = 2x - 3$ and $g(x) = x^2 + 2$ we have

    $$(f \circ g)(-3) = f(g(-3)) = f((-3)^2 + 2) = f(11) = 2(11) - 3 = 19.$$

5. For $f(x) = 2x - 3$ and $g(x) = x^2 + 2$ we have

    $$(f \circ f)(-2) = f(f(-2)) = f(2(-2) - 3) = f(-7) = 2(-7) - 3 = -17.$$

7. For $f(x) = 2x + 1$ and $g(x) = 3x - 1$ we have

    $$(f \circ g)(x) = f(g(x)) = f(3x - 1) = 2(3x - 1) + 1 = 6x - 1;$$
    $$(g \circ f)(x) = g(f(x)) = g(2x + 1) = 3(2x + 1) - 1 = 6x + 2.$$

    Since the domain of f and g is the set of all real numbers, the domain of each of the compositions is also the set of all real numbers.

9. For $f(x) = \frac{1}{x}$ and $g(x) = x^2 + 2x$ we have

    $$(f \circ g)(x) = f(g(x)) = f(x^2 + 2x) = \frac{1}{x^2 + 2x}$$

    and

    $$(g \circ f)(x) = g(f(x)) = g\left(\frac{1}{x}\right) = \left(\frac{1}{x}\right)^2 + 2\left(\frac{1}{x}\right) = \frac{1}{x^2} + \frac{2}{x} = \frac{1 + 2x}{x^2}.$$

    Domain $f : \{x \mid x \neq 0\} = (-\infty, 0) \cup (0, \infty)$; Domain $g : (-\infty, \infty)$; Domain
    $f \circ g : \{x \mid x^2 + 2x = x(x + 2) \neq 0\} = \{x \mid x \neq 0, x \neq -2\} = (-\infty, -2) \cup (-2, 0) \cup (0, \infty)$; Domain
    $g \circ f : \{x \neq 0 \mid 1/x \text{ is defined}\} = \{x \mid x \neq 0\} = (-\infty, 0) \cup (0, \infty)$.

11. For $f(x) = \sqrt{x - 1}$ and $g(x) = x^2 - 3$ we have

    $$(f \circ g)(x) = f(g(x)) = f(x^2 - 3) = \sqrt{x^2 - 3 - 1} = \sqrt{x^2 - 4};$$
    $$(g \circ f)(x) = g(f(x)) = f(\sqrt{x - 1}) = (\sqrt{x - 1})^2 - 3 = x - 4.$$

    Domain $f : \{x \mid x - 1 \geq 0\} = [1, \infty)$; Domain $g : (-\infty, \infty)$;
    Domain $f \circ g$:
    $\{x \mid x^2 - 3 \geq 1\} = \{x \mid (x - 2)(x + 2) \geq 0\} = (-\infty, -2] \cup [2, \infty)$;
    Domain $g \circ f : \{x \mid x \geq 1 \text{ and } f(x) \text{ is defined}\} = [1, \infty)$.

13. For $f(x) = \frac{1}{x}$ and $g(x) = \frac{1}{x+1}$ we have

$$(f \circ g)(x) = f(g(x)) = f\left(\frac{1}{x+1}\right) = \frac{1}{\frac{1}{x+1}} = x + 1;$$

$$(g \circ f)(x) = g(f(x)) = g\left(\frac{1}{x}\right) = \frac{1}{\frac{1}{x}+1} = \frac{1}{\frac{1+x}{x}} = \frac{x}{x+1}.$$

Domain $f : \{x \mid x \neq 0\} = (-\infty, 0) \cup (0, \infty)$;

Domain $g : \{x \mid x + 1 \neq 0\} = (-\infty, -1) \cup (-1, \infty)$;

Domain $f \circ g : \{x \neq -1 \mid \frac{1}{x+1} \neq 0\} = (-\infty, -1) \cup (-1, \infty)$;

Domain $g \circ f : \{x \neq 0 \mid \frac{1}{x} \neq -1\} = (-\infty, -1) \cup (-1, 0) \cup (0, \infty)$.

15. The inside operation of $h(x) = (3x^2 - 2)^4$ is $3x^2 - 2$ and the outside operation is raising to the fourth power, x^4. So define $g(x) = 3x^2 - 2$ and $f(x) = x^4$. This gives

$$(f \circ g)(x) = f(g(x)) = f(3x^2 - 2) = (3x^2 - 2)^4 = h(x).$$

17. The inside operation of $h(x) = \sqrt[3]{x - 4}$ is $x - 4$ and the outside operation is the cube root of a number, $\sqrt[3]{x}$. So define $g(x) = x - 4$ and $f(x) = \sqrt[3]{x}$. This gives

$$(f \circ g)(x) = f(g(x)) = f(x - 4) = \sqrt[3]{x - 4} = h(x).$$

19. Define the first operation to be $x + 2$ and then take the reciprocal, so let $g(x) = x + 2$ and $f(x) = \frac{1}{x}$. Then

$$(f \circ g)(x) = f(g(x)) = f(x + 2) = \frac{1}{x + 2} = h(x).$$

21. a. (i) $(f \circ g)(-2) = f(g(-2)) = f(1) = 2$ (ii) $(g \circ f)(-2) = g(f(-2)) = g(0) = 0$

(iii) $(g \circ f)(2) = g(f(2)) = g(0) = 0$ (iv) $(f \circ g)(2) = f(g(2)) = f(-1) = -2$

b. (i) $(f \circ g)(x) = f(g(x)) = 0$ whenever $g(x) = -2, 0, 2$, that is, $x = -4, 0, 4$

(ii) $(g \circ f)(x) = g(f(x)) = 0$ whenever $f(x) = 0$, that is, $x = -2, 0, 2$

23. The figure shows the graphs of $y = f(2x)$, $y = f(-2x)$, $y = f(\frac{1}{2}x)$, and $y = f(-\frac{1}{2}x)$.

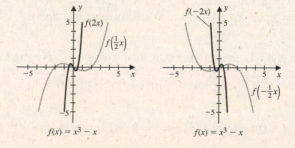

25. a. $y = x - 1$ **b.** $y = (x-1)^2$ **c.** $y = x^2 - 2x = (x-1)^2 - 1$ **d.** $y = |x^2 - 2x|$
   e. $y = \frac{1}{|x^2-2x|}$ **f.** $y = -\frac{1}{|x^2-2x|}$

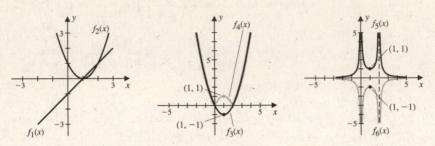

27. a. For $f(x) = \frac{1}{|x^2+2x-1|}$, completing the square on the quadratic in the denominator gives

$$f(x) = \frac{1}{|x^2 + 2x + 1 - 1 - 1|} = \frac{1}{|(x+1)^2 - 2|}.$$

Then letting $g_1(x) = (x+1)^2$, $g_2(x) = x - 2$, $g_3(x) = |x|$ and $g_4(x) = \frac{1}{x}$ we have

$$g_4(g_3(g_2(g_1(x)))) = g_4(g_3(g_2((x+1)^2))) = g_4(g_3((x+1)^2 - 2))$$

$$= g_4\left(\left|(x+1)^2 - 2\right|\right) = \frac{1}{|(x+1)^2 - 2|} = f(x).$$

b. The graphs of $y = g_1(x)$, $y = g_2(g_1(x))$, $y = g_3(g_2(g_1(x)))$, and $y = f(x)$ are shown in the figures.

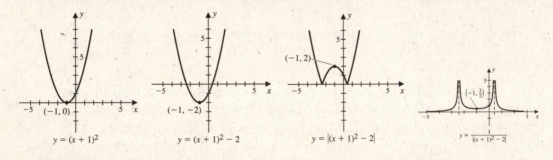

29. Let $f(x) = x$ and $g(x) = x + 1$. Then $(f \circ g)(x) = f(g(x)) = f(x+1) = x + 1$ and $(g \circ f)(x) = g(f(x)) = g(x) = x + 1$. There are many other examples.

31. If f is an odd function and g is an even function, then $f(-x) = -f(x)$ and $g(-x) = g(x)$. So

$$(f \circ g)(-x) = f(g(-x)) = f(g(x)) = (f \circ g)(x)$$

and

$$(g \circ f)(-x) = g(f(-x)) = g(-f(x)) = g(f(x)) = (g \circ f)(x).$$

33. Let $f(x) = ax + b$ and $g(x) = cx + d$. Then

$$(f \circ g)(x) = a(cx + d) + b = acx + ad + b$$
$$(g \circ f)(x) = c(ax + b) + d = acx + bc + d.$$

a. $(f \circ g)(x) = (g \circ f)(x)$ implies $ad + b = bc + d$.

b. $(f \circ g)(x) = f(x)$ implies $ac = a$ and $ad + b = b$ so $c = 1$ and $d = 0$.

c. $(f \circ g)(x) = g(x)$ implies $ac = c$ and $ad + b = d$ so $a = 1, b = 0$.

35. $g(x) = 3f(x - 1) + 2$

37. The volume of a sphere of radius r is $V(r) = \frac{4}{3}\pi r^3$. So $V(t) = \frac{4}{3}\pi(3 + 0.01t)^3$.

39. a. The volume of a sphere of radius r is $V(r) = \frac{4}{3}\pi r^3$ so

$$V(t) = V(r(t)) = \frac{4}{3}\pi\left(3\sqrt{t} + 5\right)^3 \text{ centimeters}^3.$$

b. The surface area is $S(r) = 4\pi r^2$ so

$$S(t) = S(r(t)) = 4\pi\left(3\sqrt{t} + 5\right)^2 \text{ centimeters}^2.$$

41. a. The value after 2 years is $(1.04)^2 x + P(1 + 1.04)$, and the value after 3 years is $(1.04)^3 x + P(1 + 1.04 + (1.04)^2)$.

b. Let $V_n(x)$ be the value of the account after n years. Then $V_n(x) = V^n(x)$, that is, V composed with itself n times.

. . . .

2.5 Inverse Functions

The *inverse* function, when it exists (not all functions have an inverse function), is the process that undoes the original operation of the function. For example, the function $f(x) = 2x - 1$ sends the real number $x = 2$ to the real number $f(2) = 3$. The inverse function for f sends the value 3 back to the originating value 2.

One-to-One Functions

Functions that have inverses are those that are *one-to-one*. A function is one-to-one provided no two different x values are sent to the same y value. This can be written as

$$f(x_1) \neq f(x_2), \text{ if } x_1 \neq x_2,$$

or equivalently, as

$$\text{if } f(x_1) = f(x_2), \text{ then } x_1 = x_2.$$

Geometrically, a function will be one-to-one provided that every horizontal line that crosses the graph of the function crosses it in only one place. For example, $f(x) = 2x - 1$ is one-to-one, but $g(x) = x^2$ is not one-to-one.

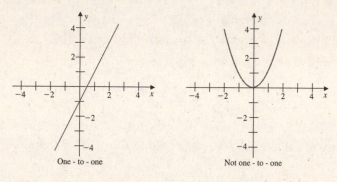

One - to - one   Not one - to - one

EXAMPLE 1

Determine if the function is one-to-one.
(a) $f(x) = x^3 - 1$ (b) $f(x) = 2 - |x|$

Solution (a) Use the second equivalent statement to verify the function is one-to-one. If

$$f(x_1) = f(x_2),$$

then

$$x_1^3 - 1 = x_2^3 - 1$$
$$x_1^3 = x_2^3$$
$$x_1 = x_2.$$

Therefore, $f(x_1) = f(x_2)$ only when $x_1 = x_2$ which means that the function is one-to-one.
(b) We can determine this function is not one-to-one by inspection. From its graph, we see for example, that

$$f(1) = 2 - |1| = 1 = 2 - |-1| = f(-1),$$

so $f(1) = f(-1)$ and the function is not one-to-one.   □

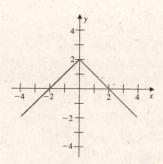

The inverse function of the function f is generally denoted f^{-1}, but some standard functions have special notations for their inverses.

EXAMPLE 2

Assume the function is one-to-one, and find the indicated value.

(a) If $f(x) = x + 3$, find $f^{-1}(1)$.   (b) If $f(x) = \dfrac{1}{x-1}$, find $f^{-1}(1)$.

Solution (a) For any function f and its inverse f^{-1},

$$f^{-1}(f(x)) = x.$$

To find $f^{-1}(1)$, we need the number x with $f(x) = 1$. If

$$f(x) = x + 3 = 1, \quad \text{then} \quad x = -2,$$

so $f^{-1}(1) = -2$.

(b) If

$$\frac{1}{x-1} = 1, \quad \text{then} \quad 1 = x - 1 \quad \text{and} \quad x = 2.$$

So $f^{-1}(1) = 2$.   □

Process for Finding an Inverse Function

When a function is one-to-one, the following process can be used to find the inverse.

Procedure for finding the Inverse of f

- Set $y = f(x)$.

- Solve for x in terms of y.

- Interchange the variables x and y.

An important relationship between the graph of a function and its inverse is they are reflections of each other across the line $y = x$.

EXAMPLE 3

Find the inverse of the one-to-one function. Sketch $y = f(x)$ and $y = f^{-1}(x)$.

(a) $f(x) = \dfrac{2x-1}{3}$ (b) $f(x) = \sqrt{x-1}$

Solution (a)

- Step 1:

$$y = \frac{2x-1}{3}$$

- Step 2:

$$y = \frac{2x-1}{3}$$
$$3y = 2x - 1$$
$$3y + 1 = 2x$$
$$x = \frac{3y+1}{2}$$

• Step 3:

$$f^{-1}(x) = \frac{3x + 1}{2}$$

Note that for any value of x

$$f^{-1}(f(x)) = f^{-1}\left(\frac{2x - 1}{3}\right) = \frac{3\left(\frac{2x-1}{3}\right) + 1}{2} = \frac{2x - 1 + 1}{2} = x.$$

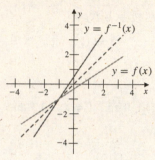

(b)

• Step 1:

$$y = \sqrt{x - 1}$$

• Step 2:

$$y^2 = x - 1$$
$$x = y^2 + 1$$

• Step 3:

$$f^{-1}(x) = x^2 + 1$$

In the figure the function is shown for $x \geq 1$ and the inverse is shown only for $x \geq 0$, because of the domain restriction. The domain of the function is $x \geq 1$ and $y = \sqrt{x - 1} \geq 0$. The range of y-values become the domain of the inverse function. The information is summarized in the table.

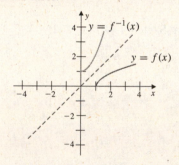

Function	Domain	Range
$f(x) = \sqrt{x - 1}$	$[1, \infty)$	$[0, \infty)$
$f^{-1}(x) = x^2 + 1$	$[0, \infty)$	$[1, \infty)$

□

Important relationships between a function and its inverse are:

- The domain of f is the range of f^{-1}.

- The range of f is the domain of f^{-1}.

EXAMPLE 4

Let $f(x) = x^2 - 4x$.

(a) Show that the function is not one-to-one.

(b) Determine a subset of the domain of the function on which it is one-to-one, and find its inverse on this restricted domain.

Solution By completing the square, we can write the function as

$$f(x) = x^2 - 4x = x^2 - 4x + 4 - 4 = (x-2)^2 - 4.$$

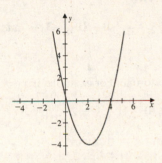
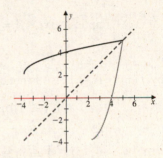

(a) The figure shows that the function does not satisfy the horizontal line test since many horizontal lines will cross the graph twice. For example, $f(0) = 0$ and $f(4) = 0$.

(b) If the domain is restricted to $[2, \infty)$, then the new function will be one-to-one. The inverse is then found using the three-step process.

- Step 1:
$$y = x^2 - 4x = (x-2)^2 - 4$$

- Step 2:
$$y + 4 = (x-2)^2$$
$$x - 2 = \sqrt{y+4}$$
$$x = \sqrt{y+4} + 2$$

- Step 3:
$$f^{-1}(x) = \sqrt{x+4} + 2$$

The domain of f^{-1}, which is the same as the range of f, is the interval $[-4, \infty)$. □

Exercise Set 2.5 (Page 119)

1. The function is one-to-one since every horizontal line crosses the curve in only one point.

3. The function is not one-to-one since many horizontal lines cross the curve in more than one point.

5. The function is not one-to-one since every horizontal line that crosses the curve intersects the curve in two points.

7. The function defined by $f(x) = 3x - 4$ is one-to-one, since

$$f(x_1) = f(x_2) \quad \text{implies} \quad 3x_1 - 4 = 3x_2 - 4 \quad \text{so} \quad 3x_1 = 3x_2 \quad \text{and} \quad x_1 = x_2.$$

9. The function defined by $f(x) = |x - 1| + 1$ is not one-to-one. For example, if $x_1 = 0$ and $x_2 = 2$, then

$$f(x_1) = f(0) = |0 - 1| + 1 = 2$$

and

$$f(x_2) = f(2) = |2 - 1| + 1 = 2.$$

So we have found x_1 and x_2, with $x_1 \neq x_2$ and $f(x_1) = f(x_2)$.

11. The function defined by $f(x) = x^4 + 1$ is not one-to-one, since, for example, $f(-1) = 2 = f(1)$.

13. The function defined by $f(x) = \begin{cases} 2x^2, & \text{if } x \geq 0 \\ x - 2 & \text{if } x < 0 \end{cases}$ is one-to-one, since the graph satisfies the horizontal line test. That is, every horizontal line that crosses the graph does so in only one point.

15. The figure shows the graph of the function and its inverse.

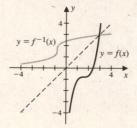

17. The figure shows the graph of the function and its inverse.

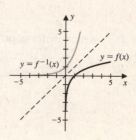

19. Since $f(x) = 2$, $f^{-1}(2) = x$. Solving $x + 1 = 2$ implies $x = 1$. So $f^{-1}(2) = 1$.

21. Since $f(x) = 1$, $f^{-1}(1) = x$. Solving $2x - 2 = 1$ implies $x = \frac{3}{2}$. So $f^{-1}(1) = \frac{3}{2}$.

23. Since $f(x) = 2$, $f^{-1}(2) = x$. Solving $\frac{1}{x} = 2$ implies $x = \frac{1}{2}$. So $f^{-1}(2) = \frac{1}{2}$.

25. Let $f(x) = 2x - 1$. Then f is one-to-one since

$$f(x_1) = f(x_2) \quad \text{implies} \quad 2x_1 - 1 = 2x_2 - 1 \quad \text{so} \quad x_1 = x_2.$$

To find f^{-1} : $y = 2x - 1$ implies $2x = y + 1$ so $x = \frac{y+1}{2}$, and $f^{-1}(x) = \frac{x+1}{2} = \frac{x}{2} + \frac{1}{2}$.

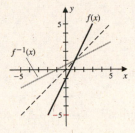

27. Let $f(x) = \sqrt{x - 3}$. Then f is one-to-one since

$$f(x_1) = f(x_2) \quad \text{implies} \quad \sqrt{x_1 - 3} = \sqrt{x_2 - 3} \quad \text{so} \quad x_1 - 3 = x_2 - 3 \quad \text{and} \quad x_1 = x_2.$$

To find f^{-1} : $y = \sqrt{x - 3}$ implies $x - 3 = y^2$ so $x = y^2 + 3$, and $f^{-1}(x) = x^2 + 3$.
Note: The domain of f is $[3, \infty)$, which is also the range of f^{-1} and the range of f is $[0, \infty)$, which is also the domain of f^{-1}.

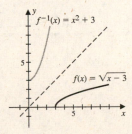

29. Let $f(x) = \frac{1}{2x}$. Then f is one-to-one since

$$f(x_1) = f(x_2) \quad \text{implies} \quad \frac{1}{2x_1} = \frac{1}{2x_2} \quad \text{so} \quad x_1 = x_2.$$

To find f^{-1}: $y = \frac{1}{2x}$ implies $2xy = 1$ so $x = \frac{1}{2y}$, and $f^{-1}(x) = \frac{1}{2x}$.

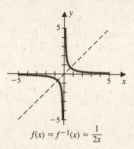

$$f(x) = f^{-1}(x) = \frac{1}{2x}$$

31. Let $f(x) = \frac{1}{\sqrt{x}}$. Then f is one-to-one since

$$f(x_1) = f(x_2) \quad \text{implies} \quad \frac{1}{\sqrt{x_1}} = \frac{1}{\sqrt{x_2}} \quad \text{so} \quad \sqrt{x_2} = \sqrt{x_1} \quad \text{and} \quad x_1 = x_2.$$

To find f^{-1}: $y = \frac{1}{\sqrt{x}}$ implies $\sqrt{x} = \frac{1}{y}$ so $x = \frac{1}{y^2}$, and $f^{-1}(x) = \frac{1}{x^2}$.

Note: The domain of f is $(0, \infty)$, which is also the range of f^{-1} and the range of f is $(0, \infty)$, which is also the domain of f^{-1}.

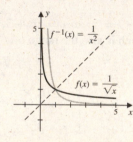

$$f^{-1}(x) = \frac{1}{x^2}$$

$$f(x) = \frac{1}{\sqrt{x}}$$

33. Let $f(x) = 1 + x^3$. Then f is one-to-one since

$$f(x_1) = f(x_2) \quad \text{implies} \quad 1 + x_1^3 = 1 + x_2^3 \quad \text{so} \quad x_1^3 = x_2^3 \quad \text{and} \quad x_1 = x_2.$$

To find f^{-1}: $y = 1 + x^3$ implies $x^3 = y - 1$ so $x = (y-1)^{\frac{1}{3}}$, and $f^{-1}(x) = (x-1)^{\frac{1}{3}}$.

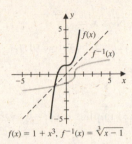

$$f(x) = 1 + x^3, \ f^{-1}(x) = \sqrt[3]{x-1}$$

35. Let $f(x) = x^2 + 1$, $x \geq 0$. Then f is one-to-one since

$$f(x_1) = f(x_2) \quad \text{implies} \quad x_1^2 + 1 = x_2^2 + 1 \quad \text{so} \quad x_1^2 = x_2^2 \quad \text{and} \quad x_1 = x_2,$$

since $x \geq 0$ (otherwise we could only conclude that $|x_1| = |x_2|$).
To find f^{-1}: $y = x^2 + 1$ implies $y - 1 = x^2$ so $x = \sqrt{y-1}$, and $f^{-1}(x) = \sqrt{x-1}$.
Note: The domain of f is $[0, \infty)$, which is also the range of f^{-1} and the range of f is $[1, \infty)$, which is also the domain of f^{-1}.

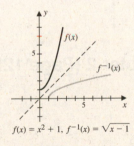

$$f(x) = x^2 + 1, \ f^{-1}(x) = \sqrt{x-1}$$

37. a. The graph of the function $f(x) = |2 - x|$ is V-shaped, opens upward and the point is at $(2, 0)$. The line $y = 1$, for example, must cross the curve two times, and $|2 - x| = 1$ implies $2 - x = 1$ or $2 - x = -1$ so $x = 1$ or $x = 3$. So the function is not one-to-one.

b. The function is one-to-one on the interval $[2, \infty)$.

$\underline{f^{-1} \text{ for } f \text{ restricted on } [2, \infty)}$: Since $2 - x \leq 0$ for $x \geq 2$,

$$y = -(2 - x) = x - 2 \text{ implies } x = y + 2, \text{ so } f^{-1}(x) = x + 2.$$

The domain of f^{-1} is the range of f, which equals $[0, \infty)$.
$f^{-1}(x) = x + 2$, for $x \geq 0$

39. a. We have $f(x) = x^2 - 2x = x(x - 2)$, and $f(x) = 0$ for both $x = 0$ and $x = 2$, so f is not one-to-one. The graph is a parabola that opens upward and has vertex $(1, -1)$.

b. The function is one-to-one on the interval $[1, \infty)$.

f^{-1} for f restricted on $[1, \infty)$:

$$y = x^2 - 2x = x^2 - 2x + 1 - 1 = (x-1)^2 - 1 \text{ implies } y + 1 = (x-1)^2$$

if and only if

$$x - 1 = \sqrt{y+1} \text{ implies } x = \sqrt{y+1} + 1, \quad \text{so } f^{-1}(x) = \sqrt{x+1} + 1.$$

The domain of f^{-1} is $[-1, \infty)$ which is also the range of f.

$f^{-1}(x) = 1 + \sqrt{x+1}$, for $x \geq -1$

41. For $f(x) = mx + b$

$$f(x_1) = f(x_2) \text{ implies } mx_1 + b = mx_2 + b \text{ so } mx_1 = mx_2$$

if and only if

$$x_1 = x_2 \text{ provided m} \neq 0.$$

So for all $m \neq 0$ the function $f(x) = mx + b$ is one-to-one. Solving for x gives

$$y = mx + b \text{ implies } y - b = mx \text{ implies } x = \frac{y-b}{m}, \quad \text{so } f^{-1}(x) = \frac{x-b}{m}.$$

....

Review Exercises for Chapter 2 (Page 120)

1. a. The graph of $y = \sqrt{x-2} + 1$ is (ii).

   b. The graph of $y = \sqrt{x} + 1$ is (i).

   c. The graph of $y = -\sqrt{x+1} + 3$ is (iv).

   d. The graph of $y = -2\sqrt{x}$ is (iii).

3. a. For $f(x) = 2x - 1$ and $g(x) = x + 2$,

$$(f+g)(x) = 3x + 1; \ (f-g)(x) = x - 3;$$

$$(fg)(x) = 2x^2 + 3x - 2; \ (f/g)(x) = \frac{2x-1}{x+2};$$

$$(f \circ g)(x) = 2x + 3; \ (g \circ f)(x) = 2x + 1;$$

$$(f \circ f)(x) = 4x - 3; \ (g \circ g)(x) = x + 4.$$

b. The domain of $f + g$, $f - g$, fg, $f \circ g$, $g \circ f$, $f \circ f$ and $g \circ g$ is $(-\infty, \infty)$. The domain of f/g is $(-\infty, -2) \cup (-2, \infty)$.

5. a. For $f(x) = x + 1$ and $g(x) = \sqrt{x-1}$,

$$(f+g)(x) = x + 1 + \sqrt{x-1}; \ (f-g)(x) = x + 1 - \sqrt{x-1};$$

$$(fg)(x) = (x+1)\sqrt{x-1}; \ (f/g)(x) = \frac{x+1}{\sqrt{x-1}};$$

$$(f \circ g)(x) = \sqrt{x-1} + 1; \ (g \circ f)(x) = \sqrt{x};$$

$$(f \circ f)(x) = x + 2; \ (g \circ g)(x) = \sqrt{\sqrt{x-1}-1};$$

b. The domain of $f+g$, $f-g$, fg and $f \circ g$ is $[1, \infty)$. The domain of f/g is $(1, \infty)$. The domain of $g \circ f$ is $[0, \infty)$. The domain of $f \circ f$ is $(-\infty, \infty)$. The domain of $g \circ g$ is $[2, \infty)$.

7. One choice is $f(x) = x^6$ and $g(x) = x^3 - 2x + 1$.
9. One choice is $f(x) = \sqrt{x}$ and $g(x) = 3x + 3$.

11. One choice is $f(x) = |x|$ and $g(x) = x^2 - 2x + 1$.

13. The graph of $f(x) = 2x + 1$ and its reciprocal are shown.

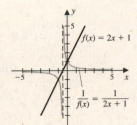

15. The graph of

$$f(x) = x^2 - 2x - 3 = (x-1)^2 - 4$$

and its reciprocal are shown.

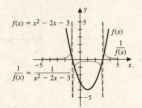

17. $f(x) = |x + 1|$

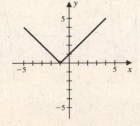

19. $f(x) = |x + 2| - 2$

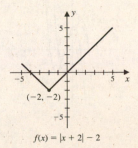

21. $g(x) = \sqrt{x+1}$

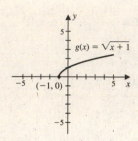

23. $g(x) = -\sqrt{x} - 2$

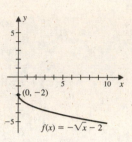

25. $f(x) = \lfloor x + 1 \rfloor$

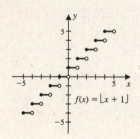

27. $f(x) = -\lfloor x \rfloor$

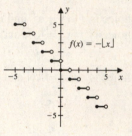

29. $f(x) = 3x - 2$

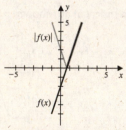

31. $f(x) = -x^2 + 3$

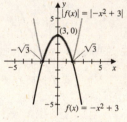

33. a. $y = f(x+1)$ **b.** $y = f(x+1) + 1$ **c.** $y = f(x-1)$ **d.** $y = f(x-1) + 2$
e. $y = f(x+1) - 1$ **f.** $y = f(x-1) - 2$ **e.** $y = f(x+1) - 1$ **f.** $y = f(x-1) - 2$
g. $y = f(2x)$ **h.** $y = f(x/2)$

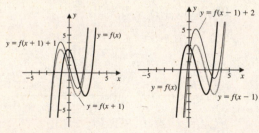

35. a. For $f(x) = 2x^2 - 3$ we have $f(-x) = 2(-x)^2 - 3 = 2x^2 - 3$;

$-f(x) = -(2x^2 - 3) = -2x^2 + 3$; $f\left(\frac{1}{x}\right) = 2\left(\frac{1}{x}\right)^2 - 3 = \frac{2}{x^2} - 3 = \frac{2-3x^2}{x^2}$;

$\frac{1}{f(x)} = \frac{1}{2x^2-3}$; $f\left(\sqrt{x}\right) = 2\left(\sqrt{x}\right)^2 - 3 = 2x - 3$; $\sqrt{f(x)} = \sqrt{2x^2 - 3}$.

b. For $f(x) = \frac{1}{x^2}$ we have $f(-x) = \frac{1}{(-x)^2} = \frac{1}{x^2}$; $-f(x) = -\frac{1}{x^2}$;

$f\left(\frac{1}{x}\right) = \frac{1}{\left(\frac{1}{x}\right)^2} = x^2$; $\frac{1}{f(x)} = \frac{1}{\frac{1}{x^2}} = x^2$;

$f\left(\sqrt{x}\right) = \frac{1}{\left(\sqrt{x}\right)^2} = \frac{1}{x}$; $\sqrt{f(x)} = \sqrt{\frac{1}{x^2}} = \frac{1}{x}$.

37. $g(x) = \frac{1}{2}f(x + 2) - 3$

39. a. For $f(x) = 2x + 1$,

$$f(x_1) = f(x_2) \text{ implies } 2x_1 + 1 = 2x_2 + 1 \text{ so } 2x_1 = 2x_2 \text{ and } x_1 = x_2.$$

The function is one-to-one. To find the inverse we have

$$y = 2x + 1 \text{ which implies } 2x = y - 1 \text{ so } x = \frac{y - 1}{2}$$

and $f^{-1}(x) = \frac{x-1}{2}$.

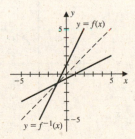

b. For $f(x) = |x - 1| + 1$, $f(0) = 2 = f(2)$, and hence the function is not one-to-one.

c. The function defined by $f(x) = 2 + x^3$ is one-to-one, since

$$f(x_1) = f(x_2) \text{ implies } 2 + x_1^3 = 2 + x_2^3 \text{ so } x_1^3 = x_2^3 \text{ and } x_1 = x_2.$$

To find the inverse we have

$$y = 2 + x^3 \text{ which implies } x^3 = y - 2 \text{ so } x = \sqrt[3]{y - 2}$$

and $f^{-1}(x) = \sqrt[3]{x - 2}$.

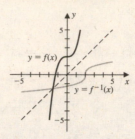

d. The graph of $f(x) = x^2 + 2x - 2$ is a parabola, so there are horizontal lines that cross the graph two times. Hence the function is not one-to-one. To show this algebraically, we have

$$f(x) = x^2 + 2x - 2 = x^2 + 2x + 1 - 1 - 2 = (x+1)^2 - 3.$$

So

$$f(x) = 0 \text{ implies } x + 1 = \pm\sqrt{3} \text{ implies } x = 1 \pm \sqrt{3}.$$

So both $f(1 + \sqrt{3}) = 0$ and $f(1 - \sqrt{3}) = 0$, which implies that the function is not one-to-one.

e. The function defined by $f(x) = \frac{1}{x^2}$ is not one-to-one, since $f(-1) = 1$ and $f(1) = 1$.

f. The function defined by $f(x) = \sqrt{x-1} + 2$ is one-to-one, since

$$f(x_1) = f(x_2) \text{ implies } \sqrt{x_1 - 1} + 2 = \sqrt{x_2 - 1} + 2 \text{ so } \sqrt{x_1 - 1} = \sqrt{x_2 - 1} \text{ implies } x_1 - 1 = x_2 - 1$$

and $x_1 = x_2$. To find the inverse we have

$$y = \sqrt{x - 1} + 2 \text{ which implies } \sqrt{x - 1} = y - 2 \text{ so } x - 1 = (y - 2)^2 \text{ implies } x = (y - 2)^2 + 1$$

and $f^{-1}(x) = (x - 2)^2 + 1$.

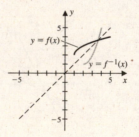

41. a. The graph of the function restricted to the subset $[2, \infty)$ is shown.

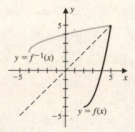

b. Since $f(x) = x^2 - 4x = x^2 - 4x + 4 - 4 = (x - 2)^2 - 4$, the vertex of the parabola is $(2, -4)$. If the domain is restricted to $[2, \infty)$, the new function is one-to-one.

c. $y = (x - 2)^2 - 4$ implies $y + 4 = (x - 2)^2$ so $x = \sqrt{y + 4} + 2$, and $f^{-1}(x) = \sqrt{x + 4} + 2$. The domain of f^{-1} is $[-4, \infty)$.

43. a. Since the cost is $C(x) = 30x + 1500$ and the selling price is $p(x) = 120 - 0.1x$, the profit function is given by

$$P(x) = xp(x) - C(x) = x(120 - 0.1x) - 30x - 1500 = -0.1x^2 + 90x - 1500.$$

b. To find the number of units sold that yields the maximum profit we find the vertex of the parabola. Then

$$-0.1x^2 + 90x - 1500 = -0.1(x^2 - 900x) - 1500$$
$$= -0.1(x^2 + 900x + 202500 - 202500) - 1500$$
$$= -0.1(x - 450)^2 + 20250 - 1500$$
$$= -0.1(x - 450)^2 + 18750.$$

So the vertex is (450, 18750) and 450 units should be produced to yield the maximum profit.

c. The maximum profit is $18,750.00.

. . . .

Chapter 2 Exercises for Calculus (Page 124)

1. The sign of $f(x)$ and the zeros of $f(x)$ give the graph of $1/f(x)$.

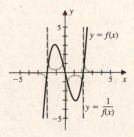

3. a. The graphs of the World population, $W(x)$, and urban population, $U(x)$, are shown.

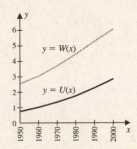

b. The graph of the non-urban population, $N(x) = W(x) - U(x)$, and the percentage of urban population, $S(x) = 100 \cdot U(x)/W(x)$, are shown.

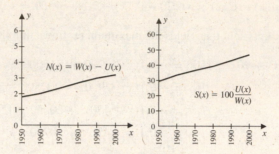

5. **a.** If x pairs of shoes are produced, the average cost for a pair of shoes is

$$\frac{C(x)}{x} = \frac{295 + 3.28x + 0.003x^2}{x} = \frac{295}{x} + 3.28 + 0.003x.$$

The graph of $y = \frac{295}{x} + 3.28 + 0.003x$ has a minimum when $x \approx 314$, so to minimize the average cost about 314 pairs of shoes should be produced.

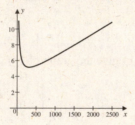

b. The revenue to the company if x pairs of shoes are sold is given by

$$R(x) = xp(x) = x\left(7.47 + \frac{321}{x}\right) = 7.47x + 321.$$

The profit is then given by

$$P(x) = R(x) - C(x) = 7.47x + 321 - (295 + 3.28x + 0.003x^2)$$
$$= 26 + 4.19x - 0.003x^2.$$

The graph of $y = 26 + 4.19x - 0.003x^2$ has a maximum when $x \approx 698$, so to maximize the profit about 698 pairs of shoes should be produced.

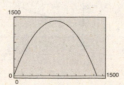

7. a. The first ten terms of the Fibonacci sequence are $1, 1, 2, 3, 5, 8, 13, 21, 34$, and 55.

    b. If d is an integer that divides both F_{n+1} and F_n, then d also divides

$$F_{n-1} = F_{n+1} - F_n.$$

Similarly, d must divide

$$F_{n-2} = F_n - F_{n-1}.$$

Proceeding in this manner, we eventually have the conclusion that d must divide $F_1 = 1$, hence $d = 1$.

9. a. We have

$$(f \circ f)(x) = f(f(x)) = \begin{cases} f(2x), & 0 \le x \le 1/2 \\ f(2 - 2x), & 1/2 < x \le 1 \end{cases}$$

$$= \begin{cases} 2(2x), & 0 \le x \le 1/4 \\ 2 - 2(2x), & 1/4 < x \le 1/2 \\ 2 - 2(2 - 2x), & 1/2 < x \le 3/4 \\ 2(2 - 2x), & 3/4 < x \le 1 \end{cases}$$

$$= \begin{cases} 4x, & 0 \le x \le 1/4 \\ 2 - 4x, & 1/4 < x \le 1/2 \\ -2 + 4x, & 1/2 < x \le 3/4 \\ 4 - 4x, & 3/4 < x \le 1 \end{cases}$$

    b. The graphs of $y = f(x)$ and $y = (f \circ f)(x) = f(f(x))$ are shown.

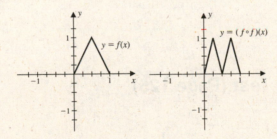

11. Let $f(x) = \frac{x+a}{x+b}$, for $a \ne b$.

    a. $\underline{f^{-1} \text{ exists}}$: First we need to show f is one-to-one. Since $a \ne b$, if $f(x_1) = f(x_2)$ then

$$\frac{x_1 + a}{x_1 + b} = \frac{x_2 + a}{x_2 + b} \text{ or } (x_1 + a)(x_2 + b) = (x_2 + a)(x_1 + b).$$

So

$$x_1 x_2 + bx_1 + ax_2 + ab = x_1 x_2 + bx_2 + ax_1 + ab$$

which implies $(b - a)x_1 = (b - a)x_2$ so $x_1 = x_2$, since $a \ne b$.

$\underline{f^{-1}(x)}$: Solving for x in the equation $y = \dfrac{x+a}{x+b}$ gives

$$y(x + b) = x + a, \quad yx - x = a - yb, \quad x(y - 1) = a - yb \quad \text{and} \quad x = \frac{a - yb}{y - 1}.$$

So

$$f^{-1}(x) = \frac{a - bx}{x - 1}.$$

b. Domain of f : $\{x \mid x \neq -b\}$; Range of f : $\{x \mid x \neq 1\}$

Domain of f^{-1} : $\{x \mid x \neq 1\}$; Range of f^{-1} : $\{x \mid x \neq -b\}$

c.

$$(f^{-1} \circ f)(x) = f^{-1}(f(x)) = f^{-1}\left(\frac{x + a}{x + b}\right) = \frac{a - b\left(\frac{x+a}{x+b}\right)}{\left(\frac{x+a}{x+b} - 1\right)}$$

$$= \frac{\frac{a(x+b)-b(x+a)}{x+b}}{\frac{x+a-(x+b)}{x+b}} = \frac{\frac{ax-bx}{x+b}}{\frac{a-b}{x+b}} = \frac{ax - bx}{a - b} = \frac{x(a - b)}{a - b} = x$$

$$(f \circ f^{-1})(x) = f(f^{-1}(x)) = f\left(\frac{a - bx}{x - 1}\right) = \frac{\frac{a-bx}{x-1} + a}{\frac{a-bx}{x-1} + b}$$

$$= \frac{\frac{a-bx+a(x-1)}{x-1}}{\frac{a-bx+b(x-1)}{x-1}} = \frac{\frac{ax-bx}{x-1}}{\frac{a-b}{x-1}} = \frac{ax - bx}{a - b} = \frac{x(a - b)}{a - b} = x$$

Chapter 2 Chapter Test (Page 125)

1. True.

2. True.

3. True.

4. False. The domain is $(-\infty, \infty)$ and the range is $(-\infty, 2]$.

5. True.

6. True.

7. False. The value of $f(2)$ is 1.

8. False. The function is constantly 2.

9. False. The function is constantly -1.

10. False. A function will have an inverse function provided every horizontal line crosses the graph of the function at most once.

11. True.

12. True.

13. True.

14. True.

15. True.

16. False. The domain is $\{x \mid x \neq -3 \text{ and } x \neq 1\}$.

17. False. $(f \circ g)(-1) = -3$

18. True.

19. False. The value of $(f \circ f)(2) = 4$.

20. True.

21. True.

22. False. It is sometimes the case that $(f \circ g)(x) - (g \circ f)(x) = 0$.

23. False. If $f(x) = x^2 - 2x + 1$ and $g(x) = \sqrt{x}$, then $(f \circ g)(4) = 1$.

24. True.

25. False. $(f + g)(2) = 6$

26. True.

27. True.

28. True.

29. True.

30. True.

31. True.

32. False. The black curve is reflected about the x-axis, and then the translations as stated are performed.

33. True.

34. False. The yellow curve could have equation $y = 2|x - 1| + 2$.

35. True.

36. True.

37. False. The green curve can be described by $y = f(x) - 2$.

38. True.

39. False. As $x \to 2^+$, $f(x) \to \infty$, but as $x \to 2^-$, $f(x) \to -\infty$.

40. True.

41. False. As $x \to 3$, the function $f(x) = \frac{2x+1}{(x-3)^2} \to \infty$.

42. False. As $x \to \infty$, the function $f(x) = \frac{1}{x} + 2 \to 2$.

CHAPTER 3: Algebraic Functions

3.1 Introduction

A *polynomial function of degree n* is a function of the form

$$P(x) = a_n x^n + a_{n-1} x^{n-1} + \cdots + a_1 x + a_0,$$

where $a_0, a_1, \ldots a_n$ are real numbers and $a_n \neq 0$. The term a_n is called the *leading coefficient*, and a_0 is called the *constant term*. The *algebraic functions* are obtained from the polynomials by any finite combination of the operations of addition, subtraction, multiplication, division, and extracting integral roots.

3.2 Polynomial Functions

EXAMPLE 1

Specify the degree, leading coefficient, and constant term of the polynomial.

(a) $P(x) = 2x^4 - x^3 + 2x^2 - x + 1$ (b) $P(x) = x^5 - 8$

(c) $P(x) = 3x - 2$ (d) $P(x) = x^{12}$

Solution

Polynomial	Degree	Leading Coefficient	Constant Term
$2x^4 - x^3 + 2x^2 - x + 1$	4	2	1
$x^5 - 8$	5	1	-8
$3x - 2$	1	3	-2
x^{12}	12	1	0

$\square$

139

Graphing Polynomial Functions

EXAMPLE 2

Use the graph of $f(x) = x^3$ shown in the figure to sketch the graph of the function.
(a) $g(x) = (x - 1)^3 + 2$ (b) $g(x) = \frac{1}{2}(x + 2)^3 - 1$

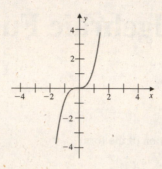

Solution (a) Because we know the general shape of the graph of $f(x) = x^3$, the graph of $y = g(x)$ can be obtained using the shifting, scaling and reflecting techniques. To sketch $y = g(x) = (x - 1)^3 + 2$, first sketch $y = (x - 1)^3$ by shifting the graph of $y = x^3$ to the right 1 unit, and then shift the resulting curve upward 2 units.

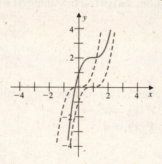

(b) First scale $y = x^3$ vertically by a factor of $\frac{1}{2}$. The new curve is below $y = x^3$ for $x > 0$, and above for $x < 0$. For example, the point $(1, 1)$ is on the graph of $y = x^3$, but the point $\left(1, \frac{1}{2}\right)$ is on the graph of $y = \frac{1}{2}x^3$. Similarly, the point $(-1, -1)$ is on the graph of $y = x^3$, but $\left(-1, -\frac{1}{2}\right)$ is on the graph of $y = \frac{1}{2}x^3$. Now shift $y = \frac{1}{2}x^3$ to the left 2 units and downward 1 unit to obtain the final graph.

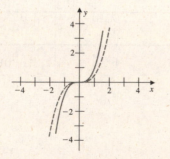 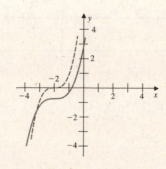

EXAMPLE 3

Use the graph of $f(x) = x^4$ shown in the figure to sketch the graph of the curve.

(a) Sketch $y = ax^4$, for $a = 2, 3, \frac{1}{2}, \frac{1}{3}$.

(b) Sketch $y = ax^4$, for $a = -1, -2, -3, -\frac{1}{2}, -\frac{1}{3}$.

(c) Sketch $y = x^4 - a$, for $a = 1, 2, -1, -2$.

(d) Sketch $y = (x - a)^4$, for $a = 1, 2, -1, -2$.

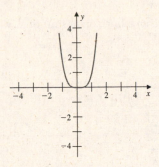

Solution (a) If the vertical line $x = 1$ is drawn to intersect the curves $y = ax^4$, for $a = 2, 3, \frac{1}{2}, \frac{1}{3}$, then it crosses the curves at the points shown in the table.

Since the curves are symmetric about the y-axis, the same relationship holds for negative x. Using one point as a model, we can draw the curves in relation to $y = x^4$, as shown in the figure.

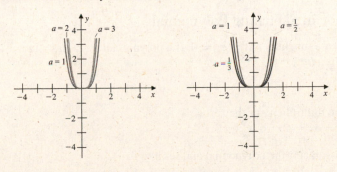

(b) The curves are the reflection about the x-axis of the curves in part (a).

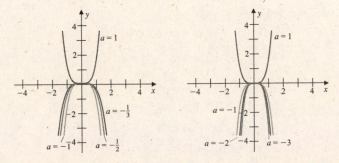

(c) If $a > 0$, then $y = x^4 - a$ is obtained from a vertical shift of $y = x^4$ downward a units. If $a < 0$, then shift upward $|a|$ units.

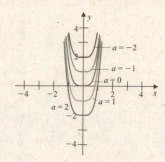

(d) If $a > 0$, then $y = (x - a)^4$ is obtained from a horizontal shift of $y = x^4$ to the right a units. If $a < 0$, then shift to the left $|a|$ units.

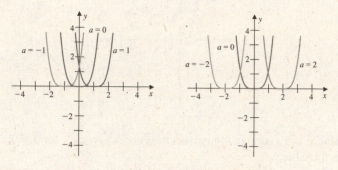

Zeros and End Behavior in Sketching Polynomials

A number c is called a *zero* of a function f if $f(c) = 0$. If a polynomial $P(x)$ has a factor of the form $(x - c)^k$, then $x = c$ is a zero of P of *multiplicity k*.

EXAMPLE 4

Find the zeros and specify the multiplicity of the zeros.
(a) $P(x) = x^3 - x$ (b) $P(x) = x^3 - 2x^2 + x$

Solution (a) To find the zeros, factor the polynomial and set it to 0.

$$P(x) = x^3 - x = x(x^2 - 1) = x(x - 1)(x + 1)$$
$$x(x - 1)(x + 1) = 0$$
$$x = 0, \ x = 1, \ x = -1$$

(b) If $P(x) = 0$, then

$$x^3 - 2x^2 + x = 0$$
$$x(x^2 - 2x + 1) = 0$$
$$x(x - 1)(x - 1) = 0$$
$$x(x - 1)^2 = 0$$
$$x = 0 \quad \text{or} \quad x = 1.$$

The zero $x = 1$ has multiplicity 2.

The zeros of a polynomial give the x-intercepts of the graph of the function.

- If the multiplicity of a zero is an even number, the graph flattens and just touches the x-axis without crossing the axis.

- If the multiplicity of the zero is an odd number greater than one, the graph flattens and then crosses the x-axis at the zero.

The information about the zeros of a polynomial together with its end behavior, gives enough information to provide a rough sketch of the graph.

The *end behavior* is what happens to the graph of the polynomial as $|x|$ becomes large, that is, at very large values of x on the positive or negative x-axis. The end behavior of a polynomial of degree n depends only on the x^n term. This is written as

$$a_n x^n + a_{n-1} x^{n-1} + \cdots + a_0 \approx a_n x^n.$$

EXAMPLE 5

Test the end behavior of the polynomial.

(a) $P(x) = x^3 + x^2 + 1$ (b) $P(x) = 5x^4 + 10x^3 + 100x^2 - x + 2$
(c) $P(x) = -3x^5 + x - 1$ (d) $P(x) = -10x^8 + x^7 + 500x^5 + x - 2$

Solution (a)

$$P(x) = x^3 + x^2 + 1 \approx x^3$$

As $|x|$ gets large, the graph behaves like the graph $y = x^3$. So, as x goes to ∞, $y = P(x) = x^3 + x^2 + 1$ also goes to ∞, and as x goes to $-\infty$, $y = P(x) = x^3 + x^2 + 1$ goes to $-\infty$. This is written as

$$P(x) \to \infty \quad \text{as} \quad x \to \infty$$
$$P(x) \to -\infty \quad \text{as} \quad x \to -\infty.$$

(b)

$$P(x) = 5x^4 + 10x^3 + 100x^2 - x + 2 \approx 5x^4$$

The end behavior is

$$P(x) \to \infty \quad \text{as} \quad x \to \infty$$
$$P(x) \to \infty \quad \text{as} \quad x \to -\infty.$$

(c)

$$P(x) = -3x^5 + x - 1 \approx -3x^5$$

The end behavior is

$$P(x) \to -\infty \quad \text{as} \quad x \to \infty$$
$$P(x) \to \infty \quad \text{as} \quad x \to -\infty.$$

(d)

$$P(x) = -10x^8 + x^7 + 500x^5 + x - 2 \approx -10x^8$$

The end behavior is

$$P(x) \to -\infty \quad \text{as} \quad x \to \infty$$
$$P(x) \to -\infty \quad \text{as} \quad x \to -\infty.$$

□

EXAMPLE 6

Use the zeros of the polynomial and the end behavior to sketch the graph.
  (a) $P(x) = x^3 - x^2 - 2x$ (b) $P(x) = x^3 - 2x^2 + x$
  (c) $P(x) = (x - 1)(x - 2)(x + 1)^3$ (d) $P(x) = (x + 1)^5(x - 1)$

Solution (a)

  • Zeros: Set $P(x) = 0$ and factor.

$$x^3 - x^2 - 2x = 0$$
$$x(x^2 - x - 2) = 0$$
$$x(x - 2)(x + 1) = 0$$
$$x = 0, \; x = 2, \; x = -1$$

  • End Behavior:

$$P(x) = x^3 - x^2 - 2x \approx x^3$$

$$P(x) \to \infty \quad \text{as} \quad x \to \infty$$
$$P(x) \to -\infty \quad \text{as} \quad x \to -\infty$$

There is only one way that the curve can cross the x-axis at $-1, 0$, and 2 and satisfy the end behavior. The curve must come *up* through $x = -1$, rather than down through the point. The curve turns somewhere between $x = -1$ and $x = 0$, passing through $x = 0$ from above, turns again between $x = 0$ and $x = 2$, and goes up through $x = 2$. The curve is shown in the figure. Without the zeros *and* the end behavior, we can not be certain how to sketch the curve. We still can *not* determine exactly where the turning points are; this is left to calculus. However, we can use a graphing device to approximate the highs and lows, called *local maximums* and *local minimums*. For example, if after plotting the curve using a graphing device, we click on the local maximum between $x = -1$ and $x = 0$, the point is approximately $(-0.6, 0.6)$. The local minimum is approximately $(1.2, -2.1)$.

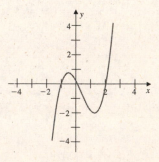

(b)

- Zeros: Set $P(x) = 0$ and factor.

$$x^3 - 2x^2 + x = 0$$
$$x(x^2 - 2x + 1) = 0$$
$$x(x - 1)(x - 1) = 0$$
$$x(x - 1)^2 = 0$$
$$x = 0, \ x = 1$$

The zero $x = 1$ has multiplicity 2.

- End Behavior:

$$P(x) = x^3 - 2x^2 + x \approx x^3$$

$$P(x) \to \infty \quad \text{as} \quad x \to \infty$$
$$P(x) \to -\infty \quad \text{as} \quad x \to -\infty.$$

The polynomial has a zero of multiplicity 2 at $x = 1$, so the curve flattens and just touches, without crossing, the x-axis at $x = 1$. Applying the end behavior produces the curve as shown in the figure.

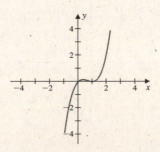

(c)

- Zeros: Since the polynomial is given in factored form, the zeros are $x = 1, x = 2$, and $x = -1$. The zero $x = -1$ has multiplicity 3.

- End Behavior: The factors of $P(x)$ are of degree 1, 1 and 3, so if the factors are expanded, and the terms collected, the degree of the polynomial is 5, and the leading coefficient is 1.

$$P(x) = (x - 1)(x - 2)(x + 1)^3 = x^5 - 4x^3 - 2x^2 + 3x + 2 \approx x^5$$

$$P(x) \to \infty \quad \text{as} \quad x \to \infty$$
$$P(x) \to -\infty \quad \text{as} \quad x \to -\infty.$$

The curve at the zero $x = -1$ flattens and also crosses the x-axis . Putting this information together, the curve is as shown in the figure.

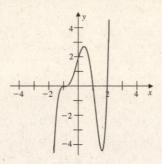

(d)

- Zeros: The factored form shows that the zeros are $x = 1$, and $x = -1$. The zero $x = -1$ has multiplicity 5.
- End Behavior: The degree of $P(x)$ is 6, and the leading coefficient is 1.

$$P(x) = (x + 1)^5(x - 1) = x^6 + 4x^5 + 5x^4 - 5x^2 - 4x - 1 \approx x^6$$

$$P(x) \to \infty \quad \text{as} \quad x \to \infty$$
$$P(x) \to \infty \quad \text{as} \quad x \to -\infty$$

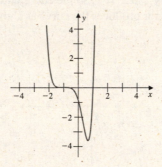

☐

EXAMPLE 7

Match the equation with the correct curve.

(a) $y = (x - 1)(x + 1)(x - 2)$ (b) $y = x(x - 1)^2$

(c) $y = (x - 1)(x + 2)^3$ (d) $y = x(x - 1)(x + 2)$

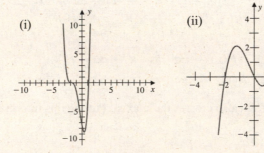

(i)   (ii)

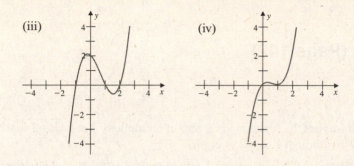

(iii)   (iv)

Solution (a) The only curve with zeros at $x = 1$, $x = -1$ and $x = 2$ is in (iii).
(b) The factor $(x - 1)^2$ implies the function has a zero of multiplicity 2 at $x = 1$. Therefore, the curve just touches at $x = 1$ without crossing through it. The only curve that satisfies this condition at $x = 1$ is in (iv).
(c) The function has a zero of multiplicity 3 at $x = -2$, so the curve flattens and crosses the x-axis at -2. It also crosses the x-axis at $x = 1$. The graph of the equation matches (i).
(d) The only function with zeros at $x = -2$, $x = 0$, and $x = 2$ is in (ii). $\square$

EXAMPLE 8

The cubic polynomial $P(x)$ has a zero of multiplicity one at $x = 1$, a zero of multiplicity two at $x = 2$, and $P(3) = 6$. Determine $P(x)$, and sketch the graph.

Solution A polynomial with the specified zeros is

$$Q(x) = (x - 1)(x - 2)^2 = x^3 - 5x^2 + 8x - 4,$$

but

$$Q(3) = (3)^3 - 5(3)^2 + 8(3) - 4 = 2.$$

The problem asks for a polynomial which when evaluated at 3 has a value of 6. Since multiplying a polynomial by a constant does not change the zeros, the polynomial $3Q(x)$ has the specified zeros and also has the valus 6 when $x = 3$. Let

$$P(x) = 3(x - 1)(x - 2)^2 = 3(x^3 - 5x^2 + 8x - 4) = 3x^3 - 15x^2 + 24x - 12.$$

Since $P(x) \approx 3x^3$, the end behavior is

$$P(x) \to \infty \quad \text{as} \quad x \to \infty \quad \text{and} \quad P(x) \to -\infty \quad \text{as} \quad x \to -\infty.$$

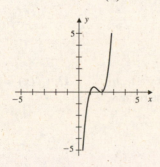

$\square$

Exercise Set 3.2 (Page 142)

1. a. The factor $(x + 2)^2$ implies the function has a zero of multiplicity 2 at $x = -2$ so the curve just touches at $x = -2$ without crossing through $x = -2$ as in (iii).

b. The factor $(x - 2)^2$ implies the function has a zero of multiplicity 2 at $x = 2$ so the curve just touches at $x = 2$ without crossing through $x = 2$ as in (iv).

c. Curve (i) is the only curve that does not exhibit the behavior of zeros of multiplicity greater than 1, and has zeros at $x = 0$, $x = 2$, and $x = -2$.

d. The factor x^2 implies the function has a zero of multiplicity 2 at the origin so the curve just touches at the origin without crossing through the origin as in (ii).

3. The graph shows a zero of at least multiplicity 2 at the origin and another positive zero so the lowest possible degree for the polynomial is 3.

5. The graph can be shifted downward so that the shifted graph has 4 distinct zeros so the lowest possible degree for the polynomial is 4.

7. Vertically stretch the graph of $y = x^3$ by a factor of 3, and then shift the resulting graph downward 2 units.

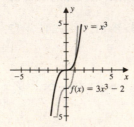

9. Shift the graph of $y = x^4$ to the right 2 units, and then shift the result upward 1 unit.

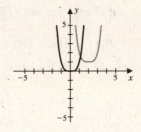

11. Vertically shrink the graph of $y = x^3$ by a factor of 2, reflect about the x-axis, shift to the right 3 units, and then shift downward 3 units.

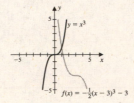

13. $f(x) = (x - 2)(x + 2)(x - 3)$

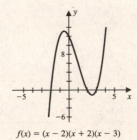

15. $f(x) = x(x+1)(x-1)(x-2)$

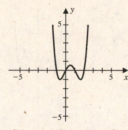

$f(x) = x(x+1)(x-1)(x-2)$

17. $f(x) = -x^3(x-1)$

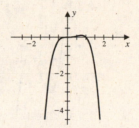

19. $f(x) = -x^2(x+1)^2$

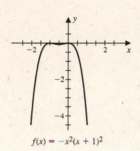

$f(x) = -x^2(x+1)^2$

21.
$$f(x) = x^3 + x^2 - 2x$$
$$= x(x^2 + x - 2)$$
$$= x(x-1)(x+2)$$

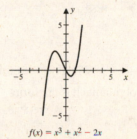

$f(x) = x^3 + x^2 - 2x$

23. $y = f(x-2)$

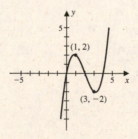

25. $y = f(x-2) + 1$

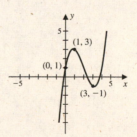

27. $y = f(-x)$

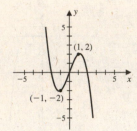

29. $y = |f(x)|$

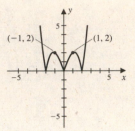

31. a. The graph of the polynomial is shown.

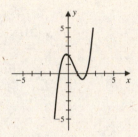

b. The least possible degree of the polynomial is 3.

33.   The polynomial has the form

$$P(x) = a(x - 1)^2(x - 2),$$

and

$$4 = P(-1) = a(-2)^2(-3) = -12a$$

implies $a = -\frac{1}{3}$.
So

$$P(x) = -\frac{1}{3}(x - 1)^2(x - 2)$$
$$= -\frac{1}{3}x^3 + \frac{4}{3}x^2 - \frac{5}{3}x + \frac{2}{3}.$$

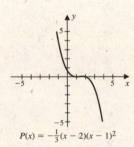

$P(x) = -\frac{1}{3}(x - 2)(x - 1)^2$

35. If the data in the table is shifted downward 2 units, then the curve through the data contains the points $(0, 0)$, $(1, 0)$, and $(2, 0)$. A polynomial that has these three zeros is $x(x - 1)(x - 2)$. Therefore a polynomial that fits the data points is

$$f(x) = 2x(x - 1)(x - 2) + 2.$$

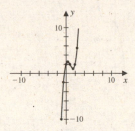

37. First, we must have for some constant a,

$$P(x) = a(x + 2)x^2(x - 2) = a(x^4 - 4x^2),$$

and since $(1, 2)$ is on the graph,

$$2 = P(1) = a(1 - 4) = -3a \text{ implies } a = -\frac{2}{3}.$$

The polynomial is

$$P(x) = -\frac{2}{3}x^4 + \frac{8}{3}x^2.$$

39. For $f_n(x) = x^n$, if $0 < x < 1$, $f_n(x) = x^n > x^{n+1} = f_{n+1}(x)$ and if $x > 1$, $f_n(x) = x^n < x^{n+1} = f_{n+1}(x)$.

41. a. $f(x) = x^4 + x^3 - 2x^2$ is increasing on $(-1.4, 0) \cup (0.7, \infty)$ and is decreasing on $(-\infty, -1.4) \cup (0, 0.7)$.
b. Local minima: $(-1.4, -2.8)$, $(0.7, -0.4)$; local maximum: $(0, 0)$

43. a. $f(x) = \frac{1}{5}x^5 - \frac{5}{4}x^4 + \frac{5}{3}x^3 + \frac{5}{2}x^2 - 6x + 1$ is increasing on $(-\infty, -1) \cup (1, 2) \cup (3, \infty)$ and is decreasing on $(-1, 1) \cup (2, 3)$.
b. Local minima: $(1, -1.9)$, $(3, -2.2)$; local maxima: $(-1, 6.4)$, $(2, -1.3)$

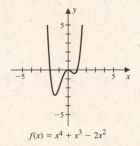

$f(x) = x^4 + x^3 - 2x^2$

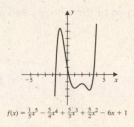

$f(x) = \frac{1}{5}x^5 - \frac{5}{4}x^4 + \frac{5}{3}x^3 + \frac{5}{2}x^2 - 6x + 1$

45. For $f(x) = x^3 - 2x^2 + x + 2$ with $a = -3$ and $b = 2$.
  a. $y = f(x)$ **b.** $y = -f(x)$ **c.** $y = f(-x)$ **d.** $y = |f(x)|$ **e.** $y = f(x-3) + 2$

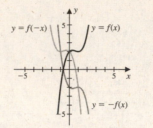 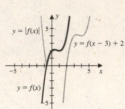

47. For $f(x) = x^4 - 400x^2$ with $a = -10$ and $b = 30000$.
  a. $y = f(x)$ **b.** $y = -f(x)$ **c.** $y = f(-x)$ **d.** $y = |f(x)|$ **e.** $y = f(x-10) + 30000$

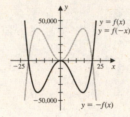 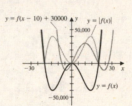

49. a. $V(x) = x(20 - 2x)(20 - 2x) = 400x - 80x^2 + 4x^3$

  b. The graph shows $y = -500 + 400x - 80x^2 + 4x^3$. The graph crosses the x-axis at $x = 5$ and $x \approx 1.9$, which are values of x that produce a volume of 500 in^3. Note the graph also crosses the x-axis at $x \approx 13$. However this value is ignored since $0 \le x \le 10$.

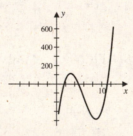

51. The graphs of the velocity and acceleration are shown below.

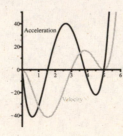

3.3   Finding Factors and Zeros of Polynomials

A polynomial $D(x)$ is a *factor* of another polynomial $P(x)$ provided there is a third polynomial $Q(x)$ with

$$P(x) = D(x) \cdot Q(x).$$

In general, $D(x)$ may not divide evenly into $P(x)$, in which case there is a remainder term $R(x)$, and

$$P(x) = D(x) \cdot Q(x) + R(x).$$

For this case, we can also write

$$\frac{P(x)}{D(x)} = Q(x) + \frac{R(x)}{D(x)}.$$

Note that this is similar to the situation of division of positive real numbers. When $0 < d \leq a$, we can divide d into a to obtain

$$a = dq + r, \quad \text{or} \quad \frac{a}{d} = q + \frac{r}{d},$$

for some number $0 < r < d$.

   The standard terminology used is given in the table.

	Divisor	Dividend	Quotient	Remainder
Real Numbers a/d	d	a	q	r
Polynomials $P(x)/D(x)$	$D(x)$	$P(x)$	$Q(x)$	$R(x)$

Division of Polynomials

EXAMPLE 1

Divide the polynomial $P(x) = x^3 - 2x^2 + x + 1$ by $x - 1$, and specify the quotient and remainder.

Solution   The division process ends when the degree of the remainder is less than the degree of the divisor. The key terms in the division process are highlighted in boxes.

$$
\begin{array}{r}
x^2 - x \\
x - 1 \overline{\smash{)}\boxed{x^3} - 2x^2 + x + 1} \\
\underline{x^3 - x^2} \\
\boxed{-x^2} + x + 1 \\
\underline{-x^2 + x} \\
1
\end{array}
$$

At each step, we determine what has to be multiplied by the leading term in the divisor, $x - 1$, to yield the term in the box. So for the first step, $x \cdot x^2 = x^3$, and at the second step, $x \cdot (-x) = -x^2$. Then we multiply each term of the divisor by the same amount and subtract the result. For example, after the first step, perform the subtraction

$$x^3 - 2x^2$$
$$-\ \underline{x^3 - x^2}.$$

The result is

$$x^3 - 2x^2 - (x^3 - x^2) = x^3 - 2x^2 - x^3 + x^2 = -x^2.$$

The quotient is

$$Q(x) = x^2 - x,$$

and the remainder is

$$R(x) = 1,$$

so

$$P(x) = x^3 - 2x^2 + x + 1 = (x - 1)(x^2 - x) + 1.$$

□

EXAMPLE 2
Let $P(x) = 5x^4 - 2x^2 + x + 3$, and $D(x) = x^2 + x - 2$. Find the quotient $Q(x)$ and remainder $R(x)$ when $P(x)$ is divided by $D(x)$.

Solution The polynomial $P(x)$ does not contain an x^3 term, and division is easier to organize when missing powers are included with a 0 coefficient, so we write $P(x) = 5x^4 + 0x^3 - 2x^2 + x + 3$. Then

$$
\begin{array}{r}
5x^2 - 5x + 13 \\
x^2 + x - 2\overline{\smash{\big)}\,5x^4 + 0x^3 - 2x^2 + x + 3} \\
\underline{5x^4 + 5x^3 - 10x^2} \\
-5x^3 + 8x^2 + x + 3 \\
\underline{-5x^3 - 5x^2 + 10x} \\
13x^2 - 9x + 3 \\
\underline{13x^2 + 13x - 26} \\
-22x + 29
\end{array}
$$

The quotient is

$$Q(x) = 5x^2 - 5x + 13,$$

the remainder is

$$R(x) = -22x + 29,$$

so

$$5x^4 - 2x^2 + x + 3 = (5x^2 - 5x + 13)(x^2 + x - 2) - 22x + 29.$$

□

The Factor Theorem

To factor a polynomial we need to find the zeros. When a polynomial $P(x)$ is divided by $x - c$ the remainder is $P(c)$.

$$P(x) = (x - c)Q(x) + R(x) \text{ so } P(c) = 0 + R(c) = R(c)$$

In the special case when $P(c) = 0$, we have $x - c$ as a factor of $P(x)$ if and only if c is a zero of $P(x)$.

EXAMPLE 3
Let $P(x) = 3x^4 - 14x^3 + 19x^2 - 4x - 4$. Use the Factor Theorem to show that $x - 1$ and $x - 2$ are factors of $P(x)$, and factor $P(x)$ completely.

Solution It is not enough to simply verify that $x = 1$ and $x = 2$ are zeros, that is $P(1) = 0$ and $P(2) = 0$. We need to perform division so we can completely factor the polynomial.

$$
\begin{array}{r}
3x^3 - 11x^2 + 8x + 4 \\
x - 1 \overline{\smash{)}3x^4 - 14x^3 + 19x^2 - 4x - 4} \\
\underline{3x^4 - 3x^3 } \\
-11x^3 + 19x^2 - 4x - 4 \\
\underline{-11x^3 + 11x^2 } \\
8x^2 - 4x - 4 \\
\underline{8x^2 - 8x } \\
4x - 4 \\
\underline{4x - 4} \\
0
\end{array}
$$

Since the remainder is 0,
$$P(x) = (x - 1)(3x^3 - 11x^2 + 8x + 4).$$

To show that 2 is a zero of $P(x)$, it is enough to show it is a zero of $3x^3 - 11x^2 + 8x + 4$. After dividing the factor $3x^3 - 11x^2 + 8x + 4$ by $x - 2$, we have

$$
\begin{aligned}
P(x) &= (x - 1)(3x^3 - 11x^2 + 8x + 4) \\
&= (x - 1)(x - 2)(3x^2 - 5x - 2).
\end{aligned}
$$

Factoring the quadratic term into $3x^2 - 5x - 2 = (3x + 1)(x - 2)$ gives the complete factored form of $P(x)$ as
$P(x) = (x - 1)(x - 2)(3x + 1)(x - 2) = (x - 1)(x - 2)^2(3x + 1)$.   □

Synthetic Division

Synthetic division is a process which simplifies the division of polynomials in cases where the divisor has the form $x - c$. It can be used for finding quotients and remainders, for solving polynomial equations $P(x) = 0$, and for evaluating polynomials at specific values.

   To introduce the technique of synthetic division, divide $x^5 - 3x^4 + 3x^3 - x^2 + x - 4$ by $x - 2$ using standard division.

$$\begin{array}{r}
x^4 - x^3 + x^2 + x + 3 \\
x - 2 \overline{\smash{\big)}\ x^5 - 3x^4 + 3x^3 - x^2 + x - 4} \\
\underline{x^5 - 2x^4} \\
-x^4 + 3x^3 - x^2 + x - 4 \\
\underline{-x^4 + 2x^3} \\
x^3 - x^2 + x - 4 \\
\underline{x^3 - 2x^2} \\
x^2 + x - 4 \\
\underline{x^2 - 2x} \\
3x - 4 \\
\underline{3x - 6} \\
2
\end{array}$$

The quotient is

$$Q(x) = x^4 - x^3 + x^2 + x + 3,$$

and the remainder is

$$R(x) = 2.$$

Recall that by the Factor Theorem, if a polynomial $P(x)$ is divided by the linear term $x - c$, then the remainder is $P(c)$. So, $P(2) = 2$.

At each step of the division process, the first columns always cancel on subtraction, so only the second columns are important for the result. If terms for every power of x are included, adding 0 for any missing powers, only the coefficients are needed and the variable terms can be dropped. The column indicates the power of x.

In the synthetic division process the coefficients of the terms of the dividend and the constant term, $-c$, of the divisor, $x - c$ are recorded in the first step. In the division example given above, we start with the form

$$\begin{array}{r|rrrrrr}
-2 & 1 & -3 & 3 & -1 & 1 & -4 \\
\hline
 & 1
\end{array}$$

which records the coefficients of the dividend $x^5 - 3x^4 + 3x^3 - x^2 + x - 4$, and the constant term $-c = -2$ from the divisor, $x - 2$. To start the process, drop the first coefficient 1 down to the third row.

Now multiply the first term in row 3 by -2, and place it in column 2 of row 2. Subtract the terms in column 2 recording the result in column 2 of row 3.

$$\begin{array}{r|rrrrrr}
-2 & 1 & -3 & 3 & -1 & 1 & -4 \\
 & & -2 \\
\hline
 & 1 & -1
\end{array}$$

Repeat the process with each successive column.

$$\begin{array}{r|rrrrrr} -2 & 1 & -3 & 3 & -1 & 1 & -4 \\ & & -2 & 2 & -2 & -2 & -6 \\ \hline & 1 & -1 & 1 & 1 & 3 & 2 \end{array}$$

The coefficients of the quotient and remainder are read from row 3. The remainder is the constant in the far right column, so $R(x) = 2$ and the quotient, in decreasing powers of x is, $Q(x) = x^4 - x^3 + x^2 + x + 3$.

In the procedure outlined, the second term in each column is subtracted from the one above. To avoid possible subtraction errors, we make the final modification in the synthetic division process. We change the -2 in the divisor to 2, and **add**, rather than subtract, the columns. The results are the same, as shown below.

$$\begin{array}{r|rrrrrr} 2 & 1 & -3 & 3 & -1 & 1 & -4 \\ & & 2 & -2 & 2 & 2 & 6 \\ \hline & 1 & -1 & 1 & 1 & 3 & 2 \end{array}$$

EXAMPLE 4

Find the quotient and remainder when $x^4 + 2x^3 - 3x^2 + 2x - 5$ is divided by $x + 3$.

Solution First note the divisor has the form

$$x + 3 = x - (-3).$$

When setting up synthetic division so that we are able to add, rather than subtract columns, we use (-3) for the multiplier.

$$\begin{array}{r|rrrrr} -3 & 1 & 2 & -3 & 2 & -5 \\ & & -3 & 3 & 0 & -6 \\ \hline & 1 & -1 & 0 & 2 & -11 \end{array}$$

The quotient is

$$Q(x) = x^3 - x^2 + 2,$$

and the remainder is

$$R(x) = -11.$$

□

EXAMPLE 5

Let $P(x) = x^6 - 2x^4 + x^3 - 3x + 5$. Find $P(3)$.

Solution To find $P(3)$, we can substitute 3 for x in the expression for $P(x)$ and do the arithmetic. But the remainder on division of $P(x)$ by $x - 3$ is also $P(3)$, so synthetic division can be used to organize the work involved. The polynomial is missing an x^5 and an x^2 term, so remember to use zeros for the coefficients of these terms in the synthetic division process. The synthetic division gives $P(3) = 590$.

$$
\begin{array}{r|rrrrrr}
3 & 1 & 0 & -2 & 1 & 0 & -3 & 5 \\
  &   & 3 & 9 & 21 & 66 & 198 & 585 \\
\hline
  & 1 & 3 & 7 & 22 & 66 & 195 & 590
\end{array}
$$

$\square$

EXAMPLE 6

Is $x - 3$ a factor of the polynomial $P(x) = x^5 - x^4 - 6x^3 + 4x - 12$?

Solution The linear factor $x - 3$ is a factor of the polynomial $P(x)$ if when divided into $P(x)$, the remainder is 0.

$$
\begin{array}{r|rrrrrr}
3 & 1 & -1 & -6 & 0 & 4 & -12 \\
  &   & 3 & 6 & 0 & 0 & 12 \\
\hline
  & 1 & 2 & 0 & 0 & 4 & 0
\end{array}
$$

So the remainder is 0, which implies that $x - 3$ is a factor, and

$$P(x) = (x - 3)(x^4 + 2x^3 + 4).$$

$\square$

The Rational Zero Test

The rational numbers that are possible zeros of the polynomial

$$P(x) = a_n x^n + a_{n-1} x^{n-1} + \cdots + a_0$$

must be of the form

$$\frac{\text{factors of } a_0}{\text{factors of } a_n}.$$

EXAMPLE 7

Let

$$P(x) = 6x^5 - 8x^3 + 4x^2 + 5x - 4.$$

Determine all possibilities for rational zeros of $P(x)$.

Solution The factors of the constant term are $\pm 1, \pm 2$, and ± 4, and the factors of the leading coefficient are $\pm 1, \pm 2, \pm 3$, and ± 6. The possible rational zeros are

$$\frac{\pm 1, \pm 2, \pm 4}{\pm 1, \pm 2, \pm 3, \pm 6} = \pm 1, \pm \frac{1}{2}, \pm \frac{1}{3}, \pm \frac{1}{6}, \pm 2, \pm \frac{2}{3}, \pm 4, \pm \frac{4}{3}.$$

$\square$

Graphing devices are useful in eliminating possibilities from the list of possible rational zeros, which can often be a very extensive list.

EXAMPLE 8

Find all rational and irrational zeros of the polynomial $P(x) = x^3 - 2x^2 - 5x + 6$, and factor the polynomial completely.

Solution The possible rational zeros are

$$\frac{\text{factors of } 6}{\text{factors of } 1} = \pm 1, \pm 2, \pm 3, \pm 6.$$

By inspection we can see that $P(1) = 1 - 2 - 5 + 6 = 0$, so $x - 1$ is a factor of the polynomial. This can also be seen from the graph shown in the figure. The graph also indicates there are zeros at $x = -2$ and $x = 3$. First divide the polynomial by $x - 1$.

$$
\begin{array}{r}
x^2 - x - 6 \\
x - 1 \overline{)\, x^3 - 2x^2 - 5x + 6} \\
\underline{x^3 - x^2} \\
-x^2 - 5x + 6 \\
\underline{-x^2 + x} \\
-6x + 6 \\
\underline{-6x + 6} \\
0
\end{array}
$$

The division can also be done synthetically.

$$
\begin{array}{r|rrrr}
1 & 1 & -2 & -5 & 6 \\
  &   & 1 & -1 & -6 \\
\hline
  & 1 & -1 & -6 & 0
\end{array}
$$

The polynomial in factored form is

$$P(x) = (x - 1)(x^2 - x - 6) = (x - 1)(x - 3)(x + 2),$$

and the zeros are $x = 1$, $x = 3$, and $x = -2$. There are no irrational zeros for this polynomial.

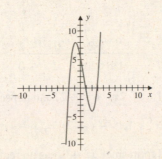

EXAMPLE 9

Find all rational and irrational zeros of the polynomial $P(x) = x^5 - 2x^3 + 2x^2 - 3x + 2$, and factor the polynomial completely.

Solution   The possible rational zeros are $\pm 1, \pm 2$. The graph indicates a zero at $x = -2$ and another zero, of multiplicity at least 2, at $x = 1$.

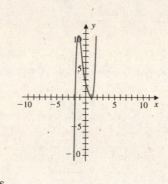

Dividing the polynomial by $(x - 1)^2$ gives

$$P(x) = (x - 1)^2(x^3 + 2x^2 + x + 2).$$

We use synthetic division to quickly show that $x = 1$ is a zero of multiplicity 2. First divide $P(x)$ by $x - 1$.

$$
\begin{array}{r|rrrrr}
1 & 1 & 0 & -2 & 2 & -3 & 2 \\
  &   & 1 & 1 & -1 & 1 & -2 \\
\hline
  & 1 & 1 & -1 & 1 & -2 & 0
\end{array}
$$

Now use synthetic division to divide the quotient $Q(x) = x^4 + x^3 - x^2 + x - 2$ again by $x - 1$.

$$
\begin{array}{r|rrrrr}
1 & 1 & 1 & -1 & 1 & -2 \\
  &   & 1 & 2 & 1 & 2 \\
\hline
  & 1 & 2 & 1 & 2 & 0
\end{array}
$$

So $P(x) = (x - 1)^2(x^3 + 2x^2 + x + 2)$. Next divide $x^3 + 2x^2 + x + 2$ by $x + 2$.

$$
\begin{array}{r|rrrr}
-2 & 1 & 2 & 1 & 2 \\
   &   & -2 & 0 & -2 \\
\hline
   & 1 & 0 & 1 & 0
\end{array}
$$

Since the remainder is 0, $x + 2$ is a factor of $x^3 + 2x^2 + x + 2$, and consequently of $P(x)$, with

$$P(x) = (x - 1)^2(x^3 + 2x^2 + x + 2) = (x - 1)^2(x + 2)(x^2 + 1).$$

The polynomial is now in complete factored form since $x^2 + 1$ has no real factors.

$\square$

EXAMPLE 10

Find all rational and irrational zeros of the polynomial $P(x) = 2x^5 - 10x^4 + 7x^3 + 28x^2 - 45x + 18$, and factor the polynomial completely.

Solution　The possible rational zeros are

$$\pm 1, \pm 2, \pm 3, \pm 6, \pm 9, \pm 18, \pm \frac{1}{2}, \pm \frac{3}{2}, \pm \frac{9}{2}.$$

The list is extensive, but we can simply substitute each value into the polynomial until we find a zero, if there is a rational zero, perform division and continue until the last factor is a quadratic, which can be solved by the quadratic formula. A graphing device reduces the process in this example very quickly.

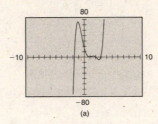

(a)

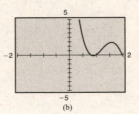

(b)

The graph in figure (a) indicates that there is a zero between -2 and -1, a zero of multiplicity 2 at $x = 1$, and zeros at $x = 2$ and $x = 3$. The close up view near $x = 1$ in figure (b) indicates $x = 1$ is *not* a root of multiplicity 2. We have

$$(x - 1)(x - 2)(x - 3) = x^3 - 6x^2 + 11x - 6,$$

which when divided into $P(x)$ gives

$$P(x) = (x^3 - 6x^2 + 11x - 6)(2x^2 + 2x - 3).$$

To factor the quadratic, we use the quadratic formula. The zeros are

$$x = \frac{-2 \pm \sqrt{4 - 4(2)(-3)}}{4} = \frac{-2 \pm \sqrt{28}}{4}$$

$$= \frac{-2 \pm 2\sqrt{7}}{4} = \frac{-1 \pm \sqrt{7}}{2}.$$

The final factorization is

$$P(x) = (x^3 - 6x^2 + 11x - 6)(2x^2 + 2x - 3)$$

$$= (x - 1)(x - 2)(x - 3)\left(x - \left(\frac{-1 + \sqrt{7}}{2}\right)\right)\left(x - \left(\frac{-1 - \sqrt{7}}{2}\right)\right).$$

Note that the Rational Zero test is of no help in finding the last two zeros since they are irrational numbers.　□

EXAMPLE 11

Sketch the graphs of $f(x) = x^3 - x^2$ and $g(x) = x^2 + x - 2$, and determine all the points of intersection.

Solution To graph the cubic $y = f(x) = x^3 - x^2$, we factor to obtain

$$f(x) = x^2(x - 1),$$

and see that f has a zero of multiplicity 2 at $x = 0$ and a zero of multiplicity 1 at $x = 1$. To graph $y = g(x)$, place the quadratic in standard form.

$$\begin{aligned}
g(x) &= x^2 + x - 2 \\
&= \left(x^2 + x\right) - 2 \\
&= \left(x^2 + x + \frac{1}{4} - \frac{1}{4}\right) - 2 \\
&= \left(x + \frac{1}{2}\right)^2 - 2 - \frac{1}{4} \\
&= \left(x + \frac{1}{2}\right)^2 - \frac{9}{4}
\end{aligned}$$

The graphs of f and g are shown below.

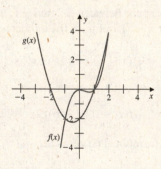

To find the points of intersection determine the values of x with $f(x) = g(x)$, that is, solve

$$x^3 - x^2 = x^2 + x - 2,$$

or

$$x^3 - 2x^2 - x + 2 = 0.$$

- Possible rational zeros: $\pm 1, \pm 2$

The figure shows a point of intersection at $x = 1$ and is verified by division.

$$
\begin{array}{r}
x^2 - x - 2 \\
x - 1 \overline{\smash{)}x^3 - 2x^2 - x + 2} \\
\underline{x^3 - x^2} \\
-x^2 - x + 2 \\
\underline{-x^2 + x}
\end{array}
$$

$$-2x + 2$$
$$\underline{-2x + 2}$$
$$0$$

The division can also be done synthetically.

$$\underline{1\,|}\,1 \quad -2 \quad -1 \quad\;\; 2$$
$$\underline{\quad\quad 1 \quad -1 \quad -2}$$
$$\quad\;\; 1 \quad -1 \quad -2 \quad\;\; 0$$

Finally

$$x^3 - 2x^2 - x + 2 = (x-1)(x^2 - x - 2) = (x-1)(x-2)(x+1),$$

and the points of intersection are

$$(1, 0), (2, 4), (-1, -2).$$

EXAMPLE 12

Find a third degree polynomial $P(x)$ that has zeros at $x = -2$, $x = -1$, and $x = 2$, and whose x^2-term has coefficient 3.

Solution　A polynomial with the desired zeros is

$$Q(x) = (x+2)(x+1)(x-2) = x^3 + x^2 - 4x - 4.$$

Multiplying a polynomial by a constant does not change the zeros of the polynomial. So to make the coefficient of the x^2-term 3, multiply $Q(x)$ by 3. Therefore,

$$P(x) = 3(x^3 + x^2 - 4x - 4) = 3x^3 + 3x^2 - 12x - 12.$$

Descarte's Rule of Signs

This is a simple test that provides information about the possible *number* of positive real zeros and negative real zeros of a polynomial $P(x)$ with real coefficients.

- The number of positive zeros is either the number of variations in sign of $P(x)$ or less than this by an even number.

- The number of negative zeros is either the number of variations in sign of $P(-x)$ or less than this by an even number.

EXAMPLE 13

Use Descarte's Rule of Signs to determine the maximum number of positive and negative zeros of the polynomial $P(x) = 3x^4 - 4x^3 - 7x^2 + x + 4$.

Solution

- Variations in sign of $P(x)$: $P(x) = \underbrace{3x^4 - 4x^3}_{+\text{ to }-} - \underbrace{7x^2 + x}_{-\text{ to }+} + 4$. There are two variations in sign, so the number

of positive real zeros is either 2 or 0.

- Variations in sign of $P(-x)$:

$$P(-x) = 3(-x)^4 - 4(-x)^3 - 7(-x)^2 + (-x) + 4$$
$$= 3x^4 + \underbrace{4x^3 - 7x^2}_{+\text{ to }-} - \underbrace{x + 4}_{-\text{ to }+}$$

The number of negative real zeros is also either 2 or 0. $\square$

Exercise Set 3.3 (Page 154)

1. The quotient is $Q(x) = 2x + 7$ and remainder is $R(x) = 11$.

$$
\begin{array}{r}
2x \quad + \quad 7 \\
x-2 \overline{\smash{\big)}\ 2x^2 \ + \ 3x \ - \ 3} \\
\underline{2x^2 \ - \ 4x } \\
7x \ - \ 3 \\
\underline{7x \ - \ 14} \\
11
\end{array}
$$

3. The quotient is $Q(x) = x^2 + 2x + 2$ and remainder is $R(x) = 0$.

$$
\begin{array}{r}
x^2 \ + \ 2x \ + \ 2 \\
x-1 \overline{\smash{\big)}\ x^3 \ + \ x^2 \ + \ 0x \ - \ 2} \\
\underline{x^3 \ - \ x^2 } \\
2x^2 \ + \ 0x \\
\underline{2x^2 \ - \ 2x } \\
2x \ - \ 2 \\
\underline{2x \ - \ 2} \\
0
\end{array}
$$

5. The quotient is $Q(x) = 2x^2 + 6$ and remainder is $R(x) = 7x + 4$.

$$
\begin{array}{r}
2x^2 \phantom{{}+{}} + 6 \\
x^2 - x - 1 \overline{\smash{\big)}\, 2x^4 - 2x^3 + 4x^2 + x - 2} \\
\underline{2x^4 - 2x^3 - 2x^2} \\
6x^2 + x - 2 \\
\underline{6x^2 - 6x - 6} \\
7x + 4
\end{array}
$$

7. We have $P(x) = x^3 - 5x^2 + 8x - 4$ and $P(1) = 1^3 - 5(1)^2 + 8(1) - 4 = 0$, so $x - 1$ is a factor. Dividing gives $P(x) = (x - 1)(x^2 - 4x + 4) = (x - 1)(x - 2)(x - 2) = (x - 1)(x - 2)^2$.

$$
\begin{array}{r}
x^2 - 4x + 4 \\
x - 1 \overline{\smash{\big)}\, x^3 - 5x^2 + 8x - 4} \\
\underline{x^3 - x^2} \\
-4x^2 + 8x \\
\underline{-4x^2 + 4x} \\
4x - 4 \\
\underline{4x - 4} \\
0
\end{array}
$$

9. We have $P(x) = 2x^4 + 3x^3 - 12x^2 - 7x + 6$ and $P(-1) = 2 - 3 - 12 + 7 + 6 = 0$,

so $x + 1$ is a factor. Dividing gives $P(x) = (x + 1)(2x^3 + x^2 - 13x + 6)$.

$$
\begin{array}{r}
2x^3 + x^2 - 13x + 6 \\
x + 1 \overline{\smash{\big)}\, 2x^4 + 3x^3 - 12x^2 - 7x + 6} \\
\underline{2x^4 + 2x^3} \\
x^3 - 12x^2 \\
\underline{x^3 + x^2} \\
-13x^2 - 7x \\
\underline{-13x^2 - 13x} \\
6x + 6 \\
\underline{6x + 6} \\
0
\end{array}
$$

If $x = -3$ is a zero of $P(x)$, then it will also be a zero of $Q(x) = 2x^3 + x^2 - 13x + 6$. Since

$$Q(-3) = 2(-3)^3 + (-3)^2 - 13(-3) + 6 = 0,$$

$x + 3$ is a factor of $Q(x)$ and also of $P(x)$. Then

$$P(x) = (x+1)(x+3)(2x^2 - 5x + 2) = (x+1)(x+3)(2x-1)(x-2).$$

$$
\begin{array}{r}
2x^2 \; - \; 5x \; + \; 2 \\
x+3 \; \overline{)\; 2x^3 \; + \; x^2 \; - \; 13x \; + \; 6} \\
\underline{2x^3 \; + \; 6x^2} \\
- \; 5x^2 \; - \; 13x \\
\underline{- \; 5x^2 \; - \; 15x} \\
2x \; + \; 6 \\
\underline{2x \; + \; 6} \\
0
\end{array}
$$

11. We have $P(x) = 3x^3 + x^2 - 8x + 4$ and

$$P\left(\frac{2}{3}\right) = 3\left(\frac{8}{27}\right) + \frac{4}{9} - \frac{16}{3} + 4 = \frac{8}{9} + \frac{4}{9} - \frac{48}{9} + \frac{36}{9} = 0,$$

so $x - \frac{2}{3}$ is a factor. Dividing gives

$$P(x) = \left(x - \frac{2}{3}\right)(3x^2 + 3x - 6) = \left(x - \frac{2}{3}\right)3(x^2 + x - 2)$$

$$= 3\left(x - \frac{2}{3}\right)(x+2)(x-1) = 3\frac{1}{3}(3x-2)(x+2)(x-1)$$

$$= (3x-2)(x+2)(x-1).$$

$$
\begin{array}{r}
3x^2 \; + \; 3x \; - \; 6 \\
x-\frac{2}{3} \; \overline{)\; 3x^3 \; + \; x^2 \; - \; 8x \; + \; 4} \\
\underline{3x^3 \; - \; 2x^2} \\
3x^2 \; - \; 8x \\
\underline{3x^2 \; - \; 2x} \\
- \; 6x \; + \; 4 \\
\underline{- \; 6x \; + \; 4} \\
0
\end{array}
$$

13. The Rational Zero Test gives the possible rational zeros

$$\frac{\text{factors of } 4}{\text{factors of } 1} = \frac{\pm 1, \pm 2, \pm 4}{\pm 1} = \pm 1, \pm 2, \pm 4.$$

15. Since the rational zeros will be the same if we divide the equation by 2, consider the equation $5x^5 - 7x^3 + 9x^2 + 3x - 2 = 0$. The possible rational zeros are

$$\frac{\text{factors of } 2}{\text{factors of } 5} = \frac{\pm 1, \pm 2}{\pm 1, \pm 5} = \pm 1, \pm 2, \pm \frac{1}{5}, \pm \frac{2}{5}.$$

17. Since the rational zeros will be the same if we multiply the equation by 2, consider the equation $3x^3 - 5x^2 + 14x - 8 = 0$. The possible rational zeros are

$$\frac{\text{factors of 8}}{\text{factors of 3}} = \frac{\pm 1, \pm 2, \pm 4, \pm 8}{\pm 1, \pm 3}$$

$$= \pm 1, \pm 2, \pm 4, \pm 8, \pm \frac{1}{3}, \pm \frac{2}{3}, \pm \frac{4}{3}, \pm \frac{8}{3}.$$

19. For the polynomial $P(x) = x^3 - 3x^2 + 4$, the possible rational zeros are $\pm 1, \pm 2, \pm 4$. By substitution, the polynomial has zeros at $x = -1$, and 2 with the zero at $x = 2$ of multiplicity 2. Dividing $(x + 1)$ into $P(x)$ gives

$$P(x) = (x + 1)(x^2 - 4x + 4) = (x + 1)(x - 2)^2.$$

21. Since

$$P(x) = 2x^4 - 3x^3 - 3x^2 + 2x = x(2x^3 - 3x^2 - 3x + 2),$$

$P(x)$ has a zero at $x = 0$. The possible rational zeros of $Q(x) = 2x^3 - 3x^2 - 3x + 2$ are

$$\frac{\pm 1, \pm 2}{\pm 1, \pm 2} = \pm 1, \pm 2, \pm \frac{1}{2}.$$

By substitution, the polynomial $Q(x)$ has zeros at $x = -1, \frac{1}{2}, 2$. Dividing $(x + 1)$ into $Q(x)$ gives

$$P(x) = x(x + 1)(2x^2 - 5x + 2) = x(x + 1)(2x - 1)(x - 2).$$

23. For the polynomial

$$P(x) = 2x^5 + x^4 - 12x^3 + 10x^2 + 2x - 3,$$

the possible rational zeros are

$$\frac{\pm 1, \pm 2, \pm 3}{\pm 1, \pm 2} = \pm 1, \pm 2, \pm 3, \pm \frac{1}{2}, \pm \frac{3}{2}.$$

By substitution, the polynomial has zeros at $x = 1$, of multiplicity 3, and zeros at $x = -3, -\frac{1}{2}$. So

$$P(x) = (x - 1)^3(2x + 1)(x + 3).$$

25. For the polynomial $P(x) = x^3 + 2x^2 - 4x + 1$, the possible rational zeros are ± 1. By substitution, the polynomial has a zero at $x = 1$. Dividing $(x - 1)$ into $P(x)$ gives

$$P(x) = (x - 1)(x^2 + 3x - 1).$$

To factor the quadratic requires the quadratic formula to find the roots of $x^2 + 3x - 1 = 0$. That is,

$$x = \frac{-3 \pm \sqrt{9 - 4(1)(-1)}}{2} = \frac{-3 \pm \sqrt{13}}{2}$$

and

$$P(x) = (x - 1)\left(x - \left(\frac{-3 + \sqrt{13}}{2}\right)\right)\left(x - \left(\frac{-3 - \sqrt{13}}{2}\right)\right).$$

27. For the polynomial $P(x) = 4x^3 - 12x^2 + 9x - 1$, the possible rational zeros are $\pm 1, \pm\frac{1}{2}, \pm\frac{1}{4}$. By substitution, the polynomial has a zero at $x = 1$. Dividing $(x - 1)$ into $P(x)$ gives

$$P(x) = (x - 1)(4x^2 - 8x + 1).$$

To factor the quadratic requires the quadratic formula to find the zeros of $4x^2 - 8x + 1 = 0$. That is,

$$x = \frac{8 \pm \sqrt{64 - 4(4)(1)}}{8} = \frac{8 \pm \sqrt{48}}{8} = \frac{8 \pm 4\sqrt{3}}{8} = 1 \pm \frac{\sqrt{3}}{2},$$

and

$$P(x) = 4(x - 1)\left(x - \left(1 + \frac{\sqrt{3}}{2}\right)\right)\left(x - \left(1 - \frac{\sqrt{3}}{2}\right)\right).$$

29. The curves $y = f(x) = x^3$ and $y = g(x) = 2x^2 + x - 2$ intersect if and only if $f(x) = g(x)$ which implies $x^3 = 2x^2 + x - 2$ so $x^3 - 2x^2 - x + 2 = 0$. For the polynomial $P(x) = x^3 - 2x^2 - x + 2$, the possible rational zeros are $\pm 1, \pm 2$. By substitution, the polynomial has a zero at $x = 1$. Dividing $(x - 1)$ into $P(x)$ gives

$$P(x) = (x - 1)(x^2 - x - 2) = (x - 1)(x + 1)(x - 2).$$

The points of intersection are $(-1, -1)$, $(1, 1)$, and $(2, 8)$. To sketch the parabola, the quadratic in standard form is

$$2x^2 + x - 2 = 2\left(x^2 + \frac{1}{2}x\right) - 2 = 2\left(x^2 + \frac{1}{2}x + \frac{1}{16} - \frac{1}{16}\right) - 2 = 2\left(x + \frac{1}{4}\right)^2 - \frac{17}{8}.$$

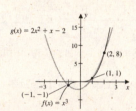

31. The curves $y = f(x) = x^3 - 1$ and $y = g(x) = 6x + 3$ intersect if and only if $f(x) = g(x)$ which implies $x^3 - 1 = 6x + 3$ so $x^3 - 6x - 4 = 0$. For the polynomial $P(x) = x^3 - 6x - 4$, the possible rational zeros are $\pm 1, \pm 2, \pm 4$. By substitution, the polynomial has a zero at $x = -2$. Dividing $(x + 2)$ into $P(x)$ gives

$$P(x) = (x + 2)(x^2 - 2x - 2).$$

To factor the quadratic requires the quadratic formula to find the zeros of $x^2 - 2x - 2 = 0$. That is,

$$x = \frac{2 \pm \sqrt{4 - 4(1)(-2)}}{2} = \frac{2 \pm \sqrt{12}}{2} = \frac{2 \pm 2\sqrt{3}}{2} = 1 \pm \sqrt{3}.$$

and

$$P(x) = (x + 2)\left(x - \left(1 + \sqrt{3}\right)\right)\left(x - \left(1 - \sqrt{3}\right)\right).$$

The points of intersection are $(-2, -9)$, $\left(1 + \sqrt{3}, \left(1 + \sqrt{3}\right)^3 - 1\right)$, and $\left(1 - \sqrt{3}, \left(1 - \sqrt{3}\right)^3 - 1\right)$.

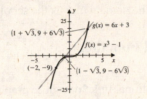

33. Since the polynomial has zeros of multiplicity 1 at $x = -3$, $x = -1$, $x = 1$, and $x = 4$, we have

$$P(x) = a(x + 3)(x + 1)(x - 1)(x - 4).$$

Since the graph passes through $(2, -3)$, we have

$$-3 = P(2) = a(5)(3)(1)(-2) = -30a \text{ so } a = \frac{1}{10}.$$

Thus

$$P(x) = \frac{1}{10}(x + 3)(x + 1)(x - 1)(x - 4) = \frac{1}{10}x^4 - \frac{1}{10}x^3 - \frac{13}{10}x^2 + \frac{1}{10}x + \frac{6}{5}.$$

35. A fourth-degree polynomial with zeros $x = -2, -1, 0, 1$ and hence, the factors $(x + 2)$, $(x + 1)$, x, $(x - 1)$, is

$$Q(x) = (x + 1)x(x - 1)(x + 2) = x^4 + 2x^3 - x^2 - 2x.$$

To have the coefficient of the x^2 term be 5, multiply the polynomial by -5. So

$$P(x) = -5(x^4 + 2x^3 - x^2 - 2x) = -5x^4 - 10x^3 + 5x^2 + 10x.$$

37. We have $P(x) = (x - 2)^3 x^2 (x - 1) = x^6 - 7x^5 + 18x^4 - 20x^3 + 8x^2$.

39. a. $x = -1$, $x = 1$, and $x = 4$.

   b. $[-1, 1]$ and $[4, \infty)$.

41. a. $x = -1$, $x = 1$, and $x = 4$.

   b. $[-1, 1]$ and $[4, \infty)$.

· ·

3.4   Rational Functions

A *rational function* has the form

$$f(x) = \frac{P(x)}{Q(x)},$$

where $P(x)$ and $Q(x)$ are polynomials. The domain of f is the set of all real numbers with $Q(x) \neq 0$.

Domains and Ranges of Rational Functions

EXAMPLE 1
Determine the domain and range of the rational function whose graph is shown in the figure.

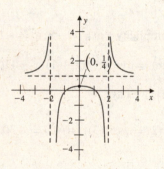

Solution

- Domain: Any vertical line, except $x = -2$ and $x = 2$, crosses the graph of the function in one point, so the domain is $(-\infty, -2) \cup (-2, 2) \cup (2, \infty)$.
- Range: Horizontal lines between the line $y = \frac{1}{4}$ and the line $y = 1$, including $y = 1$, do not intersect the graph, and are excluded from the range. The range is $\left(-\infty, \frac{1}{4}\right] \cup (1, \infty)$.

                                                  □

EXAMPLE 2
Find the domain and any x- and y-intercepts of the rational function.

(a) $f(x) = \dfrac{(x-1)(x+2)}{(x+3)(x-4)}$ (b) $f(x) = \dfrac{x^2 - x - 6}{x^2 - 4}$

Solution (a)

- Domain: All real numbers for which the denominator is *not* zero. To exclude all real numbers that make the denominator 0 solve

$$(x + 3)(x - 4) = 0 \text{ implies } x = -3 \quad \text{or} \quad x = 4.$$

The domain is $(-\infty, -3) \cup (-3, 4) \cup (4, \infty)$.

- x-intercepts: Solve $y = f(x) = 0$. The fraction will be 0 provided the numerator is 0, so

$$(x - 1)(x + 2) = 0 \text{ implies } x = 1 \quad \text{or} \quad x = -2,$$

and the graph crosses the x-axis at the points $(1, 0)$ and $(-2, 0)$.

- y-intercept: Set $x = 0$, so $y = \dfrac{(-1)(2)}{(3)(-4)} = \dfrac{1}{6}$ and the graph crosses the y-axis at $\left(0, \frac{1}{6}\right)$.

(b) First factor the numerator and denominator,

$$f(x) = \frac{x^2 - x - 6}{x^2 - 4} = \frac{(x - 3)(x + 2)}{(x - 2)(x + 2)} = \frac{x - 3}{x - 2}, \quad \text{when} \quad x \neq 2.$$

- Domain: Setting the denominator to 0 gives ,

$$x - 2 = 0, \quad \text{which implies} \quad x = 2,$$

Hence $x = 2$ must be eliminated from the domain. The value $x = -2$ must also be eliminated from the domain since the original equation is not defined at $x = -2$. As a consequence, the domain is $(-\infty, -2) \cup (-2, 2) \cup (2, \infty)$. The graph has a hole at the point $\left(-2, \frac{5}{4}\right)$.

- x-intercepts: Solve for $y = f(x) = 0$.

$$x - 3 = 0$$
$$x = 3$$

The graph crosses the x-axis at the point $(3, 0)$. Do not set the original numerator to 0 since this gives $x = -2$ as an x-intercept. However this value is not in the domain of the function.

- y-intercept: Set $x = 0$. Then $y = \frac{3}{2}$, and the graph crosses the y-axis at $\left(0, \frac{3}{2}\right)$.

$\square$

Horizontal and Vertical Asymptotes

An *asymptote* of a graph is a line that the graph approaches. A horizontal line $y = a$ is a *horizontal asymptote* to the graph of f, if

$$f(x) \to a \text{ as } x \to \infty \quad \text{or} \quad f(x) \to a \text{ as } x \to -\infty.$$

The vertical line $x = a$ is a *vertical asymptote* to the graph of f, provided

$$f(x) \to \infty \text{ as } x \to a^+ \quad \text{or} \quad f(x) \to -\infty \text{ as } x \to a^+$$
$$f(x) \to \infty \text{ as } x \to a^- \quad \text{or} \quad f(x) \to -\infty \text{ as } x \to a^-.$$

The notation $x \to a^+$ means x approaches a with values greater than a, and $x \to a^-$ means x approaches a with values less than a.

To find horizontal asymptotes, use the end behavior of the polynomials in the numerator and denominator of the rational function. For example, when $|x|$ is large,

$$f(x) = \frac{2x^3 - 2x^2 - 8x + 2}{3x^3 + x^2 - x + 5} \approx \frac{2x^3}{3x^3} = \frac{2}{3}.$$

As x is selected further from 0, on the positive or negative x-axis, the values of $f(x)$ get closer to $\frac{2}{3}$. So the curve flattens to the horizontal line $y = \frac{2}{3}$.

Vertical asymptotes can occur only for vertical lines $x = a$ where the denominator of the rational function is 0 at $x = a$. For example,

$$f(x) = \frac{1}{(x-1)(x+2)}$$

has vertical asymptotes $x = 1$ and $x = -2$.

As x gets close to 1 from the right, $x - 1 > 0$, and gets close to zero. The factor $(x + 2)$ approaches 3, so $(x - 1)(x + 2) > 0$ approaches 0. The quotient $f(x) = \frac{1}{(x-1)(x+2)}$ then becomes arbitrarily large. This is written

$$f(x) \to \infty \quad \text{as} \quad x \to 1^+.$$

As x gets close to 1 from the left, $x - 1 < 0$, and gets close to zero. The factor $(x + 2)$ still approaches 3, so $(x - 1)(x + 2) < 0$ and approaches 0. So as x approaches 1 from the left, the quotient $f(x) = \frac{1}{(x-1)(x+1)}$ becomes arbitrarily large in magnitude but negative. This is written

$$f(x) \to -\infty \quad \text{as} \quad x \to 1^-.$$

The same analysis near the vertical line $x = -2$, gives

$$f(x) \to -\infty \quad \text{as} \quad x \to -2^+$$
$$f(x) \to \infty \quad \text{as} \quad x \to -2^-.$$

The vertical asymptotes are $x = 1$ and $x = 2$.

EXAMPLE 3

Sketch the graph of

$$f(x) = \frac{x^2 - x - 2}{x - 2},$$

labeling all horizontal and vertical asymptotes and x- and y-intercepts.

Solution   First factor the numerator, so

$$f(x) = \frac{(x - 2)(x + 1)}{x - 2} = x + 1, \quad \text{when} \quad x \neq 2.$$

The value $x = 2$ is not in the domain of f since it makes the denominator 0, so the domain of f is $(-\infty, 2) \cup (2, \infty)$. The graph has no horizontal or vertical asymptotes. The axis intercepts are $(0, 1)$ and $(-1, 0)$. The graph is the same as the graph of the line $y = x + 1$, except that the point $(2, 3)$ is removed.

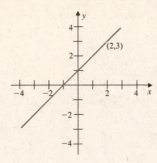

Be very careful not to assume a graph has a vertical asymptote whenever the denominator is 0. These are the only possibilities, but if the numerator and denominator have a common factor that can be cancelled, a zero of the denominator might not give a vertical asymptote. □

EXAMPLE 4

Sketch the graph of

$$f(x) = \frac{x - 1}{x^2 - 2x - 3},$$

labeling any horizontal and vertical asymptotes, and x- and y-intercepts.

Solution To determine the vertical asymptotes, first factor the denominator.

$$f(x) = \frac{x - 1}{x^2 - 2x - 3} = \frac{x - 1}{(x - 3)(x + 1)}$$

- Vertical Asymptotes: Setting the denominator to 0 gives

$$(x - 3)(x + 1) = 0 \Rightarrow x = 3 \quad \text{or} \quad x = -1,$$

so the vertical asymptotes are $x = 3$ and $x = -1$. To sketch the graph, analyze the behavior of the graph near the vertical asymptotes. The sign of the fraction tells whether the graph goes to infinity or negative infinity.

As $x \to 3^+$, the sign of the fraction is $\dfrac{+}{(+)(+)}$, so $f(x) \to \infty$.

As $x \to 3^-$, the sign of the fraction is $\dfrac{+}{(-)(+)}$, so $f(x) \to -\infty$.

As $x \to -1^+$, the sign of the fraction is $\dfrac{-}{(-)(+)}$, so $f(x) \to \infty$.

As $x \to -1^-$, the sign of the fraction is $\dfrac{-}{(-)(-)}$, so $f(x) \to -\infty$.

- Horizontal Asymptotes: The end behavior of the rational function is given by

$$f(x) = \frac{x - 1}{x^2 - 2x - 3} \approx \frac{x}{x^2} = \frac{1}{x}$$

$$f(x) \to 0 \quad \text{as} \quad x \to \infty$$
$$f(x) \to 0 \quad \text{as} \quad x \to -\infty.$$

The graph has a horizontal asymptote $y = 0$.

- *x*-intercepts: Solve $y = f(x) = 0$.

$$f(x) = \frac{x-1}{x^2 - 2x - 3} = 0$$
$$x - 1 = 0$$
$$x = 1$$

- *y*-intercept: Set $x = 0$, then $y = \frac{1}{3}$.

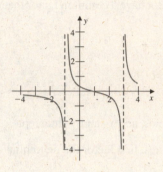

$\square$

EXAMPLE 5

Sketch the graph of

$$f(x) = \frac{x^2 - 4}{x^2 - 9},$$

labeling all horizontal and vertical asymptotes, and *x*- and *y*-intercepts.

Solution

$$f(x) = \frac{x^2 - 4}{x^2 - 9} = \frac{(x-2)(x+2)}{(x-3)(x+3)}$$

- Vertical Asymptotes: $x = -3, x = 3$

	Sign of Fraction	Behavior Near Asymptote
$x \to -3^+$	$\frac{(-)(-)}{(-)(+)}$	$f(x) \to -\infty$
$x \to -3^-$	$\frac{(-)(-)}{(-)(-)}$	$f(x) \to \infty$
$x \to 3^+$	$\frac{(+)(+)}{(+)(+)}$	$f(x) \to \infty$
$x \to 3^-$	$\frac{(+)(+)}{(-)(+)}$	$f(x) \to -\infty$

- Horizontal Asymptote: $y = 1$

$$f(x) = \frac{x^2 - 4}{x^2 - 9} \approx \frac{x^2}{x^2} = 1$$

- x-intercepts: $(2, 0), (-2, 0)$

$$\frac{(x-2)(x+2)}{(x-3)(x+3)} = 0$$
$$(x-2)(x+2) = 0$$
$$x = 2, x = -2$$

- y-intercept: $\left(0, \frac{4}{9}\right)$

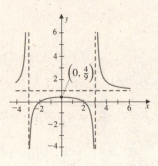

Slant Asymptotes

If the degree of the denominator of a rational function is exactly 1 greater than the degree of the numerator, then the graph approaches a non-horizontal line as $x \to \infty$ and as $x \to -\infty$. Such a line is called a *slant asymptote* to the graph. The equation of the slant asymptote comes from the fact that in this case

$$f(x) = \frac{P(x)}{Q(x)} = (ax + b) + \frac{R(x)}{Q(x)},$$

where a and b are constants and the degree of $R(x)$ is less than the degree of $Q(x)$. The quotient has the form $ax + b$, since the degree of $P(x)$ is exactly one greater than the degree of $Q(x)$.

EXAMPLE 6

Use a graphing device to sketch the graph of

$$f(x) = \frac{2x^2 + 3x - 1}{x + 2}.$$

Show any vertical and slant asymptotes, and x- and y-intercepts.

Solution There is a vertical asymptote at $x = -2$, and since the degree of the numerator is one greater than the degree of the denominator the graph also has a slant asymptote. We will use division to find the slant asymptote.

$$\begin{array}{r}
2x - 1 \\
x + 2 \overline{\smash{)}\,2x^2 + 3x - 1} \\
\underline{2x^2 + 4x} \\
-x - 1 \\
\underline{-x - 2} \\
1
\end{array}$$

Synthetic division can also be used.

$$\begin{array}{r}
-2 \underline{\,\big|\,2 \quad\quad 3 \quad -1} \\
-4 \quad\quad 2 \\
\hline
2 \quad -1 \quad\quad 1
\end{array}$$

The function can be written as

$$f(x) = 2x - 1 + \frac{1}{x + 2}.$$

Therefore,

$$\frac{1}{x+2} \to 0 \text{ as } x \to \infty \quad \text{so} \quad f(x) \to 2x - 1 \text{ as } x \to \infty$$

$$\frac{1}{x+2} \to 0 \text{ as } x \to -\infty \quad \text{so} \quad f(x) \to 2x - 1 \text{ as } x \to -\infty.$$

The slant asymptote is $y = 2x - 1$.

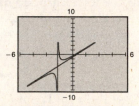

EXAMPLE 7

Determine a rational function that satisfies all of the following conditions.

(i) The graph has the vertical asymptotes $x = 3$ and $x = -2$.

(ii) The graph has the horizontal asymptote $y = 2$.

(iii) The graph has x-intercepts at 2 and 4.

Solution For the rational function to have vertical asymptotes $x = 3$ and $x = -2$, the simplified form of the function must have a denominator that contains the factors $(x - 3)$ and $(x + 2)$. The graph has the horizontal asymptote $y = 2$, so the numerator and the denominator have the same degree, and the leading coefficient of the numerator is twice that of the denominator. The graph crosses the x-axis at 2 and 4, so the numerator contains the factors $x - 2$ and $x - 4$. A possible function satisfying all these conditions is

$$f(x) = \frac{2(x-2)(x-4)}{(x-3)(x+2)} = \frac{2x^2 - 12x + 16}{x^2 - x - 6}.$$

$\square$

Applications

EXAMPLE 8

A can in the shape of a right cylinder with radius r and height h is to have a volume of 60 inches^3. The cost of the top and the bottom of the can is 2 cents per square inch, and the cost for the sides is 3 cents per square inch. Express the cost of the box in terms of
(a) the variables r and h,
(b) the variable r only, and
(c) the variable h only.
(d) Approximate the dimensions of the can that minimizes the cost.

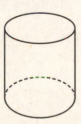

Solution The cost of a piece of material is the area times the price per square foot. The area of the top and bottom of the can is

$$A_{T,B} = 2(\pi r^2).$$

The area of the side of the can is the product of the height and the circumference of the can,

$$A_S = 2\pi rh.$$

(a) The cost of the can is

$$C = (2\pi r^2)(0.02) + (2\pi rh)(0.03) = 0.04\pi r^2 + 0.06\pi rh \text{ dollars.}$$

(b) To eliminate the variable h from the equation for cost, we need an equation relating h and r. We have yet to use the information that the volume of the can is 60 inches^3. The volume is

$$60V = (\text{area of the base}) \times (\text{height})$$
$$= \pi r^2 h.$$

Solving for h gives,

$$h = \frac{60}{\pi r^2},$$

and the cost in terms of the variable r is

$$C(r) = 0.04\pi r^2 + 0.06\pi r \left(\frac{60}{\pi r^2}\right) = 0.04\pi r^2 + \frac{3.6}{r} \text{ dollars.}$$

(c) Using the volume to solve for r in terms of h gives

$$\pi r^2 h = 60, \quad r^2 = \frac{60}{\pi h}, \quad \text{and} \quad r = \sqrt{\frac{60}{\pi h}}.$$

So

$$C(h) = 0.04\pi \left(\sqrt{\frac{60}{\pi h}}\right)^2 + 0.06\pi h \sqrt{\frac{60}{\pi h}} = \frac{2.4}{h} + 0.06\sqrt{60\pi h} \text{ dollars.}$$

(d) To approximate the dimensions of the can that minimizes the cost, graph the cost function and locate the minimum point on the graph. The minimum appears to occur when

$$r \approx 2.4 \quad \text{and} \quad h \approx \frac{60}{\pi (2.4)^2} \approx 3.3.$$

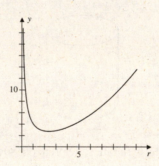

Exercise Set 3.4 (Page 168)

1. **a.** Domain$(f) = \{x \mid x \neq -1\} = (-\infty, -1) \cup (-1, \infty)$
    b. Range$(f) = \{x \mid x \neq 0\} = (-\infty, 0) \cup (0, \infty)$
    c. Vertical asymptote: $x = -1$; horizontal asymptote: $y = 0$

3. **a.** Domain$(f) = \{x \mid x \neq -1, x \neq 1\} = (-\infty, -1) \cup (-1, 1) \cup (1, \infty)$
    b. Range$(f) = (-\infty, 1) \cup [2, \infty)$
    c. Vertical asymptotes: $x = -1, x = 1$; horizontal asymptote: $y = 1$

5. a. $\text{Domain}(f) = (-\infty, 1)$

   b. $\text{Range}(f) = (-\infty, \infty)$

   c. Vertical asymptote: $x = 1$; horizontal asymptote: none

7. For $f(x) = \frac{(x-1)(x+1)}{(x-2)(x+2)}$, we have:

$x \to -2^{-}$	$\frac{(-)}{(-)(-)}$	$f(x) \to -\infty$
$x \to -2^{+}$	$\frac{(-)}{(-)(+)}$	$f(x) \to \infty$
$x \to 2^{-}$	$\frac{(+)}{(-)(+)}$	$f(x) \to -\infty$
$x \to 2^{+}$	$\frac{(+)}{(+)(+)}$	$f(x) \to \infty$
$x \to -\infty$	$\frac{(-)}{(-)(-)}$	$f(x) \to -1$
$x \to \infty$	$\frac{(+)}{(+)(+)}$	$f(x) \to 1$

9. For $f(x) = \frac{x^2-1}{x^2-2x+1}$, we have:

$x \to 1^{-}$	$\frac{(+)}{(-)}$	$f(x) \to 0$
$x \to 1^{+}$	$\frac{(+)}{(+)}$	$f(x) \to 0$
$x \to -1^{-}$	$\frac{(+)}{(+)}$	$f(x) \to -\infty$
$x \to -1^{+}$	$\frac{(+)}{(-)}$	$f(x) \to \infty$
$x \to -\infty$	$\frac{(-)}{(-)}$	$f(x) \to 1$
$x \to \infty$	$\frac{(+)}{(+)}$	$f(x) \to 1$

11. For $f(x) = \frac{x-3}{x+1}$, we have:

   a. $\text{Domain}(f) = \{x \mid x \neq -1\} = (-\infty, -1) \cup (-1, \infty)$

   b. Solve

$$\frac{x-3}{x+1} = 0 \text{ which implies } x - 3 = 0, \text{ so } x = 3.$$

x-intercepts: $(3, 0)$

Set $x = 0$ to get

$$y = \frac{0-3}{0+1} = -3.$$

y-intercept: $(0, -3)$

   c. Since for x large in magnitude

$$\frac{x-1}{x+1} \approx \frac{x}{x} = 1.$$

Vertical asymptote: $x = -1$; horizontal asymptote: $y = 1$.

13. For

$$f(x) = \frac{3x^2 - 11x - 4}{x^2 - 1} = \frac{(3x+1)(x-4)}{(x-1)(x+1)},$$

we have:

a. Domain(f) = $\{x \mid x \neq -1, x \neq 1\} = (-\infty, -1) \cup (-1, 1) \cup (1, \infty)$

b. Solve

$$\frac{(3x+1)(x-4)}{(x-1)(x+1)} = 0 \text{ which implies } 3x + 1 = 0, x - 4 = 0, \text{ so } x = 4, x = -\frac{1}{3}.$$

x-intercepts: $(4, 0)$, $\left(-\frac{1}{3}, 0\right)$

Set $x = 0$ to get

$$y = \frac{(1)(-4)}{(-1)(1)} = 4.$$

y-intercept: $(0, 4)$;

c. Since for x large in magnitude

$$\frac{(3x+1)(x-4)}{(x-1)(x+1)} \approx \frac{3x^2}{x^2} = 3.$$

Vertical asymptotes: $x = -1, x = 1$; horizontal asymptote: $y = 3$.

15. For

$$f(x) = \frac{x^3 - 2x^2 - x + 2}{x^2} = \frac{(x+1)(x-1)(x-2)}{x^2},$$

we have:

a. Domain(f) = $\{x \mid x \neq 0\} = (-\infty, 0) \cup (0, \infty)$

b. Solve

$$\frac{(x+1)(x-1)(x-2)}{x^2} = 0 \text{ which implies } (x+1)(x-1)(x-2) = 0, \text{ so } x = -1, 1, 2.$$

x-intercepts: $(-1, 0), (1, 0),$ and $(2, 0)$

y-intercept: None, since $x = 0$ is not in the domain of the function.

c. For x large in magnitude

$$\frac{(x+1)(x-1)(x-2)}{x^2} \approx \frac{x^3}{x^2} = x.$$

Vertical asymptotes: $x = 0$; horizontal asymptote: None

17. For $f(x) = \frac{3}{x-2}$. Vertical asymptotes: $x = 2$; horizontal asymptote: $y = 0$; x-intercepts: none; y-intercept: $\left(0, -\frac{3}{2}\right)$

19. For $f(x) = \frac{x}{x-3}$. Vertical asymptotes: $x = 3$; horizontal asymptote: $y = 1$; x-intercepts: $(0, 0)$; y-intercept: $(0, 0)$

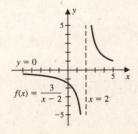

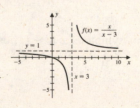

21. For $f(x) = \frac{(2x-1)(x+2)}{(3x+1)(x-3)}$. Vertical asymptotes: $x = -\frac{1}{3}$, $x = 3$; horizontal asymptote: $y = \frac{2}{3}$; x-intercepts: $\left(\frac{1}{2}, 0\right)$, $(-2, 0)$; y-intercept: $\left(0, \frac{2}{3}\right)$

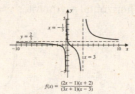

23. For

$$f(x) = \frac{x^2 - 9}{x^2 - 16} = \frac{(x-3)(x+3)}{(x-4)(x+4)}.$$

Vertical asymptotes: $x = 4$, $x = -4$; horizontal asymptote: $y = 1$; x-intercepts: $(-3, 0)$, $(3, 0)$; y-intercept: $(0, 9/16)$

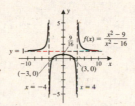

25. For

$$f(x) = \frac{2x - 3}{x^2 - x - 6} = \frac{2x - 3}{(x-3)(x+2)}.$$

Vertical asymptotes: None; horizontal asymptote: None; x-intercept: $(-3, 0)$; y-intercept: $(0, 3)$

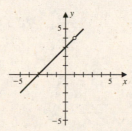

27For

$$f(x) = \frac{x^2 - x - 2}{x^2 - 2x - 3} = \frac{(x-2)(x+1)}{(x-3)(x+1)}$$
$$= \frac{x - 2}{x - 3}, \text{ for } x \neq -1.$$

Vertical asymptote: $x = 3$; horizontal asymptote: $y = 1$; x-intercepts: $(2, 0)$; y-intercept: $(0, 2/3)$PA240

29. For

$$f(x) = \frac{3x - 2}{x^2 + x - 6} = \frac{3x - 2}{(x - 2)(x + 3)}.$$

Vertical asymptotes: $x = -3$, $x = 2$; horizontal asymptote: $y = 0$; x-intercepts: $(2/3, 0)$; y-intercept: $(0, 1/3)$

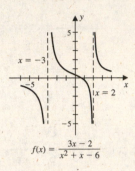

$$f(x) = \frac{3x - 2}{x^2 + x - 6}$$

31. One example is $f(x) = \dfrac{1}{x - 1}$.

33. One example is

$$f(x) = \frac{(x - 3)(x + 4)}{(x - 2)(x + 3)}.$$

35. One example is

$$f(x) = \frac{3(x - 2)}{x} = \frac{3x - 6}{x}.$$

37. $f(x) = \frac{x^2}{x-1} = x + 1 + \dfrac{1}{x - 1}$

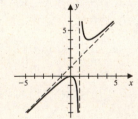

39. $f(x) = \dfrac{x^2 + x - 2}{x + 1} = x - \dfrac{2}{x + 1}$

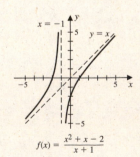

$$f(x) = \frac{x^2 + x - 2}{x + 1}$$

41. $f(x) = \dfrac{x}{x^2 - 3x + 2} = \dfrac{x}{(x-1)(x-2)}$

43. $f(x) = \dfrac{x^2}{x^2 - 5x + 6} = \dfrac{x^2}{(x-2)(x-3)}$

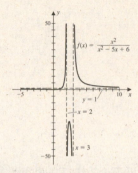

45. We have $A = \pi r^2 + 2\pi rh$ and $318 = \pi r^2 h$ which implies $h = \frac{318}{\pi r^2}$, so

$$A(r) = \pi r^2 + 2\pi r\left(\frac{318}{\pi r^2}\right) = \pi r^2 + \frac{636}{r}.$$

From the graph of $y = A(r)$, the minimum point can be estimated to be

$$r \approx 4.66, \quad \text{so } h = \frac{318}{\pi r^2} \approx 4.66.$$

47. The number of bacteria at time t is $n = 10{,}000\left(\dfrac{3t^2 + 1}{t^2 + 1}\right)$.

a. For t large,

$$n = 10{,}000\left(\frac{3t^2 + 1}{t^2 + 1}\right) \approx 10{,}000\left(\frac{3t^2}{t^2}\right) = 30{,}000$$

so the bacteria colony stabilizes.

b. The bacteria colony stabilizes to the level 30,000.

c. To find when the number of bacteria exceeds 22,000, solve

$$10{,}000\left(\frac{3t^2 + 1}{t^2 + 1}\right) > 22{,}000 \text{ or } 3t^2 + 1 > 2.2t^2 + 2.2$$

This implies that

$$0.8t^2 > 1.2, \text{ so } t^2 - 1.5 > 0 \text{ and } t > \sqrt{1.5} = \frac{\sqrt{6}}{2} \approx 1.2.$$

49. a. For $P(t) = \dfrac{at^2}{t^2 + 1}$, the parameter a determines the height of the curve.

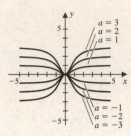

   b. As t increases the population stabilizes to the value a.

· ·

3.5   Other Algebraic Functions

Power Functions

The *rational power functions* have the form

$$f(x) = x^{\frac{m}{n}} = \left(\sqrt[n]{x}\right)^m = \sqrt[n]{x^m},$$

where $\dfrac{m}{n}$ is a rational number, and n is an integer greater than 1.

EXAMPLE 1
Use the graph of $y = x^{\frac{1}{3}}$, to sketch the graph of the function.
(a) $f(x) = (x - 2)^{\frac{1}{3}} - 1$ (b) $f(x) = -2(x - 1)^{\frac{1}{3}} + 2$ (c) $f(x) = (x + 1)^{\frac{2}{3}}$

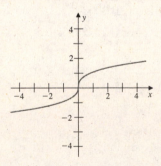

Solution (a) Shift the graph of $y = x^{\frac{1}{3}}$ to the right 2 units and downward 1 unit.

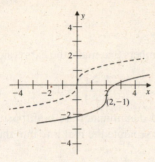

(b) Vertically scale the graph of $y = x^{\frac{1}{3}}$ by a factor of 2, so for $x > 0$ the graph of $y = 2x^{\frac{1}{3}}$ is above the graph of $y = x^{\frac{1}{3}}$, and it is below for $x < 0$. Then reflect the resulting graph about the x-axis and shift it to the right 1 unit and upward 2 units to obtain the graph of $f(x) = -2(x - 1)^{\frac{1}{3}} + 2$.

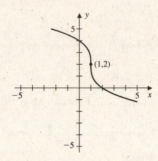

(c) The square of a positive number less than 1 is less than the number itself, and the square of a number greater than 1 is greater than the number itself. We use this to compare the graphs of $y = x^{2/3}$ and $y = x^{1/3}$.

- For $0 < x < 1$, we have $x^{\frac{2}{3}} = \left(x^{\frac{1}{3}}\right)^2 < x^{\frac{1}{3}}$ and the graph of $y = x^{2/3}$ is below the graph of $y = x^{1/3}$.

- For $x > 1$, we have $x^{\frac{2}{3}} = \left(x^{\frac{1}{3}}\right)^2 > x^{\frac{1}{3}}$ and the graph of $y = x^{2/3}$ is above the graph of $y = x^{1/3}$.

- For $x = 1$, we have $x^{\frac{2}{3}} = x^{\frac{1}{3}}$, and the graphs intersect.

Since

$$(-x)^{\frac{2}{3}} = \left(\sqrt[3]{-x}\right)^2 = \left(-\sqrt[3]{x}\right)^2 = \left(\sqrt[3]{x}\right)^2 = x^{\frac{2}{3}},$$

the function $g(x) = x^{\frac{2}{3}}$ is even and the graph is symmetric with respect to the y-axis. The graph of $f(x) = (x + 1)^{\frac{2}{3}}$ is obtained by shifting the graph of $y = x^{\frac{2}{3}}$ to the left 1 unit.

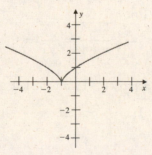

$\square$

EXAMPLE 2

Determine the domain of the function $\sqrt{\dfrac{2 - x}{x + 1}}$.

Solution The domain is the set of all real numbers that make the expression under the radical greater than or equal to 0. Solve

$$\frac{2-x}{x+1} \ge 0,$$

which is greater than 0 when numerator and denominator are both positive, or are both negative. It is zero when $2 - x = 0$, that is, $x = 2$. The linear factors separate the real line into the intervals $(-\infty, -1)$, $(-1, 2)$ and $(2, \infty)$, as shown on the sign chart.

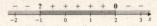

The domain is $(-1, 2]$.   □

EXAMPLE 3

Sketch the graph of $f(x) = \sqrt{\dfrac{x-3}{x+1}}$, and label any axis intercepts and asymptotes.

Solution

- Domain: $\left\{ x \mid \dfrac{x-3}{x+1} \ge 0 \right\} = (-\infty, -1) \cup [3, \infty)$. This can be verified from the sign chart.

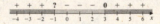

- x-intercepts:

$$\sqrt{\frac{x-3}{x+1}} = 0$$
$$\frac{x-3}{x+1} = 0$$
$$x - 3 = 0$$
$$x = 3$$

The graph crosses the x-axis at $(3, 0)$.

- y-intercepts: None, since $x = 0$ is not in the domain of the function.
- Horizontal asymptote: $y = 1$
  For $|x|$ large the end behavior is

$$\sqrt{\frac{x-3}{x+1}} \approx \frac{\sqrt{x}}{\sqrt{x}} = 1.$$

- Vertical asymptote: $x = -1$.
  Because of the domain of f, x can only approach -1 from the left, and

$$f(x) \to \infty \quad \text{as} \quad x \to -1^-.$$

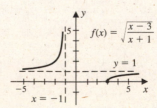

□

Exercise Set 3.5 (Page 175)

1. The domain of $f(x) = \sqrt{x^2 + 2x - 15}$ consists of all x satisfying $x^2 + 2x - 15 \geq 0$ which implies $(x + 5)(x - 3) \geq 0$, so $x \leq -5$ or $x \geq 3$, that is, $(-\infty, -5] \cup [3, \infty)$.

3. The domain of $f(x) = \sqrt{(x - 1)^2(x + 2)}$ consists of all x satisfying $(x - 1)^2(x + 2) \geq 0$ which implies $(x + 2) \geq 0$, so $x \geq -2$, that is, $[-2, \infty)$.

5. The domain of $f(x) = \sqrt{(x + 1)(x - 3)(x + 4)}$ consists of all x satisfying $(x + 1)(x - 3)(x + 4) \geq 0$. The solution to this inequality can be found from the following sign chart. The domain is where the last inequality is positive or 0, that is, $[-4, -1] \cup [3, \infty)$.

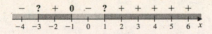

7. The domain of
$$f(x) = \sqrt{\frac{1 - x}{x + 3}}$$
consists of all x satisfying $\dfrac{1 - x}{x + 3} \geq 0$ which implies $\dfrac{x - 1}{x + 3} \leq 0$ so $-3 < x \leq 1$, that is, $(-3, 1]$.

9. The domain of
$$f(x) = \sqrt{\frac{x + 1}{x^2 + 2x - 3}} = \sqrt{\frac{x + 1}{(x + 3)(x - 1)}}$$
consists of all x satisfying $\dfrac{x + 1}{(x + 3)(x - 1)} \geq 0$. The solution to this inequality can be found from the

following sign chart. The domain is the where the last inequality is positive or 0, that is, $(-3, -1] \cup (1, \infty)$.

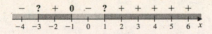

11. **a.** iii   **b.** vi   **c.** i   **d.** v   **e.** iv   **f.** ii

13. $f(x) = x^{3/2} - 2$

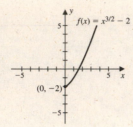

15. $f(x) = (x - 1)^{1/3} + 1$

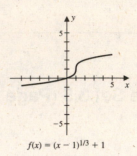

17. $f(x) = -2(x + 1)^{2/3} - 2$

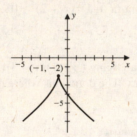

19.
$$f(x) = (1 - x)^{4/3} - 1$$
$$= (-(x - 1))^{4/3} - 1$$
$$= (x - 1)^{4/3} - 1$$

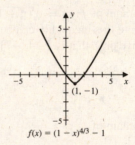

21. $f(x) = x^{-1/2} - 2$

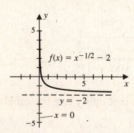

23. $f(x) = \sqrt{\dfrac{x + 2}{x - 1}}$

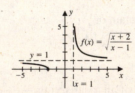

25. $f(x) = \sqrt{\dfrac{2-x}{x+2}}$

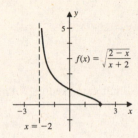

27. $f(x) = \dfrac{x}{\sqrt{x^2+1}}$

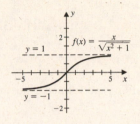

29. $f(x) = \dfrac{x-1}{\sqrt{(x+1)(x-2)}}$

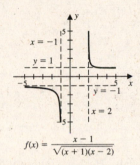

31. $f(x) = \dfrac{x}{\sqrt{4-x^2}} = \dfrac{x}{\sqrt{(2-x)(2+x)}}$

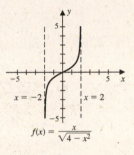

33. $w \approx 0.398$ kg.

. . . .

3.6   Complex Roots of Polynomials

A *complex number* is an expression of the form $a + bi$, where a and b are real numbers, and $i = \sqrt{-1}$, that is, $i^2 = -1$. The number a is called the *real part*, and b is called the *imaginary part*. The *conjugate* of the complex number $a + bi$ is $\overline{a + bi} = a - bi$.

Arithmetic Operations on Complex Numbers

EXAMPLE 1

Write the complex number in standard form $a + bi$.

  (a) $(1 - 3i) + (2 + 4i)$ (b) $(3 - 2i) - (7 + 9i)$

  (c) $(2 - 3i) \cdot (3 + 5i)$ (d) $(4 - 3i) \cdot \overline{(7 - 2i)}$

  (e) $\dfrac{2}{3 - i}$ (f) $\dfrac{4 - i}{2 + 3i}$

Solution (a) Addition and subtraction of complex numbers is performed simply by adding and subtracting real and imaginary parts separately.

$$(1 - 3i) + (2 + 4i) = (1 + 2) + (-3 + 4)i = 3 + i$$

(b)

$$(3 - 2i) - (7 + 9i) = (3 - 7) + (-2 - 9)i = -4 - 11i$$

(c) Multiplication is performed using the distributive law for multiplying real numbers, keeping in mind that $i^2 = -1$.

$$
\begin{aligned}
(2 - 3i) \cdot (3 + 5i) &= 2(3 + 5i) - 3i(3 + 5i) \\
&= 6 + 10i - 9i - 15i^2 \\
&= 6 + i - 15(-1) \\
&= 21 + i
\end{aligned}
$$

(d)

$$
\begin{aligned}
(4 - 3i) \cdot \overline{(7 - 2i)} &= (4 - 3i) \cdot (7 + 2i) \\
&= 28 + 8i - 21i - 6i^2 \\
&= 28 - 13i + 6 \\
&= 34 - 13i
\end{aligned}
$$

(e) Complex division is similar to rationalizing a fraction containing radicals. Here we multiply the numerator and denominator of the complex fraction by the conjugate of the denominator.

$$
\begin{aligned}
\frac{2}{3 - i} &= \frac{2}{3 - i} \cdot \frac{\overline{3 - i}}{\overline{3 - i}} \\
&= \frac{2}{3 - i} \cdot \frac{3 + i}{3 + i} \\
&= \frac{6 + 2i}{9 - i^2} \\
&= \frac{6 + 2i}{10} = \frac{3}{5} + \frac{1}{5}i
\end{aligned}
$$

(f) The conjugate of $2 + 3i$ is $2 - 3i$, so

$$
\begin{aligned}
\frac{4 - i}{2 + 3i} &= \frac{4 - i}{2 + 3i} \cdot \frac{2 - 3i}{2 - 3i} \\
&= \frac{8 - 12i - 2i + 3i^2}{4 - 9i^2} \\
&= \frac{5 - 14i}{13} = \frac{5}{13} - \frac{14}{13}i.
\end{aligned}
$$

$\square$

EXAMPLE 2

Show that $1 - i$ is a solution to the equation $f(x) = x^2 - 2x + 2 = 0$.

Solution Substituting $1 - i$ for x gives

$$
\begin{aligned}
f(1 - i) &= (1 - i)^2 - 2(1 - i) + 2 \\
&= (1 - i)(1 - i) - 2 + 2i + 2 \\
&= 1 - 2i + i^2 + 2i \\
&= 1 + i^2 \\
&= 1 - 1 = 0.
\end{aligned}
$$

□

Complex Zeros of Polynomials

EXAMPLE 3

Find the zeros of the quadratic function $f(x) = x^2 - 2x + 4$, and write the function in factored form.

Solution Since

$$f(x) = x^2 - 2x + 4 = (x^2 - 2x + 1) + 3 = (x - 1)^2 + 3 > 0$$

for all x, there are no real factors of $f(x)$, so we use the quadratic formula to first find the zeros. The quadratic equation

$$x^2 - 2x + 4 = 0$$

has

$$a = 1, \ b = -2, \ c = 4,$$

so

$$
\begin{aligned}
x &= \frac{-b \pm \sqrt{b^2 - 4ac}}{2a} \\
&= \frac{2 \pm \sqrt{4 - 4(1)(4)}}{2(1)} \\
&= \frac{2 \pm \sqrt{-12}}{2} \\
&= \frac{2 \pm 2\sqrt{-3}}{2} = 1 \pm \sqrt{-3}.
\end{aligned}
$$

The discriminant, which is the term under the radical, is negative, so the zeros are complex numbers. Since

$$\sqrt{-3} = \sqrt{3}\sqrt{-1} = \sqrt{3}i,$$

the zeros are

$$x = 1 \pm \sqrt{3}i,$$

and the quadratic can be factored as

$$f(x) = x^2 - 2x + 4$$
$$= \left(x - \left(1 + \sqrt{3}i\right)\right)\left(x - \left(1 - \sqrt{3}i\right)\right).$$

□

EXAMPLE 4

Find the zeros of

$$f(x) = x^4 + 2x^2 - 8.$$

Solution The expression $x^4 + 2x^2 - 8$ is quadratic in the variable x^2. Let $u = x^2$. Then

$$x^4 + 2x^2 - 8 = u^2 + 2u - 8 = 0$$
$$(u - 2)(u + 4) = 0$$
$$u = 2, u = -4.$$

Since $u = x^2$,

$$x^2 = 2, x^2 = -4$$
$$x = \pm\sqrt{2}, x = \pm\sqrt{-4} = \pm 2i.$$

So the zeros of the polynomial are $\sqrt{(2)}$, $-\sqrt{(2)}$, $2i$, and $-2i$. □

Polynomials with real coefficients have the property that if $a + bi$ is a zero with multiplicity k, then its conjugate $a - bi$ is also a zero with multiplicity k.

EXAMPLE 5

Show that $x = -1 + 2i$ is a zero of the function $f(x) = x^3 + x - 10$, find all zeros of $f(x)$, and factor $f(x)$ completely.

Solution We have

$$f(-1 + 2i) = (-1 + 2i)^3 + (-1 + 2i) - 10$$
$$= (-1 + 2i)(-1 + 2i)^2 - 11 + 2i$$
$$= (-1 + 2i)(1 - 4i + 4i^2) - 11 + 2i$$
$$= (-1 + 2i)(-3 - 4i) - 11 + 2i$$
$$= 3 + 4i - 6i - 8i^2 - 11 + 2i$$
$$= 3 + 6i - 6i + 8 - 11 = 0.$$

Since $-1 + 2i$ is a zero of the function $f(x)$, its conjugate $-1 - 2i$ is also a zero. To find the last solution, divide the polynomial

$$(x - (-1 + 2i))(x - (-1 - 2i)) = x^2 + 2x + 5,$$

into $x^3 + x - 10$.

$$
\begin{array}{r}
x - 2 \\
x^2 + 2x + 5 \overline{\smash{\big)}\, x^3 + 0x^2 + x - 10} \\
\underline{x^3 + 2x^2 + 5x} \\
-2x^2 - 4x - 10 \\
\underline{-2x^2 - 4x - 10} \\
0
\end{array}
$$

Thus

$$
x^3 + x - 10 = (x - (-1 + 2i))(x - (-1 - 2i))(x - 2),
$$

and the zeros of $f(x) = x^3 + x - 10$ are

$$
x = -1 + 2i, \ x = -1 - 2i, \ \text{and} \ x = 2.
$$

$\square$

EXAMPLE 6

Find all zeros and factor completely the function $f(x) = x^4 + 2x^3 - 2x^2 - 8x - 8$.

Solution

- Possible rational zeros: $\pm 1, \pm 2, \pm 4, \pm 8$

We can substitute these values and check whether $f(x) = 0$. For example, $f(1) = -15 \neq 0$, $f(-1) = -3 \neq 0$, and so on. However, the computer generated graph of $y = f(x) = x^4 + 2x^3 - 2x^2 - 8x - 8$ indicates likely zeros at $x = 2$ and $x = -2$.

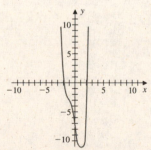

Use synthetic division to verify these are zeros and to factor $f(x)$.

$$
\begin{array}{r|rrrrr}
2 & 1 & 2 & -2 & -8 & -8 \\
  &   & 2 & 8 & 12 & 8 \\
\hline
  & 1 & 4 & 6 & 4 & 0
\end{array}
$$

$$
\begin{array}{r|rrrr}
-2 & 1 & 4 & 6 & 4 \\
   &   & -2 & -4 & -4 \\
\hline
   & 1 & 2 & 2 & 0
\end{array}
$$

So $x = 2$ and $x = -2$ are zeros and

$$f(x) = (x - 2)(x + 2)(x^2 + 2x + 2).$$

The solutions to $x^2 + 2x + 2 = 0$ are

$$x = \frac{-2 \pm \sqrt{4 - 4(1)(2)}}{2} = \frac{-2 \pm \sqrt{-4}}{2} = \frac{-2 \pm 2i}{2} = -1 \pm i.$$

- Zeros: $x = 2, x = -2, x = -1 + i, x = -1 - i$
- Factored: $f(x) = (x - 2)(x + 2)(x - (-1 + i))(x - (-1 - i))$

□

EXAMPLE 7

Find a polynomial of degree 5 with real coefficients, leading coefficient 1, zeros i, and $3 - i$, and whose graph passes through the origin.

Solution Since i and $3 - i$ are zeros, their conjugates $-i$ and $3 + i$ are also zeros. Since the graph passes through the origin, 0 is another zero. The only fifth-degree polynomial with leading coefficient 1 that satisfies these conditions is

$$f(x) = x(x + i)(x - i)(x - (3 - i))(x - (3 + i))$$
$$= x^5 - 6x^4 + 11x^3 - 6x^2 + 10x.$$

□

Exercise Set 3.6 (Page 184)

1. $(-3+i) + (2-3i) = (-3+2) + (1-3)i$
$= -1 - 2i$

3. $(-3+5i) - (2-3i) = (-3-2) + (5+3)i$
$= -5 + 8i$

5. $3i \cdot (2+i) = 6i + 3i^2 = -3 + 6i$

7. $(2-i) \cdot (3+i) = 6 + 2i - 3i - i^2 = 7 - i$

9. $(6+5i) \cdot (-3-2i) = -18 - 12i - 15i - 10i^2$
$= -8 - 27i$

11. $(2-3i) \cdot (2+3i) = 4 + 6i - 6i - 9i^2$
$= 13$

13. $(3-8i) \cdot \overline{(2+i)} = (3-8i)(2-i)$
$= 6 - 3i - 16i + 8i^2 = -2 - 19i$

15. $i^5 = i^4 \cdot i = i^2 \cdot i^2 \cdot i =$
$(-1)(-1)i = i$

17. $i^{104} = (i^4)^{26}$
$= 1^{26} = 1$

19. $\sqrt{-9} = 3\sqrt{-1}$
$= 3i$

21. $\left(2+\sqrt{-5}\right)\left(1+\sqrt{-1}\right) = \left(2+\sqrt{5}i\right)\left(1+i\right)$
which reduces to
$2 + 2i + \sqrt{5}i + \sqrt{5}i^2 = \left(2-\sqrt{5}\right) + \left(2+\sqrt{5}\right)i.$

23. $\sqrt{-\dfrac{16}{9}} = \dfrac{4}{3}\sqrt{-1} = \dfrac{4}{3}i$

25. $\dfrac{1}{1-i} = \dfrac{1}{1-i} \cdot \dfrac{1+i}{1+i} = \dfrac{1+i}{2} = \dfrac{1}{2} + \dfrac{1}{2}i$

27. $\dfrac{1}{2+3i} = \dfrac{1}{2+3i} \cdot \dfrac{2-3i}{2-3i} = \dfrac{2-3i}{13} = \dfrac{2}{13} - \dfrac{3}{13}i$

29. $\dfrac{1-4i}{1+4i} = \dfrac{1-4i}{1+4i} \cdot \dfrac{1-4i}{1-4i} = \dfrac{1-8i+16i^2}{17} = \dfrac{-15-8i}{17} = -\dfrac{15}{17} - \dfrac{8}{17}i$

31. a. Solving for the zeros we have

$$x^2 + 4 = 0 \text{ implies } x^2 = -4 \text{ so } x = \pm\sqrt{-4} = \pm 2i.$$

b. $f(x) = x^2 + 4 = (x - 2i)(x + 2i)$

33. a. Solving for the zeros we have

$$x^2 - 2x + 2 = 0 \text{ implies } x = \frac{2 \pm \sqrt{(-2)^2 - 4(1)(2)}}{2} = \frac{2 \pm \sqrt{-4}}{2} = \frac{2 \pm 2i}{2} = 1 \pm i.$$

b. $f(x) = (x - (1 - i))(x - (1 + i))$

35. a. Solving for the zeros we have

$$2x^2 - x + 2 = 0 \text{ implies } x = \frac{1 \pm \sqrt{(-1)^2 - 4(2)(2)}}{2(2)} = \frac{1 \pm \sqrt{-15}}{4} = \frac{1 \pm \sqrt{15}i}{4}.$$

b.
$$f(x) = \left(x - \left(\frac{1 - \sqrt{15}i}{4} \right) \right) \left(x - \left(\frac{1 + \sqrt{15}i}{4} \right) \right)$$

37. We have
$$x^4 - 4 = (x^2 - 2)(x^2 + 2) = (x - \sqrt{2})(x + \sqrt{2})(x - \sqrt{2}i)(x + \sqrt{2}i) = 0,$$
and the roots are $-\sqrt{2}, \sqrt{2}, -\sqrt{2}i, \sqrt{2}i$.

39. We have
$$x^4 - x^2 - 6 = (x^2 - 3)(x^2 + 2) = \left(x - \sqrt{3} \right) \left(x + \sqrt{3} \right) \left(x - \sqrt{2}i \right) \left(x + \sqrt{2}i \right) = 0,$$
and the roots are $-\sqrt{3}, \sqrt{3}, -\sqrt{2}i, \sqrt{2}i$.

41. By inspection, $x = -2$ is a solution to $x^3 + 8 = 0$, so $x + 2$ is a factor. Dividing $x^3 + 1$ by $x + 2$ gives $x^3 + 1 = (x + 2)(x^2 - 2x + 4)$. Then solving for the roots of $x^2 - 2x + 4 = 0$ gives

$$x = \frac{2 \pm \sqrt{4 - 4(1)(4)}}{2} = \frac{2 \pm \sqrt{-12}}{2} = \frac{2 \pm 2\sqrt{3}i}{2} = 1 \pm \sqrt{3}i.$$

The roots are $-2, 1 + \sqrt{3}i, 1 - \sqrt{3}i$.

43. For $f(x) = x^3 - 2x^2 + 9x - 18$, we have

$$f(3i) = (3i)^3 - 2(3i)^2 + 9(3i) - 18 = -27i + 18 + 27i - 18 = 0.$$

Since $3i$ is a root, the conjugate $-3i$ is also a root, so $(x - 3i)(x + 3i) = x^2 + 9$ is a factor. Dividing $x^2 + 9$ into $f(x)$ gives
$$f(x) = (x^2 + 9)(x - 2).$$

The third solution is $x = 2$.

45. For $f(x) = x^4 - 2x^3 - 2x^2 - 2x - 3$, we have

$$f(i) = (i)^4 - 2(i)^3 - 2(i)^2 - 2(i) - 3 = 1 + 2i + 2 - 2i - 3 = 0.$$

Since i is a zero, the conjugate $-i$ is also a zero, so

$$(x - i)(x + i) = x^2 + 1$$

is a factor. Dividing $x^2 + 1$ into $f(x)$ gives

$$f(x) = (x^2 + 1)(x^2 - 2x - 3) = (x^2 + 1)(x - 3)(x + 1).$$

The third and fourth solutions are $x = -1, x = 3$.

47. The possible rational roots of $f(x) = x^3 - 3x^2 + 9x - 27$ are

$$\pm 1, \pm 3, \pm 9, \pm 27,$$

and by substitution, $x = 3$ is a root. Then

$$f(x) = (x - 3)(x^2 + 9) = (x - 3)(x - 3i)(x + 3i).$$

49. Since $f(x) = x^4 - x^2 - 2x + 2$ has a zero of multiplicity 2 at $x = 1$, we have

$$f(x) = (x - 1)^2(x^2 + 2x + 2) = (x - 1)^2(x - (-1 - i))(x - (-1 + i)).$$

51. Since $2i$ is a root, the conjugate $-2i$ is also a root and

$$f(x) = (x - 2)(x - 2i)(x + 2i) = (x - 2)(x^2 + 4) = x^3 - 2x^2 + 4x - 8.$$

53. Since $\sqrt{3}i$ and $3i$ are roots, the conjugates $-\sqrt{3}i$ and $-3i$ are also roots, so a polynomial with the specified roots is

$$f(x) = (x - \sqrt{3}i)(x + \sqrt{3}i)(x - 3i)(x + 3i) = (x^2 + 3)(x^2 + 9) = x^4 + 12x^2 + 27.$$

55. Since $x = -2$ is a zero of multiplicity 2, $(x + 2)^2$ is a factor, and since $1 + 2i$ is a root, the conjugate $1 - 2i$ is also a root and since $x = 0$ is a root x is a factor. So

$$f(x) = x(x + 2)^2(x - (1 + 2i))(x - (1 - 2i))$$
$$= x(x^2 + 4x + 4)(x^2 - 2x + 5) = x^5 + 2x^4 + x^3 + 12x^2 + 20x.$$

57. To have a real number solution it must be true that $a \geq 2b$.

Review Exercises for Chapter 3 (Page 185)

1. a. (ii) The graph crosses the x-axis at $x = 1$ and $x = -2$. Since the zero at $x = -1$ is of multiplicity 2, the curve just touches and turns without crossing the x-axis at $x = -1$.

b. (iii) The graph crosses the x-axis at $x = 0$ and $x = 3$. Since the zero at $x = 2$ is of multiplicity 2, the curve just touches and turns without crossing the x-axis at $x = 2$.

c. (iv) The graph crosses the x-axis at $x = 1$ and $x = -2$. Since the zero at $x = 1$ is of multiplicity 3, the curve flattens and crosses the x-axis at $x = 1$.

d. (i) The graph touches without crossing the x-axis at $x = 1$ and $x = -2$ since both zeros are of multiplicity 2.

3. $f(x) = -2(x-1)^2 + 2$

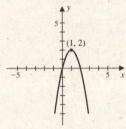

$$f(x) = -2(x-1)^2 + 2$$

5. $f(x) = -x^4 - 3$

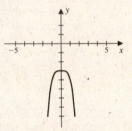

7. $f(x) = (x+1)(x+2)(x-3)$

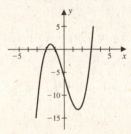

$$f(x) = (x+1)(x+2)(x-3)$$

9. $f(x) = \frac{1}{2}(x-2)^3(x+1)$

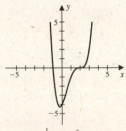

$$f(x) = \frac{1}{2}(x-2)^3(x+1)$$

11.

$$f(x) = x^3 - \frac{1}{2}x^2 - \frac{1}{2}x$$

$$= \frac{1}{2}x(2x^2 - x - 1)$$

$$= \frac{1}{2}x(2x + 1)(x - 1)$$

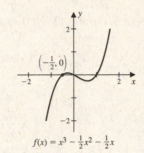

$$f(x) = x^3 - \frac{1}{2}x^2 - \frac{1}{2}x$$

13. Degree 3, leading coefficient positive. The graph shows a zero of at least multiplicity 2, and one other zero of multiplicity 1, so the degree is at least 3. The end behavior is the same as that for $y = x^3$, so the leading coefficient is positive.

15. Degree 4, leading coefficient negative. Adding or subtracting a positive constant from a polynomial shifts the curve upward or downward but does not change the degree. The curve can be shifted downward so as to appear to have 4 zeros of multiplicity 1, so the degree of the original polynomial is at least 4. Since the end behavior is the opposite of that of $y = x^4$ the leading coefficient is negative.

17. $Q(x) = 2x + 9$ and $R(x) = 23$

19. $Q(x) = 3x^2 - 4x + 6$ and $R(x) = -11$

21. $P(x) = 3x^4 - 9x^3 - 2x^2 + 5x + 3$

a.
$$\frac{\pm 1, \pm 3}{\pm 1, \pm 3} = \pm 1, \pm \frac{1}{3}, \pm 3$$

b. Since
$$P(3) = 3(3)^4 - 9(3)^3 - 2(3)^2 + 5(3) + 3 = 3^5 - 3^5 - 18 + 15 + 3 = 0,$$

$x - 3$ is a factor.

c. $P(x) = (x - 3)(3x^3 - 2x - 1) = (x - 3)(x - 1)(3x^2 + 3x + 1)$

23. $P(x) = x^5 - 3x^4 - 5x^3 + 27x^2 - 32x + 12$

a.
$$\frac{\pm 1, \pm 2, \pm 3, \pm 4, \pm 6, \pm 12}{\pm 1} = \pm 1, \pm 2, \pm 3, \pm 4, \pm 6, \pm 12$$

b. $P(-3) = (-3)^5 - 3(-3)^4 - 5(-3)^3 + 27(-3)^2 - 32(-3) + 12 = 0$, so $x + 3$ is a factor.

c. $P(x) = (x + 3)(x^4 - 6x^3 + 13x^2 - 12x + 4) = (x + 3)(x - 1)^2(x - 2)^2$

25. $P(x) = x^4 - 5x^3 + 2x^2 + 22x - 20$

Since $3 + i$ is a zero, so is $3 - i$ and

$$(x - (3 + i))(x - (3 - i)) = x^2 - 6x + 10.$$

Then

$$P(x) = (x^2 - 6x + 10)(x^2 + x - 2) = (x - (3 + i))(x - (3 - i))(x + 2)(x - 1).$$

27. $P(x) = x^5 - x^4 + 10x^3 - 10x^2 + 9x - 9$

Since i is a zero, so is $-i$ and

$$(x - i)(x + i) = x^2 + 1.$$

Then

$$P(x) = (x^2 + 1)(x^3 - x^2 + 9x - 9) = (x^2 + 1)(x - 1)(x^2 + 9)$$
$$= (x - i)(x + i)(x - 1)(x - 3i)(x + 3i).$$

29. For $f(x) = \dfrac{x - 4}{x - 1}$.

a. Domain$(f) = \{x \mid x \neq 1\} = (-\infty, 1) \cup (1, \infty)$

b. x-intercepts: Solve

$$\frac{x - 4}{x - 1} = 0 \text{ which implies } x - 4 = 0, \text{ so } x = 4;$$

y-intercept: Set $x = 0$ to get $y = 4$.

c. Vertical asymptotes: Solve $x - 1 = 0$ implies $x = 1$; horizontal asymptotes: For x large in magnitude

$$\frac{x - 4}{x - 1} \approx \frac{x}{x} = 1, \quad \text{so} \quad y = 1.$$

31. For $f(x) = \dfrac{x^2 - 2x + 1}{2x^2 - 18} = \dfrac{(x - 1)^2}{2(x - 3)(x + 3)}$.

a. Domain$(f) = \{x \mid x \neq -3, x \neq 3\} = (-\infty, -3) \cup (-3, 3) \cup (3, \infty)$

b. x-intercepts: Solve

$$\frac{(x - 1)^2}{2(x - 3)(x + 3)} = 0 \text{ which implies } (x - 1)^2 = 0, \text{ so } x = 1;$$

y-intercept: Set $x = 0$ to give $y = -\frac{1}{18}$.

c. Vertical asymptotes: Solve $(x - 3)(x + 3) = 0$ implies $x = -3, x = 3$; horizontal asymptotes: For x large in magnitude,

$$\frac{(x - 1)^2}{2(x - 3)(x + 3)} \approx \frac{x^2}{2x^2} = \frac{1}{2}, \quad \text{so} \quad y = \frac{1}{2}.$$

33. For $f(x) = \dfrac{x^3 - x^2 - 4x + 4}{x^3} = \dfrac{(x + 2)(x - 2)(x - 1)}{x^3}$.

a. Domain$(f) = \{x \mid x \neq 0\} = (-\infty, 0) \cup (0, \infty)$

b. x-intercepts: Solve

$$\frac{(x + 2)(x - 2)(x - 1)}{x^3} = 0$$

implies

$$x + 2 = 0, x - 2 = 0, x - 1 = 0, \text{ so } x = -2, x = 2, x = 1;$$

y-intercept: None, since 0 is not in the domain.

c. Vertical asymptotes: Solve $x^3 = 0$ to get $x = 0$;

horizontal asymptotes: For x large in magnitude

$$\frac{(x + 2)(x - 2)(x - 1)}{x^3} \approx \frac{x^3}{x^3} = 1, \quad \text{so} \quad y = 1.$$

35. For $f(x) = \dfrac{3}{x - 2}$.

Horizontal asymptotes: $y = 0$; vertical asymptotes: $x = 2$; x-intercepts: none; y-intercept: $(0, -3/2)$

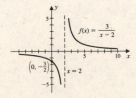

37. For $f(x) = -\dfrac{2}{(x - 1)(x - 2)}$.

Horizontal asymptotes: $y = 0$; vertical asymptotes: $x = 1, x = 2$; x-intercepts: none; y-intercept: $(0, -1)$

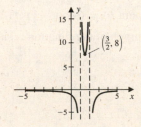

39. For

$$f(x) = \frac{4}{x^2 - 4} = \frac{4}{(x-2)(x+2)}.$$

Horizontal asymptotes: $y = 0$; vertical asymptotes: $x = -2$, $x = 2$; x-intercepts: none; y-intercept: $(0, -1)$

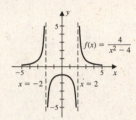

41. For

$$f(x) = \frac{x^2 - 4}{x^2 + 5x} = \frac{(x-2)(x+2)}{x(x+5)}.$$

Horizontal asymptotes: $y = 1$; vertical asymptotes: $x = -5$, $x = 0$; x-intercepts: $(2, 0)$, $(-2, 0)$; y-intercept: none

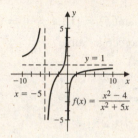

43. For

$$f(x) = \frac{x^2 - 2x + 1}{x + 1} = x - 3 + \frac{4}{x + 1},$$

we have the following graph.

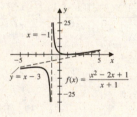

45. For $f(x) = \sqrt{\dfrac{x - 2}{x + 1}}$.

a. Domain$(f) = (-\infty, -1) \cup [2, \infty)$

b. The graph of the function is shown.

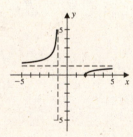

c. Horizontal asymptotes: $y = 1$; vertical asymptotes: $x = -1$; x-intercepts: $(2, 0)$; y-intercept: none

47. For $f(x) = \sqrt{\dfrac{4-x}{x+4}}$.

 a. Domain$(f) = (-4, 4]$

 b. The graph of the function is shown.

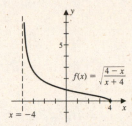

Horizontal asymptotes: none; vertical asymptotes: $x = -4$; x-intercepts: $(4, 0)$; y-intercept: $(0, 1)$

49. For $f(x) = \dfrac{x^2}{\sqrt{9-x^2}} = \dfrac{x^2}{\sqrt{(3-x)(3+x)}}$.

 a. Domain$(f) = (-3, 3)$

 b. The graph of the function is shown.

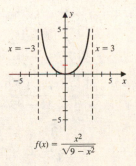

Horizontal asymptotes: none; vertical asymptotes: $x = 3$, $x = -3$; x-intercept: $(0, 0)$; y-intercept: $(0, 0)$

51. $f(x) = x^3 - 2x^2 - x + 2$
   a. increasing: $(-\infty, -0.2) \cup (1.5, \infty)$;
   decreasing: $(-0.2, 1.5)$
   b. local maximum: $(-0.2, 2.1)$; local minimum:
   $(1.5, -0.6)$

53. $f(x) = \dfrac{2}{3}x^3 + \dfrac{7}{2}x^2 - 12x$
   a. increasing: $(-\infty, -4.7) \cup (1.3, \infty)$;
   decreasing: $(-4.7, 1.3)$
   b. local maximum: $(-4.7, 64.5)$; local minimum:
   $(1.3, -8.2)$

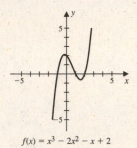

$f(x) = x^3 - 2x^2 - x + 2$

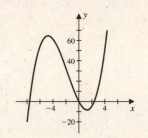

55. $\dfrac{2 + i\sqrt{2}}{4} = \dfrac{1}{2} + \dfrac{\sqrt{2}}{4}i$

57. $(2 - i) - (3 - 2i) = (2 - 3) + (-1 + 2)i = -1 + i$

59. $(2 - i) \cdot \overline{(2 + i)} = (2 - i) \cdot (2 - i) = 4 - 4i + i^2 = 3 - 4i$

61. $i^{20} = (i^4)^5 = 1^5 = 1$

63. $\quad \cdot \dfrac{2 + 3i}{4 - 7i} = \dfrac{2 + 3i}{4 - 7i} \cdot \dfrac{4 + 7i}{4 + 7i} = \dfrac{8 + 14i + 12i + 21i^2}{16 + 49} = \dfrac{-13 + 26i}{65} = -\dfrac{1}{5} + \dfrac{2}{5}i$

65. **a.** $y = f(x - 1)$ **b.** $y = f(x - 1) - 1$ **c.** $y = -f(x + 1) + 1$ **d.** $y = |f(x)|$

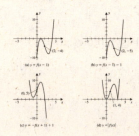

67. A possible graph is shown.

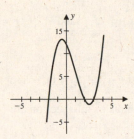

69. First we have

$$P(x) = a(x - 1)(x - 3)(x + 1) = a(x^3 - 3x^2 - x + 3).$$

If $P(0) = 1$, then

$$1 = a((-1)(-3)(1)) \text{ implies } 3a = 1 \text{ so } a = \frac{1}{3}.$$

And

$$P(x) = \frac{1}{3}x^3 - x^2 - \frac{1}{3}x + 1.$$

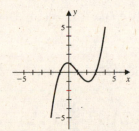

71. a. The graph of the function is shown.

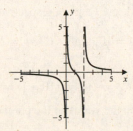

b. The graph will have vertical asymptotes $x = 0$ and $x = 2$ if the denominator of the function has factors x and $x - 2$. To have a horizontal asymptote $y = 0$, the degree of the numerator must be less than the degree of the denominator. If the numerator is $x - 1$, then $f(1) = 0$, and a possible definition for the function is

$$f(x) = \frac{x - 1}{x(x - 2)}.$$

73. To find the points of intersection of the curves

$$f(x) = x^3 \quad \text{and} \quad g(x) = 2x^2 + x - 2,$$

solve $x^3 = 2x^2 + x - 2$ which implies

$$x^3 - 2x^2 - x + 2 = (x-1)(x-2)(x+1) = 0,$$

so $x = -1, x = 1,$ and $x = 2$. The points of intersection are $(-1, -1), (1, 1), (2, 8)$.

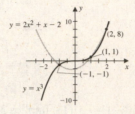

75. If the polynomial has integer coefficients and a zero $-i$, then the conjugate i is also a zero. The polynomial is of degree three with zeros, $1, i, -i$, so

$$P(x) = (x-1)(x-i)(x+i) = (x-1)(x^2+1) = x^3 - x^2 + x - 1.$$

77. If the polynomial has integer coefficients and zeros at $\sqrt{2}i$ and $2i$, then the conjugates $-\sqrt{2}i$ and $-2i$ are also zeros. The polynomial has the form

$$P(x) = a\left(x - \sqrt{2}i\right)\left(x + \sqrt{2}i\right)(x - 2i)(x + 2i) = a(x^2 + 2)(x^2 + 4) = a(x^4 + 6x^2 + 8),$$

and if the constant term is to be 8, then $a = 1$, so $P(x) = x^4 + 6x^2 + 8$.

79. If the length and width of the rectangle are l and w, respectively, and the perimeter is 20, then

$$2l + 2w = 20 \text{ implies } l + w = 10, \text{ so } w = 10 - l.$$

The area is

$$A = lw = l(10 - l) = 10l - l^2.$$

To find l so the area is a maximum find the vertex of the parabola. Completing the square gives

$$-l^2 + 10l = -(l^2 - 10l + 25 - 25) = -(l - 5)^2 + 25$$

and the vertex is $(5, 25)$. So $l = 5$ implies $w = 10 - 5 = 5$, and the rectangle with maximum area is a square of side 5ft and area 25ft^2.

· ·

Chapter 3 Exercises for Calculus (Page 188)

1. a. First we have $P(x) = 2x^3 - 3x^2 = x^2(2x - 3)$.

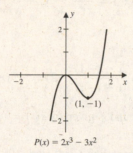

$$P(x) = 2x^3 - 3x^2$$

b. Adding a constant, C, to $P(x)$ shifts the graph upward C units, if $C > 0$, and downward $|C|$ units, if $C < 0$. If $C < 0$, $Q(x) = P(x) + C$ has 1 zero. If $0 < C < 1$, $Q(x) = P(x) + C$ has 3 zeros. If $C = 1$, $Q(x) = P(x) + C$ has 2 zeros. If $C > 1$, $Q(x) = P(x) + C$ has 1 zero. If $C = 0$, $Q(x) = P(x) + C$ has two real zeros.

3. a. Let $P(x) = mx + b$. Then

$$(P \circ P)(x) = m(mx + b) + b = m^2 x + b(m + 1),$$

which is a linear polynomial with positive slope since the slope is $m^2 > 0$.

b. Let $P(x) = ax^2 + bx + c$. Then

$$\begin{aligned}(P \circ P)(x) &= a(ax^2 + bx + c)^2 + b(ax^2 + bx + c) + c \\ &= a^3 x^4 + 2a^2 b x^3 + (2a^2 c + ab^2 + ab)x^2 + (2abc + b^2)x + (ac^2 + bc + c)\end{aligned}$$

which is a polynomial of degree 4.

5. The possible rational roots of $f(x) = x^3 + x^2 + kx - 3$ are ± 1 and ± 3. Then

$$f(1) = 0 \text{ implies } k = 1, \qquad f(-1) = 0 \text{ implies } k = -3,$$
$$f(3) = 0 \text{ implies } k = -11, \quad \text{and} \quad f(-3) = 0 \text{ implies } k = -7.$$

For

$$k = 1: \ f(x) = (x - 1)(x^2 + 2x + 3);$$
$$k = -3: \ f(x) = (x + 1)(x^2 - 3) = (x + 1)\left(x - \sqrt{3}\right)\left(x + \sqrt{3}\right);$$
$$k = -11: \ f(x) = (x - 3)(x^2 + 4x + 1) = (x - 3)\left(x - \left(-2 + \sqrt{3}\right)\right)\left(x - \left(-2 - \sqrt{3}\right)\right);$$
$$k = -7: \ f(x) = (x + 3)(x^2 - 2x - 1) = (x + 3)\left(x - \left(1 + \sqrt{2}\right)\right)\left(x - \left(1 - \sqrt{2}\right)\right).$$

So $f(x)$ has at least one rational zero for $k = 1, -3, -11,$ and -7.

7. a. Since

$$P(x_1) = \frac{x_1 - x_2}{x_1 - x_2} y_1 + \frac{x_1 - x_1}{x_2 - x_1} y_2 = y_1,$$

the point (x_1, y_1) is on the line. Similarly,

$$P(x_2) = \frac{x_2 - x_2}{x_1 - x_2} y_1 + \frac{x_2 - x_1}{x_2 - x_1} y_2 = y_2,$$

so (x_2, y_2) is on the line.

b. We have

$$y = \frac{x - 1}{-1 - 1}(6) + \frac{x - (-1)}{1 - (-1)}(-2) = -3(x - 1) - (x + 1) = -4x + 2$$

9. The height of an object above the ground is given by $s(t) = v_0 t - \frac{g}{2} t^2$.

a. The initial height of the object occurs when $t = 0$, and $s(0) = 0$.

b. To determine how long the object is in the air we need to find the time when the object strikes the ground. Solving

$$0 = s(t) = v_0 t - \frac{g}{2} t^2 = t\left(v_0 - \frac{g}{2} t\right)$$

gives $t = 0$ and $t = 2v_0/g$. Hence $t = 2v_0/g$ is the amount of time the object is in the air.

c. To determine the maximum height reached by the object find the vertex of the parabola. Completing the square we have

$$-\frac{g}{2} t^2 + v_0 t = -\frac{g}{2}\left(t^2 - \frac{2v_0}{g} t\right) = -\frac{g}{2}\left(t^2 - \frac{2v_0}{g} t + \frac{v_0^2}{g^2} - \frac{v_0^2}{g^2}\right) = -\frac{g}{2}\left(t - \frac{v_0}{g}\right)^2 + \frac{v_0^2}{2g}.$$

So the vertex is $\left(v_0/g, v_0^2/2g\right)$ and the maximum height reached is $v_0^2/2g$.

d. The object reaches the maximum height after $t = v_0/g$.

11. The dimensions on the figure imply that $V(x) = x(20 - 2x)\left(\frac{50-3x}{2}\right) = x(10 - x)(50 - 3x)$.

· ·

Chapter 3 Chapter Test (Page 190)

1. True.

2. True.

3. True.

4. True.

5. False. The polynomial

$$P(x) = x^2 + 1,$$

has no real numbers with $P(x) = 0$.

6. True.

7. False. The graph of the polynomial

$$P(x) = x^3 + x^2 - 2x$$

crosses the x-axis at $x = 1$, $x = -2$, and $x = 0$.

8. False. The graph of the polynomial

$$P(x) = x(x - 1)^3(x + 1)^2$$

crosses the x-axis twice. It touches the axis at $x = -1$, but does not cross.

9. True.

10. True.

11. False. The polynomials

$$P(x) = 2x^4 - 3x^3 + 7x - 10$$

and

$$Q(x) = 3x^4 - 5x^3 + 9x - 12$$

do not have the same zeros, since one is not a multiple of the other.

12. False. The polynomials

$$P(x) = x^4 - 2x^3 + 6x - 9$$

and

$$Q(x) = x^4 - 2x^3 + 6x - 5$$

do not have the same zeros, since one is not a multiple of the other.

13. False. The possible rational zeros of the polynomial

$$P(x) = 2x^5 - 2x^3 + x^2 - 3x + 5$$

are

$$\pm 1, \pm \frac{1}{2}, \pm 5, \pm \frac{5}{2}.$$

14. False. The possible rational zeros of the polynomial $P(x) = 3x^4 - 2x^3 + x^2 - 2x + 4$ are

$$\pm 1, \pm 2, \pm 4, \pm \tfrac{1}{3}, \pm \tfrac{2}{3}, \pm \tfrac{4}{3}.$$

15. True.

16. False. The polynomial shown in the figure has a zero of odd multiplicity at $x = 2$.

17. True.

18. False. The degree of the polynomial shown in the figure is at least 6.

19. False. The polynomial shown in the figure has zeros at $x = -1$, $x = 1$, and $x = 2$.

20. False. The end-behavior of the polynomial shown in the figure is similar to the end-behavior of $-x^3$, but the given graph approaches the end-behavior more rapidly because its degree is higher.

21. True.

22. True.

23. True.

24. False. One zero of the polynomial

$$P(x) = x^5 - 5x^4 + 6x^3 + 4x^2 - 8x$$

is $x = 0$, and another is $x = -1$. Factoring $P(x)$ using these observations gives

$$P(x) = x(x + 1)(x^3 - 6x^2 + 12x - 8)$$

and the right-most term has a zero at $x = 2$. Continuing we have

$$x^3 - 6x^2 + 12x - 8 = (x - 2)^3$$

and

$$P(x) = x(x + 1)(x - 2)^3.$$

So, $P(x)$ has a zero of multiplicity 3 at $x = 2$.

25. False. The rational function

$$f(x) = \frac{1 - x^2}{x^2 + 1}$$

has no vertical asymptotes, but it has a horizontal asymptote at $y = -1$.

26. False. Since

$$f(x) = \frac{x^2 - 3x + 2}{x - 2}$$
$$= \frac{(x - 2)(x - 1)}{x - 2} = x - 1,$$

for $x \neq 2$, the function does not have a vertical asymptote. The graph is given by the line with equation, $y = x - 1$, except that there is a hole in the graph at $(2, 1)$.

27. False. The rational function

$$f(x) = \frac{x^2 - 1}{x^2 - 4}$$

has vertical asymptotes at $x = 2$ and $x = -2$.

28. True.

29. False. The graph has vertical asymptotes $x = 2$ and $x = -2$.

30. False. The graph has a horizontal asymptote $y = 1$.

31. True.

32. True.

33. True.

34. False. The graph of the rational function has a horizontal asymptote $y = 2$.

35. False. The rational function $f(x) = \dfrac{1 - x^2}{x^2 + 1}$ has no vertical asymptotes, but it has a horizontal asymptote at $y = -1$.

36. True.

37. False. The rational function $f(x) = \dfrac{x^2 - 1}{x^2 + 4}$ has vertical asymptotes at $x = 2$ and $x = -2$.

38. False. The rational function $f(x) = \dfrac{x^2 + 5x + 4}{x^2 - 2x - 3} = \dfrac{x + 4}{x - 3}$, if $x \neq -1$, has a vertical asymptote at $x = 3$.

39. False. The function

$$f(x) = \frac{x^2 + 1}{x^2 - 3x + 2}$$

satisfies

$$f(x) \to -\infty, \text{ as } x \to 1^+$$
$$f(x) \to \infty, \text{ as } x \to 1^-.$$

40. True.

41. True.

42. False. The function

$$f(x) = \frac{2x^3 - 3x^2 + x - 1}{3x^3 - 4x^2 - 2}$$

has a horizontal asymptote $y = \frac{2}{3}$.

43. True.

44. False. For large values of x,

$$\frac{10x^2 - x + 1}{5x^2} \geq \frac{5x^3 - x^2 - 2x + 1}{3x^3 + 1},$$

since

$$\frac{10x^2 - x + 1}{5x^2} \to 2, \text{ as } x \to \infty$$

and

$$\frac{5x^3 - x - 2x + 1}{3x^3 + 1} \to \frac{5}{3}, \text{ as } x \to \infty.$$

CHAPTER 4: Trigonometric Functions

4.1 Introduction

The *trigonometric functions* $f(t) = \cos t$, and $g(t) = \sin t$ are defined as the x- and y-coordinates of points $P(t) = (\cos t, \sin t)$ on the unit circle. The point $P(t)$ is called the terminal point, and t is the length of the arc along the circle measured counterclockwise from the starting point $(1, 0)$. The measure of t is called radian measure. The sine and cosine functions are the basis for the other trigonometric functions. Master these and you have mastered most of trigonometry.

4.2 Measuring Angles

Angles are measured using both radians and degrees.

Relationship Between Degrees and Radians

- $360° = 2\pi$ radians

- $1° = \frac{\pi}{180} \approx 0.01745$ radians

- 1 radian $= \frac{180°}{\pi} \approx 57.296°$

Radian measure is the method most often used in PreCalculus and Calculus.

EXAMPLE 1
Convert from degrees to radians or from radians to degrees.
(a) $\frac{\pi}{4}$ (b) $\frac{5\pi}{6}$ (c) $240°$ (d) $315°$

Solution We have
(a) $\frac{\pi}{4} = \left(\frac{\pi}{4}\right)\left(\frac{180°}{\pi}\right) = 45°$
(b) $\frac{5\pi}{6} = \left(\frac{5\pi}{6}\right)\left(\frac{180°}{\pi}\right) = 150°$
(c) $240° = 240\left(\frac{\pi}{180}\right) = \frac{4\pi}{3}$
(d) $315° = 315\left(\frac{\pi}{180}\right) = \frac{7\pi}{4}$

$\square$

211

The Unit Circle

The *unit circle* is the circle with center at the origin $(0, 0)$ and radius 1. The equation of the unit circle is

$$x^2 + y^2 = 1.$$

EXAMPLE 2

Verify that the point is on the unit circle.

(a) $\left(\frac{\sqrt{3}}{2}, \frac{1}{2}\right)$ (b) $\left(-\frac{1}{3}, \frac{2\sqrt{2}}{3}\right)$

Solution (a) Substitute the coordinates into the equation $x^2 + y^2 = 1$, and if the equation is satisfied, the point will lie on the unit circle. Substituting $x = \frac{\sqrt{3}}{2}$ and $y = \frac{1}{2}$ gives

$$\left(\frac{\sqrt{3}}{2}\right)^2 + \left(\frac{1}{2}\right)^2 = \frac{(\sqrt{3})^2}{2^2} + \frac{1}{2^2} = \frac{3}{4} + \frac{1}{4} = 1,$$

and the point lies on the unit circle.

(b)

$$\left(-\frac{1}{3}\right)^2 + \left(\frac{2\sqrt{2}}{3}\right)^2 = \frac{1}{9} + \frac{2^2(\sqrt{2})^2}{9} = \frac{1}{9} + \frac{8}{9} = 1$$

$\square$

EXAMPLE 3

Find the point on the unit circle that satisfies the given conditions.

(a) The point has x-coordinate $\frac{4}{5}$ and negative y-coordinate.

(b) The point has y-coordinate $-\frac{5}{6}$ and is in quadrant IV.

(c) The point has x-coordinate $-\frac{\sqrt{2}}{2}$ and is in quadrant II.

Solution (a) The point lies on the unit circle, so $x^2 + y^2 = 1$ and

$$\left(\frac{4}{5}\right)^2 + y^2 = 1$$

$$y^2 = 1 - \frac{16}{25} = \frac{9}{25}$$

$$y = \pm\sqrt{\frac{9}{25}} = \pm\frac{3}{5}.$$

Since the y-coordinate is negative, the point is $\left(\frac{4}{5}, -\frac{3}{5}\right)$.

(b) Solving for the x-coordinate, we have

$$x^2 + \left(-\frac{5}{6}\right)^2 = 1$$

$$x^2 = 1 - \frac{25}{36} = \frac{11}{36}$$

$$x = \pm \frac{\sqrt{11}}{6}.$$

The quadrants of the plane are labeled I, II, III and IV counterclockwise starting in the upper right. In quadrant IV the x-coordinate is positive and y-coordinate is negative, so $x = \frac{\sqrt{11}}{6}$, and the point is $\left(\frac{\sqrt{11}}{6}, -\frac{5}{6}\right)$.

(c)

$$\left(-\frac{\sqrt{2}}{2}\right)^2 + y^2 = 1$$

$$y^2 = 1 - \frac{1}{2} = \frac{1}{2}$$

$$y = \pm \sqrt{\frac{1}{2}} = \pm \frac{1}{\sqrt{2}} = \pm \frac{1}{\sqrt{2}} \cdot \frac{\sqrt{2}}{\sqrt{2}}$$

$$y = \pm \frac{\sqrt{2}}{2}$$

In quadrant II the y-coordinate is positive, so the point is $\left(-\frac{\sqrt{2}}{2}, \frac{\sqrt{2}}{2}\right)$.   □

Terminal Points

If $t > 0$, then $P(t)$ is the point on the unit circle found by measuring t units counterclockwise along the arc of the circle starting at $(1, 0)$. If $t < 0$, then measure clockwise around the circle starting at $(1, 0)$.

EXAMPLE 4

Find the location of the terminal point $P(t)$ on the unit circle.

(a) $t = \frac{\pi}{4}$ (b) $t = \frac{7\pi}{6}$ (c) $t = 7\pi$ (d) $t = -\frac{3\pi}{2}$ (e) $t = -\frac{4\pi}{3}$

Solution (a) The circumference of the unit circle is 2π, so $\frac{\pi}{2}$ is one fourth the way around. Since $\frac{\pi}{4} = \frac{1}{2}(\frac{\pi}{2})$, and $\frac{\pi}{2}$ is one-fourth around the circle, $\frac{\pi}{4}$ is one-eighth the way around the circle.

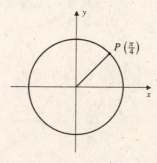

(b) Since $\frac{\pi}{6} = \frac{1}{3}(\frac{\pi}{2})$, an arc of length $\frac{\pi}{6}$ is one third of the arc from $(1, 0)$ to $(0, 1)$. Also, $P(\pi) = (-1, 0)$, since an arc of π units is half the unit circle. So $P\left(\frac{7\pi}{6}\right)$ is one third of the way between $(-1, 0)$ and $(0, -1)$.

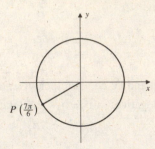

(c) An arc of length π is half the length of the unit circle, or one-half a complete revolution. Since 2π is one revolution around the circle, 7π is three and one-half revolutions. So $P(7\pi)$ is the same as $P(\pi)$ and located at the point $(-1, 0)$.

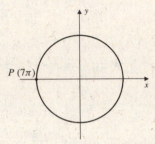

(d) The point $P\left(-\frac{3\pi}{2}\right)$ is three-fourths of the way around the circle in the clockwise direction and is located at $(0, 1)$.

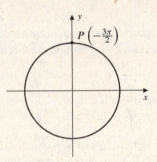

(e) Since $-\frac{\pi}{3} = \frac{2}{3}\left(-\frac{\pi}{2}\right)$, the point $P\left(-\frac{\pi}{3}\right)$ is two-thirds of the way from $(1, 0)$ to $(0, -1)$. Since $P\left(-\frac{4\pi}{3}\right)$ is in quadrant II it is two-thirds of the way between $(-1, 0)$ and $(0, 1)$.

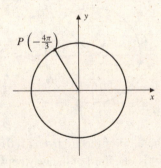

Reference Number

For any real number t the *reference number* r associated with t is the shortest distance along the unit circle from t to the x-axis. For any t, the reference number satisfies $0 \le r \le \frac{\pi}{2}$.

EXAMPLE 5

Find the reference number r for the given value of t, and show $P(t)$ and $P(r)$ on the unit circle.
(a) $t = \frac{5\pi}{6}$ (b) $t = -\frac{2\pi}{3}$ (c) $t = \frac{5\pi}{4}$

Solution

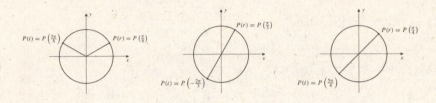

EXAMPLE 6

Suppose that $P(t)$ has coordinates $\left(\frac{12}{13}, \frac{5}{13}\right)$. Find the coordinates of the point.
(a) $P(t + \pi)$ (b) $P(-t)$ (c) $P(t - \pi)$ (d) $P(-t - \pi)$

Solution   The reference number for each of the points is t, so the coordinates of each of the points have the same magnitude as those of $P(t)$. However, the coordinates might be negative. For example, $P(t)$ lies in quadrant I, and $P(t + \pi)$ lies in quadrant III, where both coordinates are negative. So

$$P(t + \pi) = \left(-\frac{12}{13}, -\frac{5}{13}\right).$$

The four points are shown in the figure.

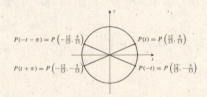

Exercise Set 4.2 (Page 200)

1. $30° = 30\frac{\pi}{180}$ radians $= \frac{\pi}{6}$ radians

3. $150° = 150\frac{\pi}{180}$ radians $= \frac{5\pi}{6}$ radians

5. $225° = 225\frac{\pi}{180}$ radians $= \frac{5\pi}{4}$ radians

7. $-72° = -72\frac{\pi}{180}$ radians $= -\frac{2\pi}{5}$ radians

9. $\frac{3\pi}{4}$ radians $= \frac{3\pi}{4}\frac{180}{\pi}$ degrees $= 135°$

11. $-\frac{11\pi}{6}$ radians $= -\frac{11\pi}{6}\frac{180}{\pi}$ degrees $= -330°$

13. $\frac{9\pi}{2}$ radians $= \frac{9\pi}{2}\frac{180}{\pi}$ degrees $= 810°$

15. The angle $405°$ is $45°$ beyond $360°$ so the angle coincides with $45°$.

17. Since $2\pi = \frac{12\pi}{6}$, the angles $\frac{\pi}{6}$ and $\frac{13\pi}{6}$ coincide.

19. The angle $150°$ is $30°$ from $180°$, but $-240°$ is $60°$ from $180°$, so the angles do not coincide.

21. a. $P\left(\frac{\pi}{2}\right)$ **b.** $P(\pi)$ **c.** $P(2\pi)$ **d.** $P\left(\frac{3\pi}{2}\right)$

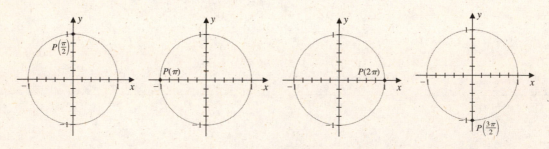

23. a. $P\left(-\frac{\pi}{4}\right)$ **b.** $P\left(-\frac{4\pi}{3}\right)$ **c.** $P\left(-\frac{37\pi}{6}\right)$ **d.** $P\left(-\frac{7\pi}{4}\right)$

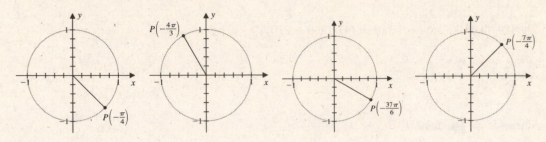

25. Let $P(t) = \left(\frac{3}{5}, \frac{4}{5}\right)$.

a. Since $P(t)$ is in quadrant I, adding π to t moves the point to quadrant III, so $P(t + \pi) = \left(-\frac{3}{5}, -\frac{4}{5}\right)$.

b. Since $P(t)$ is in quadrant I, $P(-t)$ is in quadrant IV, so $P(-t) = \left(\frac{3}{5}, -\frac{4}{5}\right)$.

c. Since $P(t)$ is in quadrant I, subtracting π from t moves the point to quadrant III, so $P(t - \pi) = \left(-\frac{3}{5}, -\frac{4}{5}\right)$.

d. Since $P(t)$ is in quadrant I, $P(-t)$ is in quadrant IV so subtracting π from $-t$ moves the point to quadrant II, so $P(-t - \pi) = \left(-\frac{3}{5}, \frac{4}{5}\right)$.

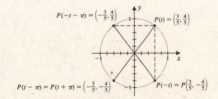

27. a. $P(t + \pi) = \left(\frac{\sqrt{5}}{3}, -\frac{2}{3}\right)$

b. $P(-t) = \left(\frac{-\sqrt{5}}{3}, -\frac{2}{3}\right)$

c. $P(t - \pi) = P(t + \pi) = \left(\frac{\sqrt{5}}{3}, -\frac{2}{3}\right)$

d. $P(-t - \pi) = \left(\frac{\sqrt{5}}{3}, \frac{2}{3}\right)$

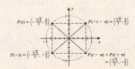

29. Since $45° = \frac{\pi}{4}$ radians, the length of the arc is $s = r\theta = (8)\frac{\pi}{4} = 2\pi$.

31. $\theta = \frac{s}{r} = \frac{6}{3} = 2$ radians $= 2\frac{180°}{\pi} = \frac{360°}{\pi}$

33. The length of the arc is $s = r\theta = (4)\left((360 - 220)\frac{\pi}{180}\right) = \frac{22\pi}{9}$.

35. $A = \frac{1}{2}r^2\theta = \frac{1}{2}(10)^2\frac{1}{2} = 25$

37. Since $45° = \frac{\pi}{4}$ radians, $r = \sqrt{\frac{2A}{\theta}} = \sqrt{\frac{(2)(2)}{\frac{\pi}{4}}} = \frac{4}{\sqrt{\pi}}$.

39. Since $65° = 65\frac{\pi}{180}$ radians $= \frac{13\pi}{36}$ radians, $A = \frac{1}{2}r^2\theta = \frac{1}{2}(16)\frac{13\pi}{36} = \frac{26\pi}{9}$.

41. Since the angle between the two cities measured north along the common meridian is
$43 - 36.5 = 6.5° = 6.5\frac{\pi}{180}$ radians, the distance between the two cities is approximately $3960\left(\frac{6.5\pi}{180}\right) \approx 449$
miles.

43. Approximately 330π miles per hour.

45. a. $4\rho = \theta$

  b. 5/4 revolutions per second

....

4.3   Right Triangle Trigonometry

Angle Measure : Radians and Degrees

The trigonometric functions for an angle θ in the right triangle can be defined using the lengths of the sides of the triangle. We have

$$\sin\theta = \frac{b}{c} = \frac{\text{opposite}}{\text{hypotenuse}}, \qquad \cos\theta = \frac{a}{c} = \frac{\text{adjacent}}{\text{hypotenuse}},$$

$$\tan\theta = \frac{b}{a} = \frac{\text{opposite}}{\text{adjacent}}, \qquad \cot\theta = \frac{a}{b} = \frac{\text{adjacent}}{\text{opposite}},$$

$$\sec\theta = \frac{c}{a} = \frac{\text{hypotenuse}}{\text{adjacent}}, \qquad \csc\theta = \frac{c}{b} = \frac{\text{hypotenuse}}{\text{opposite}}.$$

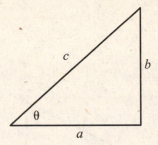

EXAMPLE 1
Find the value of the six trigonometric functions of the angle θ shown in the right triangle.

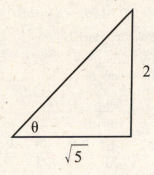

Solution The length of the hypotenuse is needed and can be found using the Pythagorean Theorem. If x denotes the length of the hypotenuse, then

$$x^2 = \left(\sqrt{5}\right)^2 + 2^2$$
$$x^2 = 9$$
$$x = 3.$$

Using the sides of the triangle gives the six trigonometric functions as

$$\sin\theta = \frac{2}{3}, \quad \cos\theta = \frac{\sqrt{5}}{3},$$

$$\tan\theta = \frac{2}{\sqrt{5}} = \frac{2\sqrt{5}}{5}, \quad \cot\theta = \frac{\sqrt{5}}{2},$$

$$\sec\theta = \frac{3}{\sqrt{5}} = \frac{3\sqrt{5}}{5}, \quad \csc\theta = \frac{3}{2}.$$

$\square$

EXAMPLE 2
Find any missing angles or sides in the triangle.

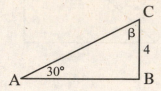

Solution Since the sum of the angles of a triangle is 180°, we have

$$\beta = 180° - 90° - 30° = 60°.$$

To find the side $\overline{AB}$, use the tangent of the 30° angle, which is

$$\tan 30° = \tan \frac{\pi}{6}$$

$$= \frac{\sin \frac{\pi}{6}}{\cos \frac{\pi}{6}} = \frac{1/2}{\sqrt{3}/2}$$

$$= \frac{1}{\sqrt{3}} = \frac{1}{\sqrt{3}} \cdot \frac{\sqrt{3}}{\sqrt{3}} = \frac{\sqrt{3}}{3}.$$

Since

$$\tan 30° = \frac{4}{AB}$$

$$AB\tan 30° = 4$$

$$AB = \frac{4}{\tan 30°}$$

$$= \frac{4}{\sqrt{3}/3} = \frac{12}{\sqrt{3}} = 4\sqrt{3}.$$

The remaining side AC can be found using the Pythagorean Theorem.

$$AC^2 = \left(4\sqrt{3}\right)^2 + 4^2 = 48 + 16 = 64 \quad \text{so} \quad AC = 8.$$

$\square$

EXAMPLE 3
From the figure determine $\sin \alpha$, $\sin \beta$, $\cos \alpha$, and $\cos \beta$. Then find The value of the trigonometric function.
(a) $\sin(\alpha + \beta)$ (b) $\cos(\alpha + \beta)$ (c) $\sin 2\alpha$
(d) $\cos 2\alpha$ (e) $\sin \frac{\alpha}{2}$ (f) $\cos \frac{\alpha}{2}$

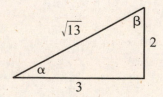

Solution Using the angle-side relationships in the figure gives

$$\sin \alpha = \frac{2}{\sqrt{13}}, \ \sin \beta = \frac{3}{\sqrt{13}}$$

$$\cos \alpha = \frac{3}{\sqrt{13}}, \ \cos \beta = \frac{2}{\sqrt{13}}.$$

(a) Since $\alpha + \beta = 90°$,

$$\sin(\alpha + \beta) = \sin(90°) = 1.$$

(b) Also,

$$\cos(\alpha + \beta) = \cos(90°) = 0.$$

(c)

$$\sin 2\alpha = 2 \sin \alpha \cos \alpha = 2 \frac{2}{\sqrt{13}} \cdot \frac{3}{\sqrt{13}} = \frac{12}{13}$$

(d)

$$\cos 2\alpha = (\cos \alpha)^2 - (\sin \alpha)^2 = \frac{9}{13} - \frac{4}{13} = \frac{5}{13}$$

(e)

$$\sin \frac{\alpha}{2} = \sqrt{\frac{1 - \cos \alpha}{2}} = \sqrt{\frac{1 - \frac{3}{\sqrt{13}}}{2}} = \sqrt{\frac{\sqrt{13} - 3}{2\sqrt{13}}}$$

(f)

$$\cos \frac{\alpha}{2} = \sqrt{\frac{1 + \cos \alpha}{2}} = \sqrt{\frac{1 + \frac{3}{\sqrt{13}}}{2}} = \sqrt{\frac{\sqrt{13} + 3}{2\sqrt{13}}}$$

$\square$

Applications

EXAMPLE 4

The angle of elevation to the top of a building is 65° from the ground when viewed 165 ft from the building. Estimate the height of the building in feet.

Solution The right triangle in the figure describes the situation.

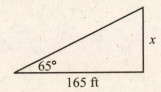

65°

x

165 ft

The sides of length x and length 165 are the opposite and adjacent sides, respectively, of the 65° angle. So we can use a calculated value for the tangent function to obtain

$$\tan 65° = \frac{x}{165}$$
$$x = 165 \tan 65° \text{ feet} \approx 354 \text{ feet.}$$

☐

EXAMPLE 5

A surveyor wants to find the distance between points A and B on opposite sides of a pond. From point A, the surveyor determines a line of sight perpendicular to AB and establishes point C along this line 150 feet from A. Angle ACB is determined to be 62°. What is the distance from A to B?

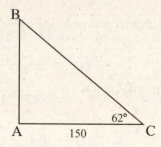

Solution The lengths AB and $AC = 150$ are the sides opposite and adjacent, respectively, to the angle of 62° and so can be related using a calculated value of the tangent function.

$$\tan 62° = \frac{AB}{AC} = \frac{AB}{150}$$
$$AB = 150 \tan 62° \approx 282 \text{ feet}$$

☐

Exercise Set 4.3 (Page 206)

1. The hypotenuse is $\sqrt{2^2 + (\sqrt{21})^2} = \sqrt{25} = 5$, so

$$\cos\theta = \frac{\sqrt{21}}{5}, \ \sin\theta = \frac{2}{5}, \ \tan\theta = \frac{2}{\sqrt{21}} = \frac{2\sqrt{21}}{21}, \cot\theta = \frac{\sqrt{21}}{2}, \ \sec\theta = \frac{5}{\sqrt{21}} = \frac{5\sqrt{21}}{21}, \ \csc\theta = \frac{5}{2}.$$

3. The hypotenuse is $\sqrt{3+1} = 2$, so

$$\cos\theta = \frac{\sqrt{3}}{2}, \ \sin\theta = \frac{1}{2}, \ \tan\theta = \frac{\sqrt{3}}{3},$$

$$\cot\theta = \frac{3}{\sqrt{3}} = \sqrt{3}, \ \sec\theta = \frac{2}{\sqrt{3}} = \frac{2\sqrt{3}}{3}, \ \csc\theta = 2.$$

5. Let x be the missing side. Then $x^2 + 25 = 169$ so $x = \sqrt{144} = 12$, and

$$\cos\theta = \frac{12}{13}, \quad \sin\theta = \frac{5}{13}, \quad \tan\theta = \frac{5}{12},$$

$$\cot\theta = \frac{12}{5}, \quad \sec\theta = \frac{13}{12}, \quad \csc\theta = \frac{13}{5}.$$

7. Since $\cos 30° = \frac{x}{16}$, it implies that $x = 16\cos 30° = 16\frac{\sqrt{3}}{2} = 8\sqrt{3}$.

9. Since $\sin 45° = \frac{7}{x}$, it implies that $x = \frac{7}{\sin 45°} = \frac{7}{\sqrt{2}/2} = \frac{14}{\sqrt{2}} = 7\sqrt{2}$.

11. Since $\tan 60° = \frac{6}{x}$, it implies that $x = \frac{6}{\tan 60°} = \frac{6}{\sqrt{3}} = \frac{6\sqrt{3}}{3} = 2\sqrt{3}$.

13. The angle $\beta = 60°$, and

$$\tan 30° = \frac{\overline{BC}}{\overline{AB}} = \frac{8}{\overline{AB}} \quad \text{so} \quad \overline{AB} = \frac{8}{\sqrt{3}/3} = 8\sqrt{3}.$$

Also,

$$\sin 30° = \frac{\overline{BC}}{\overline{AC}} = \frac{8}{\overline{AC}} \quad \text{so} \quad \overline{AC} = \frac{8}{1/2} = 16.$$

Note that once $\overline{AB}$ is computed $\overline{AC}$ can be found using the Pythagorean Theorem as well. That is, $\overline{AC} = \sqrt{192 + 64} = \sqrt{256} = 16$.

15. The angle $\alpha = 45°$, and

$$\cos 45° = \frac{\overline{AB}}{\overline{AC}} = \frac{\overline{AB}}{5} \text{ so } \overline{AB} = \frac{5\sqrt{2}}{2}.$$

Also,

$$\sin 45° = \frac{\overline{BC}}{5} \text{ so } \overline{BC} = \frac{5\sqrt{2}}{2}.$$

17. a. The missing side is $\sqrt{9-4} = \sqrt{5}$, and

$$\cos\theta = \frac{\sqrt{5}}{3}, \tan\theta = \frac{2}{\sqrt{5}} = \frac{2\sqrt{5}}{5}, \cot\theta = \frac{\sqrt{5}}{2},$$

$$\sec\theta = \frac{3}{\sqrt{5}} = \frac{3\sqrt{5}}{5}, \csc\theta = \frac{3}{2}.$$

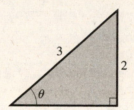

b. The missing side is $\sqrt{25-1} = \sqrt{24} = 2\sqrt{6}$, and

$$\sin\theta = \frac{2\sqrt{6}}{5}, \tan\theta = 2\sqrt{6}, \cot\theta = \frac{1}{2\sqrt{6}} = \frac{\sqrt{6}}{12},$$

$$\sec\theta = 5, \csc\theta = \frac{5}{2\sqrt{6}} = \frac{5\sqrt{6}}{12}.$$

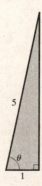

c. The hypotenuse is $\sqrt{16+9} = 5$, and

$$\sin\theta = \frac{4}{5}, \cos\theta = \frac{3}{5}, \cot\theta = \frac{3}{4}, \sec\theta = \frac{5}{3}, \csc\theta = \frac{5}{4}.$$

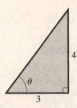

d. The missing side is $\sqrt{9-4} = \sqrt{5}$, and

$$\sin\theta = \frac{2}{3}, \cos\theta = \frac{\sqrt{5}}{3}, \tan\theta = \frac{2}{\sqrt{5}} = \frac{2\sqrt{5}}{5},$$

$$\cot\theta = \frac{\sqrt{5}}{2}, \sec\theta = \frac{3}{\sqrt{5}} = \frac{3\sqrt{5}}{5}.$$

19. Following the work in Example 6, we have

$$L = 6400\cos 45° = 3200\sqrt{2} \approx 4525.5 \text{ km.}$$

21. If h denotes the height of the building, then

$$\tan 30° = \frac{h}{1} \quad \text{so} \quad h = \frac{\sqrt{3}}{3} \approx 0.577 \text{ mi} \approx 3048 \text{ ft.}$$

23. If h denotes the height of the cliff, then

$$\tan 60° = \frac{h}{40} \quad \text{so} \quad h = 40\sqrt{3} \approx 69.3 \text{ft.}$$

25. Let x be the distance from the base of the tower to the campfire. Since

$$\cos 10.5° = \sqrt{1 - (0.182)^2} = 0.983, \quad \text{we have} \quad \cot 10.5° = \frac{0.983}{0.182} = 5.403$$

and

$$x = 80\cot 10.5° \approx 431.64.$$

So the campfire is approximately 431.64 feet from the base of the tower.

4.4   The Sine and Cosine Functions

A point on the unit circle with coordinates $P(t) = (x, y)$ has

$$\sin t = y, \quad \cos t = x.$$

These definitions are also used for the sine and cosine of an angle with radian measure t. The conversion formula from radians to degrees gives the sine and cosine in degree measure. The values of sine and cosine at some standard angles are given in the table.

t	0	$\frac{\pi}{6}$	$\frac{\pi}{4}$	$\frac{\pi}{3}$	$\frac{\pi}{2}$	π	$\frac{3\pi}{2}$	2π
	0°	30°	45°	60°	90°	180°	270°	360°
$P(t)$	$(1,0)$	$\left(\frac{\sqrt{3}}{2}, \frac{1}{2}\right)$	$\left(\frac{\sqrt{2}}{2}, \frac{\sqrt{2}}{2}\right)$	$\left(\frac{1}{2}, \frac{\sqrt{3}}{2}\right)$	$(0,1)$	$(-1,0)$	$(0,-1)$	$(1,0)$
$\sin t$	0	$\frac{1}{2}$	$\frac{\sqrt{2}}{2}$	$\frac{\sqrt{3}}{2}$	1	0	-1	0
$\cos t$	1	$\frac{\sqrt{3}}{2}$	$\frac{\sqrt{2}}{2}$	$\frac{1}{2}$	0	-1	0	1

EXAMPLE 1

Find $\sin t$ and $\cos t$ for the given value of t.
(a) $t = \frac{3\pi}{4}$ (b) $t = \frac{7\pi}{6}$ (c) $t = \frac{5\pi}{3}$ (d) $t = -\frac{7\pi}{4}$

Solution   (a) The reference number for $t = \frac{3\pi}{4}$ is $r = \frac{\pi}{4}$. To find the sine and cosine of t, use the sine and cosine of r and determine the appropriate sign for each based on the quadrant in which $P(t)$ lies. Using the table, we have

$$\sin \frac{\pi}{4} = \frac{\sqrt{2}}{2}, \quad \cos \frac{\pi}{4} = \frac{\sqrt{2}}{2}.$$

Since $P\left(\frac{3\pi}{4}\right)$ lies in quadrant II, $\sin \frac{3\pi}{4} > 0$ and $\cos \frac{3\pi}{4} < 0$, so

$$\sin \frac{3\pi}{4} = \frac{\sqrt{2}}{2}, \cos \frac{3\pi}{4} = -\frac{\sqrt{2}}{2}.$$

(b) for $t = \frac{\pi}{6}$
- Reference number: $r = \frac{\pi}{6}$
- Quadrant: The point $P(t)$ lies in quadrant III, where $\sin t < 0$ and $\cos t < 0$.
  Using the table, we have

$$\sin \frac{7\pi}{6} = -\sin \frac{\pi}{6} = -\frac{1}{2}, \cos \frac{7\pi}{6} = -\cos \frac{\pi}{6} = -\frac{\sqrt{3}}{2}.$$

(c) for $t = \frac{5\pi}{3}$
- Reference number: $r = \frac{\pi}{3}$
- Quadrant: The point $P(t)$ lies in quadrant IV, where $\sin t < 0$ and $\cos t > 0$. So

$$\sin \frac{5\pi}{3} = -\sin \frac{\pi}{3} = -\frac{\sqrt{3}}{2}, \cos \frac{5\pi}{3} = \cos \frac{\pi}{3} = \frac{1}{2}.$$

(d) for $t = \frac{-7\pi}{4}$
- Reference number: $r = \frac{\pi}{4}$
- Quadrant: The point $P(t)$ lies in quadrant I, where $\sin t > 0$ and $\cos t > 0$. So

$$\sin\left(-\frac{7\pi}{4}\right) = \sin\frac{\pi}{4} = \frac{\sqrt{2}}{2}, \cos\left(-\frac{7\pi}{4}\right) = \cos\frac{\pi}{4} = \frac{\sqrt{2}}{2}.$$

$\square$

EXAMPLE 2
(a) Find all values of t in the interval $[0, 2\pi]$ that satisfy the equation $\cos t = -\frac{1}{2}$.

(b) Find all values of t in the interval $[0, 2\pi]$ that satisfy the equation $\sin 3t = \frac{\sqrt{2}}{2}$.

Solution (a) The cosine is negative in quadrants II and III. We want a reference angle r with $0 < r < \frac{\pi}{2}$ so that

$$\cos r = \frac{1}{2} \text{ which implies } r = \frac{\pi}{3}.$$

The angle in quadrant II with reference angle $\frac{\pi}{3}$ is $\frac{2\pi}{3}$, and the angle in quadrant III is $\frac{4\pi}{3}$. So

$$t = \frac{2\pi}{3}, \frac{4\pi}{3}.$$

(b) First note that if $0 \leq t \leq 2\pi$, then $0 \leq 3t \leq 6\pi$. So we need to consider values of $3t$ between 0 and 6π. The sine is positive in quadrants I and II, so

$$\sin 3t = \frac{\sqrt{2}}{2}$$

for

$$3t = \frac{\pi}{4} \qquad\qquad 3t = \frac{3\pi}{4}$$
$$3t = \frac{\pi}{4} + 2\pi = \frac{9\pi}{4} \qquad 3t = \frac{3\pi}{4} + 2\pi = \frac{11\pi}{4}$$
$$3t = \frac{\pi}{4} + 4\pi = \frac{17\pi}{4} \qquad 3t = \frac{\pi}{4} + 4\pi = \frac{19\pi}{4}.$$

Hence

$$t = \frac{\pi}{12}, \frac{3\pi}{12}, \frac{9\pi}{12}, \frac{11\pi}{12}, \frac{17\pi}{12}, \frac{19\pi}{12}.$$

$\square$

EXAMPLE 3
Find all t in the interval $[0, 2\pi]$, satisfying $(\cos t)^2 - \cos t - 2 = 0$.

Solution If we set $x = \cos t$, then the equation is the same as

$$x^2 - x - 2 = 0$$
$$(x + 1)(x - 2) = 0$$
$$x = -1, x = 2.$$

Now solve

$$\cos t = -1, \cos t = 2.$$

The equation $\cos t = 2$ has no solutions since $-1 \leq \cos t \leq 1$ (recall that the same restriction holds for $\sin t$, that is, $-1 \leq \sin t \leq 1$). The solutions are

$$\cos t = -1 \text{ so } t = \pi.$$

$\square$

Exercise Set 4.4 (Page 219)

1. **a.** $\frac{\pi}{4}$ **b.** $\frac{\pi}{3}$ **c.** $\pi - \frac{2\pi}{3} = \frac{\pi}{3}$ **d.** $\pi - \frac{3\pi}{4} = \frac{\pi}{4}$

3. **a.** $\frac{\pi}{3}$ **b.** $\frac{7\pi}{6} - \pi = \frac{\pi}{6}$ **c.** $\pi - \frac{2\pi}{3} = \frac{\pi}{3}$ **d.** $\pi - \frac{3\pi}{4} = \frac{\pi}{4}$

5. **a.** $45°$ **b.** $180° - 150° = 30°$ **c.** $240° - 180° = 60°$ **d.** $360° - 330° = 30°$

7. $\sin \frac{\pi}{6} = \frac{1}{2}$; $\cos \frac{\pi}{6} = \frac{\sqrt{3}}{2}$

9. $\sin \frac{5\pi}{6} = \sin \frac{\pi}{6} = \frac{1}{2}$; $\cos \frac{5\pi}{6} = -\cos \frac{\pi}{6} = -\frac{\sqrt{3}}{2}$

11. $\sin \frac{4\pi}{3} = -\sin \frac{\pi}{3} = -\frac{\sqrt{3}}{2}$; $\cos \frac{4\pi}{3} = -\cos \frac{\pi}{3} = -\frac{1}{2}$

13. $\sin \frac{13\pi}{6} = \sin \frac{\pi}{6} = \frac{1}{2}$; $\cos \frac{13\pi}{6} = \cos \frac{\pi}{6} = \frac{\sqrt{3}}{2}$

15. $\sin \left(-\frac{\pi}{3}\right) = -\sin \frac{\pi}{3} = -\frac{\sqrt{3}}{2}$; $\cos \left(-\frac{\pi}{3}\right) = \cos \frac{\pi}{3} = \frac{1}{2}$

17. $\sin \left(-\frac{5\pi}{6}\right) = -\sin \frac{\pi}{6} = -\frac{1}{2}$; $\cos \left(-\frac{5\pi}{6}\right) = -\cos \frac{\pi}{6} = -\frac{\sqrt{3}}{2}$

19. $\sin \left(-\frac{5\pi}{4}\right) = \sin \frac{\pi}{4} = \frac{\sqrt{2}}{2}$; $\cos \left(-\frac{5\pi}{4}\right) = -\cos \frac{\pi}{4} = -\frac{\sqrt{2}}{2}$

21. $\sin \left(-\frac{3\pi}{2}\right) = \sin \frac{\pi}{2} = 1$; $\cos \left(-\frac{3\pi}{2}\right) = \cos \frac{\pi}{2} = 0$

23. $\sin(-7\pi) = \sin(\pi) = 0$; $\cos(-7\pi) = \cos(\pi) = -1$

25. $\sin 45° = \frac{\sqrt{2}}{2}$; $\cos 45° = \frac{\sqrt{2}}{2}$

27. $\sin 150° = \sin 30° = \frac{1}{2}$; $\cos 150° = -\cos 30° = -\frac{\sqrt{3}}{2}$

29. $\sin 240° = -\sin 60° = -\frac{\sqrt{3}}{2}$; $\cos 240° = -\cos 60° = -\frac{1}{2}$

31. $\sin 330° = -\sin 30° = -\frac{1}{2}$; $\cos 330° = \cos 30° = \frac{\sqrt{3}}{2}$

33. $\sin(-60°) = -\sin 60° = -\frac{\sqrt{3}}{2}$; $\cos(-60°) = \cos 60° = \frac{1}{2}$

35. $\sin(-240°) = \sin 60° = \frac{\sqrt{3}}{2}$; $\cos(-240°) = -\cos 60° = -\frac{1}{2}$

37. $\sin 540° = \sin 180° = 0$; $\cos 540° = \cos 180° = -1$

39. $\cos t = \frac{\sqrt{2}}{2}$ implies $t = \frac{\pi}{4}, \frac{7\pi}{4}$

41. $\sin t = -\frac{1}{2}$ implies $t = \frac{7\pi}{6}, \frac{11\pi}{6}$

43. $\cos t = 1$ implies $t = 0, 2\pi$

45. If $0 \le t \le 2\pi$, then $0 \le \frac{t}{2} \le \pi$ and $\cos \frac{t}{2} = \frac{1}{2}$ implies $\frac{t}{2} = \frac{\pi}{3}$ so $t = \frac{2\pi}{3}$.

47. The function $f(x) = (\cos x)^2$ is even since

$f(-x) = (\cos(-x))^2 = (\cos x)^2 = f(x)$.

49. The function $f(x) = |x| \sin x$ is odd since

$f(-x) = |-x| \sin(-x) = |x| (-\sin x) = -|x| \sin x = -f(x)$.

51. Since $\sin t = -\frac{2\sqrt{2}}{3}$ and $\cos t = \frac{1}{3}$, the terminal point $P(t) = \left(\frac{1}{3}, -\frac{2\sqrt{2}}{3} \right)$ and lies in quadrant IV.

 a. $P(t + \pi)$ lies in quadrant II, and $\sin(t + \pi) = \frac{2\sqrt{2}}{3}$, $\cos(t + \pi) = -\frac{1}{3}$.

 b. $P(-t)$ lies in quadrant I, and $\sin(-t) = \frac{2\sqrt{2}}{3}$, $\cos(-t) = \frac{1}{3}$.

 c. $P\left(t + \frac{\pi}{2}\right)$ lies in quadrant I and the coordinates of the point are interchanged, so
$\sin\left(t + \frac{\pi}{2}\right) = \frac{1}{3}$, $\cos\left(t + \frac{\pi}{2}\right) = \frac{2\sqrt{2}}{3}$.

d. $P\left(-t + \frac{\pi}{2}\right)$ lies in quadrant II and the coordinates of the point are interchanged, so
$\sin\left(-t + \frac{\pi}{2}\right) = \frac{1}{3}$, $\cos\left(-t + \frac{\pi}{2}\right) = -\frac{2\sqrt{2}}{3}$.

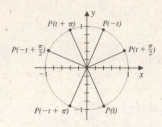

53. We have

$$2(\sin t)^2 + \sin t - 1 = (2\sin t - 1)(\sin t + 1) = 0,$$

which holds when $2\sin t - 1 = 0$ or $\sin t + 1 = 0$. Hence $\sin t = \frac{1}{2}$ or $\sin t = -1$ which implies $t = \frac{\pi}{6}, \frac{5\pi}{6}, \frac{3\pi}{2}$.

55. Squaring both sides of the equation $\sin t + \cos t = 1$ gives

$$1 = (\sin t + \cos t)^2 = (\sin t)^2 + 2\sin t \cos t + (\cos t)^2 = 1 + 2\sin t \cos t,$$

So $\sin t \cos t = 0$.

Hence, the equation $\sin t + \cos t = 1$ is satisfied precisely when one of $\cos t$ or $\sin t$ is 0 and the other is 1. So $t = 0, \frac{\pi}{2}, 2\pi$.

4.5   Graphs of the Sine and Cosine Functions

A function f is *periodic* if a positive number T exists with $f(x + T) = f(x)$, for all x in the domain of f. The smallest value of T is called the *period* of the function. Since the circumference of the unit circle is 2π, we have

$$\sin(x + 2\pi) = \sin x \quad \text{and} \quad \cos(x + 2\pi) = \cos x.$$

Since 2π is the smallest value for which this is true, the period of the sine and cosine functions is 2π.

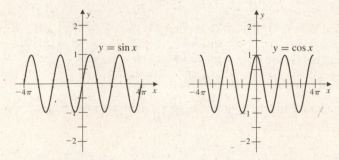

Three important features of the graphs of

$$y = A \sin(Bx + C), \text{ and } y = A \cos(Bx + C)$$

where $A \neq 0$ and $B > 0$ are the amplitude, period, and phase shift.

- Amplitude: The *amplitude* is the height of the sine or cosine wave, which is $|A|$.

- Period: The period is $\frac{2\pi}{B}$.

- Phase Shift: To determine the amount that the standard sine or cosine function is horizontally shifted, first write

$$y = A \sin B \left(x + \frac{C}{B} \right) \text{ or } y = A \cos B \left(x + \frac{C}{B} \right).$$

The phase shift is $\left| \frac{C}{B} \right|$, and the graph is horizontally shifted to the left $\frac{C}{B}$ units if $\frac{C}{B} > 0$, and horizontally shifted to the right $\left| \frac{C}{B} \right|$ if $\frac{C}{B} < 0$.

EXAMPLE 1

Use the graph of the sine or cosine to sketch one period of the graph. Specify the amplitude and the period of the graph.

(a) $y = \frac{1}{2} \cos x$ (b) $y = -2 \cos 3x$ (c) $y = 2 \sin 2x$ (d) $y = -\frac{1}{2} \sin \frac{1}{2}x$

Solution (a)

- Amplitude: $|A| = \frac{1}{2}$
- Period: 2π
- x-intercepts on $[0, 2\pi]$:

$$\frac{1}{2} \cos x = 0$$
$$\cos x = 0$$
$$x = \frac{\pi}{2}, \frac{3\pi}{2}$$

- Maximums and minimums:

$$\frac{1}{2} \cos x = \frac{1}{2}, \frac{1}{2} \cos x = -\frac{1}{2}$$
$$\cos x = 1, \cos x = -1$$
$$x = 0, 2\pi, \ x = \pi$$

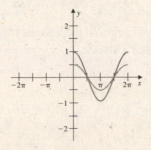

(b) The equation $y = -2\cos 3x$ has the form $y = -A\cos Bx$ with $A = -2$ and $B = 3$.
- Amplitude: $|A| = 2$
- Period: $\frac{2\pi}{B} = \frac{2\pi}{3}$

   Since the period is $\frac{2\pi}{3}$, one complete wave occurs on the interval $\left[0, \frac{2\pi}{3}\right]$.

- x-intercepts on $\left[0, \frac{2\pi}{3}\right]$: Check all values of x so that $0 \le x \le \frac{2\pi}{3} \Rightarrow 0 \le 3x \le 2\pi$.

$$-2\cos 3x = 0$$
$$\cos 3x = 0$$
$$3x = \frac{\pi}{2}, \frac{3\pi}{2}$$
$$x = \frac{\pi}{6}, \frac{\pi}{2}$$

- Maximums and minimums on $\left[0, \frac{2\pi}{3}\right]$:

$$-2\cos 3x = 2, \; -2\cos 3x = -2$$
$$\cos 3x = -1, \; \cos 3x = 1$$
$$3x = \pi, \; 3x = 0, 2\pi$$
$$x = \frac{\pi}{3}, \; x = 0, \frac{2\pi}{3}$$

The figure shows the graph of $y = \cos x$ and $y = -2\cos 3x$. Notice that on the interval $[0, 2\pi]$ the graph of $y = -2\cos 3x$ has 3 complete waves and is reflected over the x-axis. The maximums and minimums are 2 and -2, respectively.

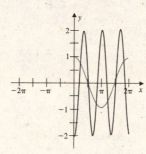

(c) The equation $y = 2\sin 2x$ has the form $y = A\sin Bx$ where $A = 2$, and $B = 2$.
- Amplitude: $|A| = 2$
- Period: $\frac{2\pi}{B} = \frac{2\pi}{2} = \pi$

   Since the period is π, one complete wave occurs on the interval $[0, \pi]$.

- x-intercepts on $[0, \pi]$: Check all values of x so that $0 \le x \le \pi \Rightarrow 0 \le 2x \le 2\pi$.

$$2\sin 2x = 0$$
$$2x = 0, \pi, 2\pi$$
$$x = 0, \frac{\pi}{2}, \pi$$

● Maximums and minimums on $[0, \pi]$:

$$2\sin 2x = 2, 2\sin 2x = -2$$
$$\sin 2x = 1, \sin 2x = -1$$
$$2x = \frac{\pi}{2}, 2x = \frac{3\pi}{2}$$
$$x = \frac{\pi}{4}, x = \frac{3\pi}{4}$$

The figure shows the graph of $y = \sin x$ and $y = 2\sin 2x$. Notice that there are 2 complete waves of the graph of $y = 2\sin 2x$ on the interval $[0, 2\pi]$. The maximums and minimums are 2 and -2, respectively.

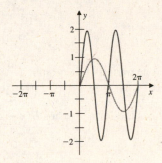

(d) The equation $y = -\frac{1}{2}\sin\frac{1}{2}x$ has the form $y = A\sin Bx$, where $A = -\frac{1}{2}$ and $B = \frac{1}{2}$. The graph of $y = -\frac{1}{2}\sin\frac{1}{2}x$ is the reflection about the x-axis of the graph of $y = \frac{1}{2}\sin\frac{1}{2}x$.

● Amplitude: $|A| = \frac{1}{2}$

● Period: $\frac{2\pi}{B} = \frac{2\pi}{1/2} = 4\pi$

● x-intercepts: Checking all x so that $0 \le x \le 4\pi \Rightarrow 0 \le \frac{1}{2}x \le 2\pi$ gives

$$-\frac{1}{2}\sin\frac{1}{2}x = 0$$
$$\frac{1}{2}x = 0, \pi, 2\pi$$
$$x = 0, 2\pi, 4\pi.$$

● Maximums and minimums:

$$-\frac{1}{2}\sin\frac{1}{2}x = \frac{1}{2}, -\frac{1}{2}\sin\frac{1}{2}x = -\frac{1}{2}$$
$$\sin\frac{1}{2}x = -1, \sin\frac{1}{2}x = 1$$
$$\frac{1}{2}x = \frac{3\pi}{2}, \frac{1}{2}x = \frac{\pi}{2}$$
$$x = 3\pi, x = \pi$$

The figure shows the graph of $y = \sin x$ and $y = -\frac{1}{2} \sin \frac{1}{2}x$. Notice that on the interval $[0, 2\pi]$ the graph of $y = -\frac{1}{2} \sin \frac{1}{2}x$ completes only half a wave. The maximums and minimums are $\frac{1}{2}$ and $-\frac{1}{2}$, respectively.

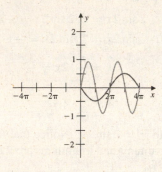

$\square$

EXAMPLE 2

Use the graph of the sine or cosine to sketch one period of the graph.

(a) $y = \cos\left(x - \frac{\pi}{4}\right)$ (b) $y = -1 + 2\sin\left(3x - \frac{\pi}{2}\right)$

Solution (a) The equation $y = \cos\left(x - \frac{\pi}{4}\right)$ has the form $y = \cos B(x - c)$, where $B = 1$ and $C = \frac{\pi}{4}$.

- Amplitude: 1
- Period: $\frac{2\pi}{B} = 2\pi$
- Phase Shift: To the right $\frac{\pi}{4}$ units.

The graph is obtained by shifting the graph of $y = \cos x$ to the right $\frac{\pi}{4}$ units.

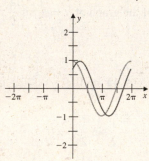

(b) First rewrite the function as

$$y = -1 + 2\sin\left(3x - \frac{\pi}{2}\right) = -1 + 2\sin 3\left(x - \frac{\pi}{6}\right).$$

- Amplitude: 2
- Period:

$$\frac{2\pi}{B} = \frac{2\pi}{3}$$

The graph is horizontally compressed and completes one entire wave on an interval of length $\frac{2\pi}{3}$.

- Phase Shift: To the right $\frac{\pi}{6}$ units.
- Vertical Shift: Adding -1 to the sine function shifts the graph downward 1 unit.

- x-intercepts of $y = 2\sin\left(3x - \frac{\pi}{2}\right)$:

$$2\sin\left(3x - \frac{\pi}{2}\right) = 0$$

$$\sin\left(3x - \frac{\pi}{2}\right) = 0$$

$$3x - \frac{\pi}{2} = 0, \pi, 2\pi, 3\pi, 4\pi, 5\pi, 6\pi$$

$$3x = \frac{\pi}{2}, \frac{3\pi}{2}, \frac{5\pi}{2}, \frac{7\pi}{2}, \frac{9\pi}{2}, \frac{11\pi}{2}, \frac{13\pi}{2}$$

$$x = \frac{\pi}{6}, \frac{\pi}{2}, \frac{5\pi}{6}, \frac{7\pi}{6}, \frac{3\pi}{2}, \frac{11\pi}{6}, \frac{13\pi}{6}$$

- Maximums and Minimums of $y = 2\sin\left(3x - \frac{\pi}{2}\right)$:

$$2\sin\left(3x - \frac{\pi}{2}\right) = 2 \text{ or } 2\sin\left(3x - \frac{\pi}{2}\right) = -2$$

$$\sin\left(3x - \frac{\pi}{2}\right) = 1 \text{ or } \sin\left(3x - \frac{\pi}{2}\right) = -1$$

$$3x - \frac{\pi}{2} = \frac{\pi}{2} \text{ or } 3x - \frac{\pi}{2} = \frac{3\pi}{2}$$

$$x = \frac{\pi}{3}, x = \frac{2\pi}{3}$$

Notice that $\frac{5\pi}{6} - \frac{\pi}{6} = \frac{4\pi}{6} = \frac{2\pi}{3}$, which is the period of the function. Also, $\frac{\pi}{3}$ is midway between $\frac{\pi}{6}$ and $\frac{\pi}{2}$, and $\frac{2\pi}{3}$ is midway between $\frac{\pi}{2}$ and $\frac{5\pi}{6}$.

To obtain the graph:

- Start with the graph of $y = \sin x$.
- Vertically stretch it by a factor of 2.
- Horizontally compress the graph so one complete wave occurs on the interval $\left[0, \frac{2\pi}{3}\right]$.
- Shift this graph to the right $\frac{\pi}{6}$ units.
- Shift the graph downward 1 unit.

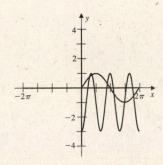

EXAMPLE 3

Find (a) a cosine function and (b) a sine function whose graph matches the curve.

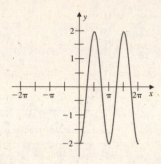

Solution (a) The curve is a cosine curve reflected over the x-axis, vertically stretched by a factor of 2, and horizontally compressed by a factor of 2. It has equation

$$y = -2\cos 2x.$$

(b) To define the curve as a sine curve is a bit more difficult. The curve has the appearance of a sine curve of the form $y = A\sin B\left(x + \frac{C}{B}\right)$, which has also been reflected about the x-axis.

- Amplitude: The height of the wave is 2, so $|A| = 2$.
- Period: One complete sine wave occurs on the interval $\left[\frac{\pi}{4}, \frac{5\pi}{4}\right]$, so the period is $\frac{5\pi}{4} - \frac{\pi}{4} = \pi$, and

$$\frac{2\pi}{B} = \pi$$

$$B = 2.$$

- Phase Shift: The curve is shifted to the left $\frac{\pi}{4}$ units.

The equation of the curve is

$$y = -2\sin 2\left(x + \frac{\pi}{4}\right).$$

The minus sign causes the reflection about the x-axis. A third equation with the same graph is $y = 2\sin\left(x - \frac{\pi}{4}\right)$.

$\square$

EXAMPLE 4

Match the equation with the graph.

(a) $y = -\cos 2\left(x - \frac{\pi}{2}\right)$ (b) $y = \cos\frac{1}{2}\left(x + \frac{\pi}{2}\right)$

(c) $y = \sin\frac{1}{2}\left(x - \frac{\pi}{2}\right)$ (d) $y = \sin 2\left(x + \frac{\pi}{2}\right)$

(i)
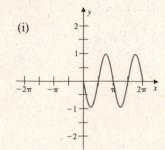

(ii)
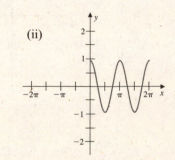

(iii) 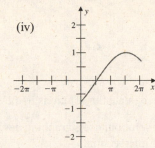 (iv)

Solution The period of the waves in (a) and (d) is $\frac{2\pi}{2} = \pi$, and the period of the waves in (b) and (c) is $\frac{2\pi}{1/2} = 4\pi$. So on an interval of length 4π, one complete wave of the curves is described in (b) or (c), but four complete waves of the curves are described in (a) or (d). As a consequence, (a) and (d) match (i) or (ii) and (b) and (c) match (iii) or (iv).
(a) If $x = 0$, $y = -\cos 2(0 - \pi/2) = -\cos(-\pi) = 1$, so (a) matches with (ii).
(b) If $x = \frac{\pi}{2}$, $y = \cos \frac{\pi}{2} = 0$, and (b) matches (iii).
(c) If $x = 0$, $y = \sin \frac{1}{2}(0 - \pi/2) = \sin(-\pi/4) = -\sqrt{2}/2 < 0$ so (c) matches with (iv).
(d) If $x = 0$, $y = \sin \pi = 0$, and (d) matches (i). □

EXAMPLE 5
Find the equation of a cosine wave that is obtained by shifting the graph of $y = \cos x$ to the left 2 units, upward 1 unit, and is horizontally compressed by a factor of 3.

Solution The cosine wave has the form
$$y = A \cos B(x + C) + D.$$

Since the graph is not vertically scaled, the amplitude remains one, and $A = 1$. If $y = \cos x$ is horizontally compressed by a factor of 3, the period is one third of $y = \cos x$, so

$$\frac{2\pi}{B} = \frac{1}{3}(2\pi) = \frac{2\pi}{3}$$
$$B = 3.$$

To shift to the left 2 units, let $C = -2$. To shift upward 1 unit, let $D = 1$. The equation of the wave is

$$y = 1 + \cos 3(x - 2).$$

□

EXAMPLE 6
Determine an appropriate viewing rectangle for the function $f(x) = \cos(200x)$, and use it to sketch the graph.

Solution Since there is no phase shift, the period of the function gives a good viewing rectangle, so compute

$$\frac{2\pi}{200} = \frac{\pi}{100} \approx 0.03.$$

Two complete waves of the function occur on the interval $\left[-\frac{\pi}{100}, \frac{\pi}{100}\right]$, and an appropriate viewing rectangle is $\left[-\frac{\pi}{100}, \frac{\pi}{100}\right] \times [-1, 1]$.

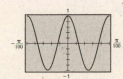

EXAMPLE 7

Use a graphing device to approximate the solutions to the equation $\cos x + x = x^3 + 1$.

Solution To approximate the solution, use a graphing device to sketch the two curves, $y = \cos x + x$ and $y = x^3 + 1$, and approximate the points of intersection.

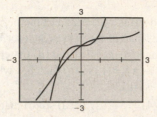

The points of intersection are

$$x = 0, \quad x \approx 0.8, \quad \text{and} \quad x \approx -1.25.$$

EXAMPLE 8

Write the function h as the composition of two functions $h(x) = f(g(x))$.
(a) $h(x) = \sqrt{\cos x}$ (b) $h(x) = 2\sin(2x + 1)$

Solution (a) The inside operation takes the cosine of the number x, and the outside operation takes the square root. So define

$$f(x) = \sqrt{x}, \text{ and } g(x) = \cos x,$$

and

$$f(g(x)) = f(\cos x) = \sqrt{\cos x}.$$

(b) The inside operation is the argument of the sine function, $2x - 1$. The outside operation is taking twice the sine. Define

$$f(x) = 2\sin x, \quad \text{and} \quad g(x) = 2x + 1,$$

so

$$f(g(x)) = f(2x + 1) = 2\sin(2x + 1).$$

Exercise Set 4.5 (Page 229)

1. a. For $y = 2\cos x$, amplitude $= 2$, period $= 2\pi$

    b. For $y = -3\cos x$, amplitude $= 3$, period $= 2\pi$

    c. For $y = \frac{1}{2}\cos 2x$, amplitude $= \frac{1}{2}$, period$= \frac{2\pi}{2} = \pi$

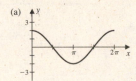 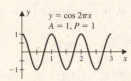 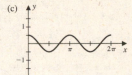

3. a. For $y = 2\cos \pi x$, amplitude $= 2$, period $= \frac{2\pi}{\pi} = 2$

    b. For $y = \cos 2\pi x$, amplitude $= 1$, period $= \frac{2\pi}{2\pi} = 1$

    c. For $y = -2\cos \frac{\pi}{2}x$, amplitude $= 2$, period $= \frac{2\pi}{\pi/2} = 4$

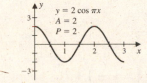 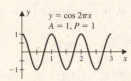 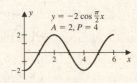

5. $y = 1 + \cos x$

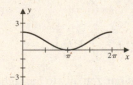

7. $y = \cos\left(x - \frac{\pi}{2}\right)$

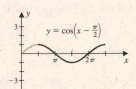

9. $y = -3 + \cos \pi x$

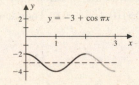

11. $y = 1 + \cos(2x + \pi) = 1 + \cos 2\left(x + \frac{\pi}{2}\right)$

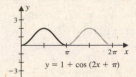

13. $y = -1 + 2\sin\left(2x - \frac{\pi}{2}\right) = -1 + 2\sin 2\left(x - \frac{\pi}{4}\right)$ **15.** $y = -2\sin(x - 1) + 3$

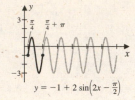

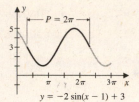

17. $y = |\cos x|$

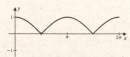

19. a. The equation of the graph has the form $y = A\cos(Bx)$, where the amplitude is 3 and the period is π. So $A = 3$, $\frac{2\pi}{B} = \pi$ which implies $B = 2$, and $y = 3\cos 2x$.

b. The equation can also be given as a sine curve shifted to the left by $\frac{\pi}{4}$ units. With amplitude 3, period π, and phase shift $-\frac{\pi}{4}$, the sine wave has equation $y = 3\sin 2\left(x + \frac{\pi}{4}\right)$.

21. a. The equation of the graph has the form

$$y = A\cos(Bx + C) + D = A\cos B\left(x + \frac{C}{B}\right) + D,$$

where the amplitude is 2 the period is 10, the curve has been shifted horizontally to the right $\frac{9}{2}$ units, and shifted upward 1 unit. So $A = 2$, $\frac{2\pi}{B} = 10$, which implies $B = \frac{\pi}{5}$, $\frac{C}{B} = -\frac{9}{2}$, $D = 1$, and $y = 2\cos\frac{\pi}{5}\left(x - \frac{9}{2}\right) + 1$.

b. The equation of the graph can also have the form

$$y = A\sin(Bx + C) + D = A\sin B\left(x + \frac{C}{B}\right) + D,$$

where the amplitude is 2, the period is 10, the curve has been shifted horizontally to the right 2 units, and the curve has been shifted upward 1 unit. So $A = 2$ and $\frac{2\pi}{B} = 10$, which implies $B = \frac{\pi}{5}$, $\frac{C}{B} = -2$, $D = 1$, and $y = 2\sin\frac{\pi}{5}(x - 2) + 1$.

23. The amplitude of the waves in (a) and (c) is 2, which could only match (iii) and (iv), and the amplitude in (b) and (d) is $\frac{1}{2}$, which could only match (i) and (ii). Setting $x = 0$ in each of the equations, the y-intercepts are $2\sin\left(0 - \frac{\pi}{2}\right) = -2$, $2\cos\left(0 - \frac{\pi}{2}\right) = 0$, $\frac{1}{2}\sin\left(0 + \frac{\pi}{2}\right) = \frac{1}{2}$, and $\frac{1}{2}\cos\left(0 + \frac{\pi}{2}\right) = 0$. The matching is then (a) iv, (b) i, (c) iii, and (d) ii.

25. $y = -1 + \frac{1}{2}\sin(x - 2)$

27. The period of $f(x) = \cos(100x)$ is $\frac{2\pi}{100} = \frac{\pi}{50} \approx 0.06$, and the amplitude is 1, so a reasonable viewing rectangle is $[-\pi/50, \pi/50] \times [-1, 1]$.

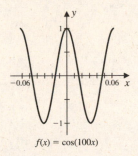

$$f(x) = \cos(100x)$$

29. The period of $f(x) = \sin\left(\frac{x}{50}\right)$ is $\frac{2\pi}{1/50} = 100\pi$, and the amplitude is 1, so a reasonable viewing rectangle is $[-100\pi, 100\pi] \times [-1, 1]$.

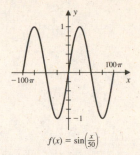

$$f(x) = \sin\left(\frac{x}{50}\right)$$

31. $\cos x = x$ so $x \approx 0.7$

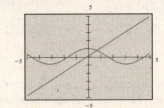

33. $\sin x + \cos x = x$ so $x \approx 1.3$

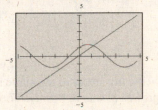

35. a. The smallest positive value of t for which

$$0 = f(t) = 2\sin\left(3t - \frac{\pi}{4}\right)$$

occurs when

$$3t - \frac{\pi}{4} = 0, \text{ so } t = \frac{\pi}{12}.$$

b. A maximum value first occurs when

$$\sin\left(3t - \frac{\pi}{4}\right) = 1,$$

so

$$3t - \frac{\pi}{4} = \frac{\pi}{2} \text{ and } t = \frac{\pi}{4}.$$

c. A minimum value first occurs when

$$\sin\left(3t - \frac{\pi}{4}\right) = -1,$$

so

$$3t - \frac{\pi}{4} = \frac{3\pi}{2} \text{ and } t = \frac{7\pi}{12}.$$

37. The amplitude is 2 and the line on which the graph is centered is $y = 1$. The period is 4, so the points on the graph are given by

$$y = 1 + 2\sin\frac{\pi}{2}x.$$

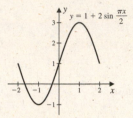

39. The graphs are shown for each part.

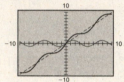 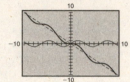 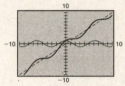 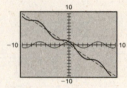

41. a. It vertically stretches the curve by a factor of 2, so the amplitude is doubled.   **b.** It horizontally compresses the curve by a factor of 2, so the period is halved.   **c.** The horizontal shift is doubled.

43. a. Since the difference between the largest data value and the smallest data value is about $70 - 26 = 44$ the amplitude is half this amount or 22. One wave appears to occur on an interval of length 12, so $\frac{2\pi}{B} = 12$ and $B = \frac{\pi}{6}$. The wave is also shifted to the right about 4.5 units and upward 48 units. A possible equation for the curve is

$$f(x) = 22\sin\left(\frac{\pi}{6}(x - 4.5)\right) + 48.$$

b.

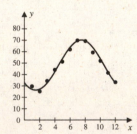

4.6   Other Trigonometric Functions

There are four additional trigonometric functions defined in terms of the sine and cosine function. They are

$$\tan x = \frac{\sin x}{\cos x}, \quad \cot x = \frac{\cos x}{\sin x},$$

$$\sec x = \frac{1}{\cos x}, \quad \csc x = \frac{1}{\sin x}.$$

The tangent and secant functions are defined whenever $\cos x \neq 0$, that is, $x \neq \frac{\pi}{2} + n\pi$ for an integer n. The cotangent and cosecant are defined whenever $\sin x \neq 0$, that is, $x \neq n\pi$ for an integer n.

Being familiar with the values of the trigonometric functions at the standard angles is essential to work in trigonometry. The next example reviews some of the standard angles.

EXAMPLE 1

Determine the six trigonometric functions for the standard angles for $-2\pi \leq t \leq -\frac{7\pi}{6}$.

Solution The table gives the values for the six trigonometric functions at t, where $-2\pi \leq t \leq -7\pi/6$. We have added the reference angles r to the table along with their sine and cosine. To find the cosine and sine at a value t, determine the quadrant where the angle lies and adjust the signs of the sine and cosine for the corresponding reference angle. The angles $t = -7\pi/6, -5\pi/4$, and $-4\pi/3$ are in quadrant II, where $\sin t > 0$ and $\cos t < 0$. The angles $t = -5\pi/3, -7\pi/4$, and $-11\pi/6$ are in quadrant I, where $\sin t > 0$ and $\cos t > 0$.

t	$-\frac{7\pi}{6}$	$-\frac{5\pi}{4}$	$-\frac{4\pi}{3}$	$-\frac{3\pi}{2}$	$-\frac{5\pi}{3}$	$-\frac{7\pi}{4}$	$-\frac{11\pi}{6}$	-2π
	$-210°$	$-225°$	$-240°$	$-270°$	$-300°$	$-315°$	$-330°$	$-360°$
r	$\frac{\pi}{6}$	$\frac{\pi}{4}$	$\frac{\pi}{3}$	$\frac{\pi}{2}$	$\frac{\pi}{3}$	$\frac{\pi}{4}$	$\frac{\pi}{6}$	0
	$30°$	$45°$	$60°$	$90°$	$60°$	$45°$	$30°$	$0°$
$\sin r$	$\frac{1}{2}$	$\frac{\sqrt{2}}{2}$	$\frac{\sqrt{3}}{2}$	1	$\frac{\sqrt{3}}{2}$	$\frac{\sqrt{2}}{2}$	$\frac{1}{2}$	0
$\cos r$	$\frac{\sqrt{3}}{2}$	$\frac{\sqrt{2}}{2}$	$\frac{1}{2}$	0	$\frac{1}{2}$	$\frac{\sqrt{2}}{2}$	$\frac{\sqrt{3}}{2}$	1
$\sin t$	$\frac{1}{2}$	$\frac{\sqrt{2}}{2}$	$\frac{\sqrt{3}}{2}$	1	$\frac{\sqrt{3}}{2}$	$\frac{\sqrt{2}}{2}$	$\frac{1}{2}$	0
$\cos t$	$-\frac{\sqrt{3}}{2}$	$-\frac{\sqrt{2}}{2}$	$-\frac{1}{2}$	0	$\frac{1}{2}$	$\frac{\sqrt{2}}{2}$	$\frac{\sqrt{3}}{2}$	1
$\tan t$	$-\frac{\sqrt{3}}{3}$	-1	$-\sqrt{3}$	$-$	$\sqrt{3}$	1	$\frac{\sqrt{3}}{3}$	0
$\cot t$	$-\sqrt{3}$	-1	$-\frac{\sqrt{3}}{3}$	0	$\frac{\sqrt{3}}{3}$	1	$\sqrt{3}$	$-$
$\sec t$	$-\frac{2\sqrt{3}}{3}$	$-\sqrt{2}$	-2	$-$	2	$\sqrt{2}$	$\frac{2\sqrt{3}}{3}$	1
$\csc t$	2	$\sqrt{2}$	$\frac{2\sqrt{3}}{3}$	1	$\frac{2\sqrt{3}}{3}$	$\sqrt{2}$	2	$-$

$\square$

EXAMPLE 2

Find the values of all the trigonometric functions, given that $\sin t = 3/5$ and $\pi/2 \le t \le \pi$.

Solution First find the $\cos t$ using the Pythagorean Identity.

$$(\cos t)^2 + (\sin t)^2 = 1$$

$$(\cos t)^2 + \left(\frac{3}{5}\right)^2 = 1$$

$$(\cos t)^2 = 1 - \frac{9}{25} = \frac{16}{25}$$

$$\cos t = \pm\sqrt{\frac{16}{25}}$$

$$\cos t = \pm\frac{4}{5}$$

Since t is in quadrant II, the cosine is negative. So

$$\cos t = -\frac{4}{5}, \quad \sin t = \frac{3}{5}$$

$$\tan t = \frac{\frac{3}{5}}{-\frac{4}{5}} = -\frac{3}{4}, \quad \cot t = \frac{-\frac{4}{5}}{\frac{3}{5}} = -\frac{4}{3}$$

$$\sec t = \frac{1}{-\frac{4}{5}} = -\frac{5}{4}, \quad \csc t = \frac{1}{\frac{3}{5}} = \frac{5}{3}.$$

□

Other Graphic Information:

	Period	x-intercepts	Vertical Asymptotes
$y = \tan x$	π	$x = \pm k\pi, k = 0, 1...$	$x = \pm(2k - 1)\frac{\pi}{2}, k = 1, 2...$
$y = \cot x$	π	$x = \pm(2k - 1)\frac{\pi}{2}, k = 1, 2...$	$x = \pm k\pi, k = 0, 1...$
$y = \sec x$	2π	none	$x = \pm(2k - 1)\frac{\pi}{2}, k = 1, 2...$
$y = \csc x$	2π	none	$x = \pm k\pi, k = 0, 1...$

EXAMPLE 3

Sketch one period of the curve.

(a) $y = 2\tan\left(x - \frac{\pi}{2}\right)$ (b) $y = \sec(3x)$ (c) $y = \cot(2x + \pi)$

Solution (a)

- Period: The same as the period of tangent, which is π.

- Phase shift: To the right $\frac{\pi}{2}$ units.

The strategy is:

- Start with the graph of $y = \tan x$.
- Vertically stretch it by a factor of 2.
- Shift the resulting graph to the right $\frac{\pi}{2}$ units.
- x-intercept:

$$2 \tan \left(x - \frac{\pi}{2} \right) = 0$$

$$\tan \left(x - \frac{\pi}{2} \right) = \frac{\sin \left(x - \frac{\pi}{2} \right)}{\cos \left(x - \frac{\pi}{2} \right)} = 0$$

$$\sin \left(x - \frac{\pi}{2} \right) = 0$$

$$x - \frac{\pi}{2} = 0$$

$$x = \frac{\pi}{2}$$

This is exactly what is expected since $y = \tan x$ crosses the x-axis at $(0, 0)$, and the graph of $y = 2 \tan \left(x - \frac{\pi}{2} \right)$ is the same, but shifted to the right $\frac{\pi}{2}$ units. The graph crosses the x-axis at $\left(\frac{\pi}{2}, 0 \right)$.

- Asymptotes: Since $y = \tan x$ has asymptotes $x = -\frac{\pi}{2}$ and $x = \frac{\pi}{2}$, $y = 2 \tan \left(x - \frac{\pi}{2} \right)$ has asymptotes $x = -\frac{\pi}{2} + \frac{\pi}{2} = 0$ and $x = \frac{\pi}{2} + \frac{\pi}{2} = \pi$.

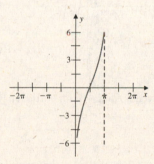

(b) To plot $y = \sec 3x$, first plot $y = \cos 3x$ and then use the reciprocal graphing technique.

- Period: $\frac{2\pi}{3}$
- Phase shift: none
- Vertical asymptotes: Since $\sec 3x = \dfrac{1}{\cos 3x}$, the vertical asymptotes occur when $\cos 3x = 0$.

$$\cos 3x = 0$$

$$3x = \pm \frac{\pi}{2}, \pm \frac{3\pi}{2}, \pm \frac{5\pi}{2}, \dots$$

$$x = \pm \frac{\pi}{6}, \pm \frac{\pi}{2}, \pm \frac{5\pi}{6}, \dots$$

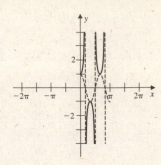

(c) Rewrite the function in form

$$y = \cot(2x + \pi) = \cot 2\left(x + \frac{\pi}{2}\right).$$

- Period: Since the period of the cotangent function is π, the period of $\cot(2x + \pi)$ is $\frac{\pi}{2}$.
- Phase shift: To the left $\frac{\pi}{2}$ units.
  One period of $y = \cot x$ occurs on the interval $(0, \pi)$, has vertical asymptotes at $x = 0$ and $x = \pi$, and crosses the x-axis at $x = \frac{\pi}{2}$. So $y = \cot 2x$ has period $\frac{\pi}{2}$, vertical asymptotes $x = 0$ and $x = \frac{\pi}{2}$, and crosses the x-axis at $x = \frac{\pi}{4}$. The graph of $\cos(2x + \pi)$ also involves a shift to the left $\frac{\pi}{2}$ units.
- Vertical asymptotes of $y = \cot 2\left(x + \frac{\pi}{2}\right) : x = -\frac{\pi}{2}, x = 0$
- x-intercept of $y = \cot 2\left(x + \frac{\pi}{2}\right) : \left(-\frac{\pi}{4}, 0\right)$

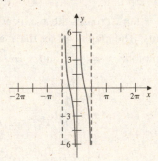

Notice that one period of the graph is also between 0 and $\frac{\pi}{2}$ with vertical asymptotes $x = 0$ and $x = \frac{\pi}{2}$, and x-intercept $\frac{\pi}{4}$.   □

EXAMPLE 4
Find all values of x in the interval $[0, 2\pi]$ that satisfy the equation $|\cot x| = 1$.

Solution We have

$$1 = |\cot x| = \left|\frac{\cos x}{\sin x}\right|, \quad \text{so } |\sin x| = |\cos x|.$$

In the interval $[0, 2\pi]$, the sine and cosine have the same magnitude, $\frac{\sqrt{2}}{2}$, at $\frac{\pi}{4}, \frac{3\pi}{4}, \frac{5\pi}{4}$, and $\frac{7\pi}{4}$. So

$$|\sin x| = |\cos x| = \frac{\sqrt{2}}{2}, \quad \text{for} \quad x = \frac{\pi}{4}, \frac{3\pi}{4}, \frac{5\pi}{4}, \frac{7\pi}{4}.$$

                                                                                                                      □

EXAMPLE 5
Find all values of x in the interval $[0, 2\pi]$ that satisfy the equation $2 \sin 2x - \sqrt{3} \tan 2x = 0$.

Solution Rewriting the equation to involve only sine and cosine gives

$$2 \sin 2x - \sqrt{3} \tan 2x = 0$$

$$2 \sin 2x - \sqrt{3} \frac{\sin 2x}{\cos 2x} = 0$$

$$\frac{2 \sin 2x \cos 2x - \sqrt{3} \sin 2x}{\cos 2x} = 0$$

$$2 \sin 2x \cos 2x - \sqrt{3} \sin 2x = 0$$

$$\sin 2x (2 \cos 2x - \sqrt{3}) = 0.$$

Solutions to this equation for $0 \le x \le 2\pi$, that is, for $0 \le 2x \le 4\pi$, occur when

$$\sin 2x = 0 \quad \text{or for} \quad \cos 2x = \frac{\sqrt{3}}{2}$$

$$2x = 0, \pi, 2\pi, 3\pi, 4\pi \quad \text{or for} \quad 2x = \frac{\pi}{6}, \frac{11\pi}{6}, \frac{13\pi}{6}, \frac{23\pi}{6},$$

$$x = 0, \frac{\pi}{12}, \frac{\pi}{2}, \frac{11\pi}{12}, \pi, \frac{13\pi}{12}, \frac{3\pi}{2}, \frac{23\pi}{12}, 2\pi.$$

$\square$

EXAMPLE 6

Determine the values of the trigonometric functions of t if $P(t)$ lies in the third quadrant and on the line $y = 3x$.

Solution To find the sine and cosine, we need to determine the x- and y-coordinates of the point $P(t)$. Since the point lies on both the line and the unit circle, it must satisfy the two equations,

$$y = 3x \quad \text{and} \quad x^2 + y^2 = 1.$$

Substituting the value of y from the first equation into the second gives

$$x^2 + (3x)^2 = 1$$

$$10x^2 = 1$$

$$x^2 = \frac{1}{10}$$

$$x = \pm\sqrt{\frac{1}{10}} = \pm\frac{\sqrt{10}}{10}.$$

Since $P(t)$ is in the third quadrant, $\cos t < 0$, and the x-coordinate of $P(t)$ is less than 0. So

$$x = -\frac{\sqrt{10}}{10} \quad \text{and} \quad y = -\frac{3\sqrt{10}}{10}.$$

The values of the trigonometric functions are

$$\sin t = -\frac{3\sqrt{10}}{10}, \quad \cos t = -\frac{\sqrt{10}}{10}$$

$$\tan t = 3, \quad \cot t = \frac{1}{3}$$

$$\sec t = -\frac{10}{\sqrt{10}} = -\sqrt{10}, \quad \csc t = -\frac{10}{3\sqrt{10}} = -\frac{\sqrt{10}}{3}.$$

$\square$

Exercise Set 4.6 (Page 237)

1. $\sin t < 0$ and $\cos t < 0$ so $P(t)$ is in quadrant III.

3. $\sin t < 0$ and $\cot t > 0$ implies $\sin t < 0$ and $\frac{\cos t}{\sin t} > 0$ so $\sin t < 0$ and $\cos t < 0$ and $P(t)$ is in quadrant III.

5. $\sin t > 0$ and $\tan t > 0$ implies $\sin t > 0$ and $\frac{\sin t}{\cos t} > 0$ so $\sin t > 0$ and $\cos t > 0$ and $P(t)$ is in quadrant I.

7. A dash in the table indicates the trigonometric function is undefined at the given value.

	$\frac{7\pi}{6}$	$\frac{5\pi}{4}$	$\frac{4\pi}{3}$	$\frac{3\pi}{2}$	$\frac{5\pi}{3}$	$\frac{7\pi}{4}$	$\frac{11\pi}{6}$	2π
$\sin t$	$-\frac{1}{2}$	$-\frac{\sqrt{2}}{2}$	$-\frac{\sqrt{3}}{2}$	-1	$-\frac{\sqrt{3}}{2}$	$-\frac{\sqrt{2}}{2}$	$-\frac{1}{2}$	0
$\cos t$	$-\frac{\sqrt{3}}{2}$	$-\frac{\sqrt{2}}{2}$	$-\frac{1}{2}$	0	$\frac{1}{2}$	$\frac{\sqrt{2}}{2}$	$\frac{\sqrt{3}}{2}$	1
$\tan t$	$\frac{\sqrt{3}}{3}$	1	$\sqrt{3}$	—	$-\sqrt{3}$	-1	$-\frac{\sqrt{3}}{3}$	0
$\cot t$	$\sqrt{3}$	1	$\frac{\sqrt{3}}{3}$	0	$-\frac{\sqrt{3}}{3}$	-1	$-\sqrt{3}$	—
$\sec t$	$-\frac{2\sqrt{3}}{3}$	$-\sqrt{2}$	-2	—	2	$\sqrt{2}$	$\frac{2\sqrt{3}}{3}$	1
$\csc t$	-2	$-\sqrt{2}$	$-\frac{2\sqrt{3}}{3}$	-1	$-\frac{2\sqrt{3}}{3}$	$-\sqrt{2}$	-2	—

9. Since $\cos t = \frac{3}{5}$ and $(\cos t)^2 + (\sin t)^2 = 1$, we have

$$(\sin t)^2 = 1 - \left(\frac{3}{5}\right)^2 = \frac{16}{25}, \text{ which implies } \sin t = \pm\sqrt{\frac{16}{25}} = \pm\frac{4}{5},$$

and since t is in quadrant I, $\sin t > 0$, so $\sin t = \frac{4}{5}$. Then

$$\tan t = \frac{4/5}{3/5} = \frac{4}{3}, \quad \cot t = \frac{3/5}{4/5} = \frac{3}{4}, \quad \sec t = \frac{1}{3/5} = \frac{5}{3}, \quad \csc t = \frac{1}{4/5} = \frac{5}{4}.$$

11. Since $\cos t = -\frac{4}{5}$, with $\pi \le t \le \frac{3\pi}{2}$ implies t is in quadrant III, so $\sin t < 0$. Then since $(\cos t)^2 + (\sin t)^2 = 1$, we have

$$(\sin t)^2 = 1 - \left(-\frac{4}{5}\right)^2 = 1 - \frac{16}{25} = \frac{9}{25}, \quad \text{which implies } \sin t = \pm\sqrt{\frac{9}{25}} = \pm\frac{3}{5},$$

and since $\sin t < 0$, $\sin t = -\frac{3}{5}$. Then

$$\tan t = \frac{-3/5}{-4/5} = \frac{3}{4}; \quad \cot t = \frac{1}{\tan t} = \frac{4}{3}, \quad \sec t = \frac{1}{-4/5} = -\frac{5}{4}, \quad \csc t = \frac{1}{-3/5} = -\frac{5}{3}.$$

13. Since $\cot t = \frac{1}{2}$, we have $\tan t = \frac{1}{1/2} = 2$. Then
$\tan t = \frac{\sin t}{\cos t} = 2$ implies $\sin t = 2\cos t$ or $\sqrt{1 - (\cos t)^2} = 2\cos t$,
so

$$1 - (\cos t)^2 = 4(\cos t)^2 \text{ implies } 5(\cos t)^2 = 1 \text{ so } \cos t = \pm\sqrt{\frac{1}{5}} = \pm\frac{\sqrt{5}}{5},$$

and since $0 < t < \frac{\pi}{2}$, $\cos t > 0$, we have $\cos t = \frac{\sqrt{5}}{5}$ and $\sin t = 2\cos t = \frac{2\sqrt{5}}{5}$. Then

$$\sec t = \frac{1}{\sqrt{5}/5} = \frac{5}{\sqrt{5}} = \sqrt{5}, \quad \csc t = \frac{1}{2\sqrt{5}/5} = \frac{5}{2\sqrt{5}} = \frac{\sqrt{5}}{2}.$$

15. Since $\sec t = 3$ implies $\cos t = \frac{1}{3}$, and since t is in quadrant IV, $\sin t < 0$. Since $(\cos t)^2 + (\sin t)^2 = 1$, we have

$$(\sin t)^2 = 1 - \left(\frac{1}{3}\right)^2 = 1 - \frac{1}{9} = \frac{8}{9} \text{ implies } \sin t = \pm\sqrt{\frac{8}{9}} = \pm\frac{2\sqrt{2}}{3},$$

and since $\sin t < 0$, $\sin t = -\frac{2\sqrt{2}}{3}$. Then

$$\tan t = \frac{-2\sqrt{2}/3}{1/3} = -2\sqrt{2}, \quad \cot t = -\frac{1}{2\sqrt{2}} = -\frac{\sqrt{2}}{4}, \quad \csc t = -\frac{1}{2\sqrt{2}/3} = -\frac{3\sqrt{2}}{4}.$$

17. $y = 3 \tan x$

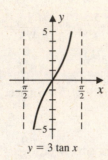

19. $y = -2 \sec x$

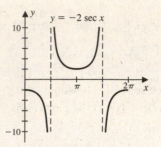

21. $y = \tan(x + \pi/2)$

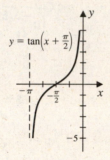

23. $y = \frac{1}{2} \sec(x + \pi/4)$

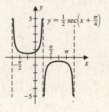

25. For $y = \tan \pi x$, the period is $\frac{\pi}{\pi} = 1$.

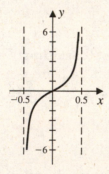

27. For $y = \tan\left(2x - \frac{\pi}{2}\right) = \tan 2\left(x - \frac{\pi}{4}\right)$, the period is $\frac{\pi}{2}$.

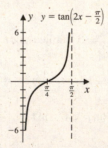

29. $\tan t + 1 = 0$ implies $\tan t = -1$ so $t = \frac{3\pi}{4}, \frac{7\pi}{4}$.

31. $(\tan t)^2 = \frac{1}{3}$ implies $\tan t = \pm \frac{\sqrt{3}}{3}$ so $t = \frac{\pi}{6}, \frac{5\pi}{6}, \frac{7\pi}{6}, \frac{11\pi}{6}$

33. $|\tan t| = 1$ implies $\tan t = 1$ or $\tan t = -1$ so $t = \frac{\pi}{4}, \frac{3\pi}{4}, \frac{5\pi}{4}, \frac{7\pi}{4}$

35. Since

$$2\sin 2t - \sqrt{2}\tan 2t = 2\sin 2t - \sqrt{2}\frac{\sin 2t}{\cos 2t} = 0 \text{ implies } \frac{2\sin 2t\cos 2t - \sqrt{2}\sin 2t}{\cos 2t} = 0,$$

we have

$$2\sin 2t\cos 2t - \sqrt{2}\sin 2t = \sin 2t(2\cos 2t - \sqrt{2}) = 0 \text{ so } \sin 2t = 0 \text{ or } \cos 2t = \frac{\sqrt{2}}{2}.$$

If $0 \le t \le 2\pi$, then $0 \le 2t \le 4\pi$, so $2t = 0, \pi, 2\pi, 3\pi, 4\pi$ or $2t = \frac{\pi}{4}, \frac{7\pi}{4}, \frac{9\pi}{4}, \frac{15\pi}{4}$, which implies that $t = 0, \frac{\pi}{2}, \pi, \frac{3\pi}{2}, 2\pi, \frac{\pi}{8}, \frac{7\pi}{8}, \frac{9\pi}{8}, \frac{15\pi}{8}$.

37. The period of $f(x) = \tan(5x)$ is $\frac{\pi}{5}$, so a reasonable viewing rectangle is $\left[-\frac{\pi}{10}, \frac{\pi}{10}\right] \times [-5, 5]$.

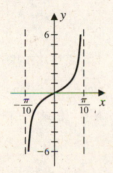

39. The period of $f(x) = \csc 100x$ is $\frac{2\pi}{100} = \frac{\pi}{50}$, so a reasonable viewing rectangle is $\left[-\frac{\pi}{50}, \frac{\pi}{50}\right] \times [-5, 5]$.

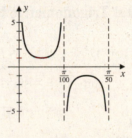

41. The period of $f(x) = \tan\left(\frac{x}{100}\right)$ is $\frac{\pi}{1/100} = 100\pi$, so a reasonable viewing rectangle is $[-50\pi, 50\pi] \times [-5, 5]$.

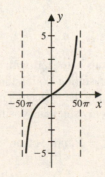

43. Let $P(t) = (x, y)$. Since $P(t)$ lies on the unit circle $x^2 + y^2 = 1$, and since the point lies on the line $y = -2x$, we have

$$x^2 + (-2x)^2 = 1 \text{ implies } 5x^2 = 1 \text{ so } x = \pm\sqrt{\frac{1}{5}} = \pm\frac{\sqrt{5}}{5}.$$

Since $P(t)$ lies in quadrant IV, $x > 0$ and $y < 0$, so we have $x = \frac{\sqrt{5}}{5}, y = -\frac{2\sqrt{5}}{5}$. Then

$$\cos t = \frac{\sqrt{5}}{5}, \quad \sin t = -\frac{2\sqrt{5}}{5}, \quad \tan t = \frac{-2\sqrt{5}/5}{\sqrt{5}/5} = -2,$$

$$\cot t = -\frac{1}{2}, \quad \sec t = \frac{5}{\sqrt{5}} = \sqrt{5}, \quad \csc t = -\frac{5}{2\sqrt{5}} = -\frac{\sqrt{5}}{2}.$$

....

4.7   Trigonometric Identities

An *identity* is an equation that is true for every value of the variable. Trigonometric identities are equations that are true for all values of the variable for which both sides are defined.

Fundamental Trigonometric Identities

$$(\sin x)^2 + (\cos x)^2 = 1, \qquad 1 + (\tan x)^2 = (\sec x)^2, \qquad 1 + (\cot x)^2 = (\csc x)^2,$$

$$\sin(-x) = -\sin x, \qquad \cos(-x) = \cos x, \qquad \tan(-x) = -\tan x.$$

EXAMPLE 1
Verify the identities.

(a) $(1 - (\cos x)^2)(\sec x)^2 = (\tan x)^2$

(b) $\tan x + \cot x = \sec x \csc x$

(c) $\dfrac{\cos x}{1 - \tan x} + \dfrac{\sin x}{1 - \cot x} = \sin x + \cos x$

(d) $(\cot x)^2 - (\cos x)^2 = (\cot x)^2 (\cos x)^2$

Solution (a) One method for verifying identities is to start with one side, replace any trigonometric functions with sines and cosines, and then try to simplify the resulting expression to that on the other side of the identity. Be on the look out for the Fundamental Identities, especially the Pythagorean Identities.

In this example rewrite the left side to involve only sines and cosines, and use the fact that

$$(\sin x)^2 + (\cos x)^2 = 1 \text{ implies that } (\sin x)^2 - 1 = (\cos x)^2.$$

$$(1 - (\cos x)^2)(\sec x)^2 = (1 - (\cos x)^2)\frac{1}{(\cos x)^2}$$

$$= (\sin x)^2 \frac{1}{(\cos x)^2}$$

$$= \frac{(\sin x)^2}{(\cos x)^2} = (\tan x)^2$$

(b) First replace the left side with sines and cosines, giving

$$\tan x + \cot x = \frac{\sin x}{\cos x} + \frac{\cos x}{\sin x}.$$

Now take the common denominator of the two fractions. Recall that for any algebraic expressions

$$\frac{a}{b} + \frac{c}{d} = \frac{ad + cb}{bd}.$$

Combining the fractions on the right side gives

$$\begin{aligned}
\tan x + \cot x &= \frac{\sin x}{\cos x} + \frac{\cos x}{\sin x} \\
&= \frac{\sin x \sin x + \cos x \cos x}{\cos x \sin x} \\
&= \frac{(\sin x)^2 + (\cos x)^2}{\cos x \sin x} \\
&= \frac{1}{\cos x \sin x} = \frac{1}{\cos x} \cdot \frac{1}{\sin x} \\
&= \sec x \csc x.
\end{aligned}$$

(c) Changing the left side to sines and cosines gives

$$\begin{aligned}
\frac{\cos x}{1 - \tan x} + \frac{\sin x}{1 - \cot x} &= \frac{\cos x}{1 - \frac{\sin x}{\cos x}} + \frac{\sin x}{1 - \frac{\cos x}{\sin x}} \\
&= \frac{\cos x}{\frac{\cos x - \sin x}{\cos x}} + \frac{\sin x}{\frac{\sin x - \cos x}{\sin x}} \\
&= \cos x \cdot \frac{\cos x}{\cos x - \sin x} + \sin x \cdot \frac{\sin x}{\sin x - \cos x} \\
&= \frac{(\cos x)^2}{\cos x - \sin x} + \frac{(\sin x)^2}{\sin x - \cos x} \\
&= \frac{(\cos x)^2}{\cos x - \sin x} - \frac{(\sin x)^2}{\cos x - \sin x} \\
&= \frac{(\cos x)^2 - (\sin x)^2}{\cos x - \sin x}.
\end{aligned}$$

The factoring formula

$$a^2 - b^2 = (a - b)(a + b),$$

with $a = \cos x$ and $b = \sin x$, allows us to simplify the final expression to

$$\begin{aligned}
\frac{\cos x}{1 - \tan x} + \frac{\sin x}{1 - \cot x} &= \frac{(\cos x - \sin x)(\cos x + \sin x)}{\cos x - \sin x} \\
&= \cos x + \sin x.
\end{aligned}$$

(d) We again change to sines and cosines, which gives

$$(\cot x)^2 - (\cos x)^2 = \frac{(\cos x)^2}{(\sin x)^2} - (\cos x)^2$$

$$= \frac{(\cos x)^2 - (\cos x)^2(\sin x)^2}{(\sin x)^2}$$

$$= \frac{(\cos x)^2(1 - (\sin x)^2)}{(\sin x)^2}$$

$$= \frac{(\cos x)^2}{(\sin x)^2} \cdot (1 - (\sin x)^2)$$

$$= (\cot x)^2(\cos x)^2.$$

□

EXAMPLE 2
Make the indicated trigonometric substitution, and simplify the expression.
(a) $\sqrt{1 - x^2}$; $x = \sin t$, for $-\frac{\pi}{2} \le t \le \frac{\pi}{2}$
(b) $\frac{\sqrt{x^2 - 1}}{x}$; $x = \sec t$, for $0 < t < \frac{\pi}{2}$
(c) $\frac{x}{(1 - x^2)^{3/2}}$; $x = \sin t$, for $-\frac{\pi}{2} < t < \frac{\pi}{2}$

Solution (a) We have

$$\sqrt{1 - x^2} = \sqrt{1 - (\sin t)^2} = \sqrt{(\cos t)^2} = |\cos t| = \cos t.$$

The absolute value can be dropped since $\cos t \ge 0$, for $-\frac{\pi}{2} \le t \le \frac{\pi}{2}$.
(b) We have

$$\frac{\sqrt{x^2 - 1}}{x} = \frac{\sqrt{(\sec t)^2 - 1}}{\sec t} = \frac{\sqrt{(\tan t)^2}}{\sec t} = \frac{\tan t}{\sec t} = \frac{\frac{\sin t}{\cos t}}{\frac{1}{\cos t}} = \sin t.$$

(c) We have

$$\frac{x}{(1 - x^2)^{3/2}} = \frac{\sin t}{(1 - (\sin t)^2)^{3/2}} = \frac{\sin t}{((\cos t)^2)^{3/2}} = \frac{\sin t}{(\cos t)^3} = \frac{\sin t}{\cos t} \cdot \frac{1}{(\cos t)^2} = \tan t(\sec t)^2.$$

□

Addition and Subtraction Identities

The fundamental identities are:

- $\sin(x \pm y) = \sin x \cos y \pm \cos x \sin y$

- $\cos(x \pm y) = \cos x \cos y \mp \sin x \sin y$

These can be used to show that

$$\tan(x \pm y) = \frac{\tan x \pm \tan y}{1 \mp \tan x \tan y},$$

as well as identities involving other trigonometric functions.

EXAMPLE 3

Determine the exact value of the trigonometric function.

(a) $\cos\left(\frac{\pi}{6} + \frac{3\pi}{4}\right)$ (b) $\sin\left(-\frac{7\pi}{12}\right)$

Solution (a) Applying the addition formula for cosine,

$$\cos\left(\frac{\pi}{6} + \frac{3\pi}{4}\right) = \cos\left(\frac{\pi}{6}\right)\cos\left(\frac{3\pi}{4}\right) - \sin\left(\frac{\pi}{6}\right)\sin\left(\frac{3\pi}{4}\right)$$

$$= \frac{\sqrt{3}}{2} \cdot \left(-\frac{\sqrt{2}}{2}\right) - \frac{1}{2} \cdot \frac{\sqrt{2}}{2}$$

$$= -\frac{\sqrt{2}}{4}\left(\sqrt{3} + 1\right).$$

(b) First express $\frac{7\pi}{12}$ as the sum of two values for which we know the exact values of the sine and cosine.

$$\frac{7\pi}{12} = \frac{\pi}{3} + \frac{\pi}{4}$$

Using the fact that $\sin(-x) = -\sin x$ and the addition formula for sine, we have

$$\sin\left(-\frac{7\pi}{12}\right) = -\sin\left(\frac{7\pi}{12}\right) = -\sin\left(\frac{\pi}{3} + \frac{\pi}{4}\right)$$

$$= -\left(\sin\left(\frac{\pi}{3}\right)\cos\left(\frac{\pi}{4}\right) + \cos\left(\frac{\pi}{3}\right)\sin\left(\frac{\pi}{4}\right)\right)$$

$$= -\left(\frac{\sqrt{3}}{2} \cdot \frac{\sqrt{2}}{2} + \frac{1}{2} \cdot \frac{\sqrt{2}}{2}\right)$$

$$= -\frac{\sqrt{2}}{4}\left(\sqrt{3} + 1\right).$$

$\square$

EXAMPLE 4

Express $\sin 3x$ in terms of $\sin x$ and $\cos x$.

Solution First write $3x = 2x + x$, and then apply the addition formula for the sine function.

$$\sin 3x = \sin(2x + x) = \sin 2x \cos x + \cos 2x \sin x$$

Next apply the addition formula a second time to $\sin 2x = \sin(x + x)$ and $\cos 2x = \cos(x + x)$.

$$\sin 3x = \sin 2x \cos x + \cos 2x \sin x$$

$$= (\sin x \cos x + \cos x \sin x) \cos x + (\cos x \cos x - \sin x \sin x) \sin x$$
$$= 2 \sin x (\cos x)^2 + \sin x (\cos x)^2 - (\sin x)^3$$
$$= 3 \sin x (\cos x)^2 - (\sin x)^3$$

□

EXAMPLE 5

Use the addition and subtraction formulas to verify the identity.

(a) $\sin\left(t + \frac{3\pi}{2}\right) = -\cos t$ (b) $\cos\left(t + \frac{3\pi}{2}\right) = \sin t$

Solution (a) Expanding the left side gives

$$\sin\left(t + \frac{3\pi}{2}\right) = \sin t \cos\left(\frac{3\pi}{2}\right) + \cos t \sin\left(\frac{3\pi}{2}\right)$$
$$= (\sin t) \cdot (0) + (\cos t) \cdot (-1)$$
$$= -\cos t.$$

(b) Using the cosine addition formula gives

$$\cos\left(t + \frac{3\pi}{2}\right) = \cos t \cos \frac{3\pi}{2} - \sin t \sin \frac{3\pi}{2}$$
$$= (\cos t) \cdot (0) - (\sin t) \cdot (-1)$$
$$= \sin t.$$

□

Double Angle Formulas

The double-angle formulas are:

- $\sin 2x = 2 \sin x \cos x$

- $\cos 2x = (\cos x)^2 - (\sin x)^2 = 2(\cos x)^2 - 1 = 1 - 2(\sin x)^2$

These can be used to derive

$$\tan 2x = \frac{2 \tan x}{1 - (\tan x)^2},$$

and identities involving other trigonometric functions.

EXAMPLE 6

Suppose that $\cos t = \frac{3}{5}$, where $0 < t < \frac{\pi}{2}$. Find $\cos 2t$, $\sin 2t$, and $\tan 2t$.

Solution First find $\sin t$, which is needed in the double-angle formulas. Since t is in the first quadrant, $\sin t > 0$ and

$$\sin t = \sqrt{1 - (\cos t)^2}$$

$$= \sqrt{1 - \left(\frac{3}{5}\right)^2} = \sqrt{\frac{16}{25}} = \frac{\sqrt{16}}{\sqrt{25}} = \frac{4}{5}.$$

Using the double angle formulas gives

$$\cos 2t = 2\left(\frac{3}{5}\right)^2 - 1 = \frac{18}{25} - 1 = -\frac{7}{25}$$

$$\sin 2t = 2\left(\frac{4}{5}\right)\left(\frac{3}{5}\right) = \frac{24}{25}$$

$$\tan 2t = \frac{\sin 2t}{\cos 2t} = \frac{\frac{24}{25}}{-\frac{7}{25}} = -\frac{24}{25} \cdot \frac{25}{7} = -\frac{24}{7}.$$

□

EXAMPLE 7

Verify the identity $\dfrac{2\tan x}{1 + (\tan x)^2} = \sin 2x$.

Solution　Replacing $\tan x$ with $\sin x / \cos x$ gives

$$\frac{2\tan x}{1 + (\tan x)^2} = \frac{2\frac{\sin x}{\cos x}}{1 + \left(\frac{\sin x}{\cos x}\right)^2}$$

$$= \frac{2\frac{\sin x}{\cos x}}{1 + \frac{(\sin x)^2}{(\cos x)^2}} = \frac{2\frac{\sin x}{\cos x}}{\frac{(\cos x)^2 + (\sin x)^2}{(\cos x)^2}}$$

$$= \frac{2\frac{\sin x}{\cos x}}{\frac{1}{(\cos x)^2}} = 2\frac{\sin x}{\cos x} \cdot (\cos x)^2$$

$$= 2\sin x \cos x = \sin 2x.$$

The graphs of $y = \dfrac{2\tan x}{1 + (\tan x)^2}$ and $y = \sin 2x$ are plotted together in the figure. Notice that the graphs coincide, which shows graphically that the equation is an identity.

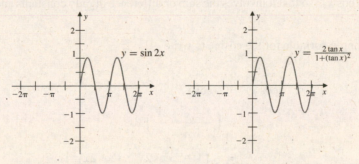

□

Half Angle Formulas

The half-angle formulas can be written either as

$$(\sin x)^2 = \frac{1 - \cos 2x}{2} \quad \text{and} \quad (\cos x)^2 = \frac{1 + \cos 2x}{2}$$

or as

$$\sin\left(\frac{x}{2}\right) = \sqrt{\frac{1 - \cos x}{2}} \quad \text{and} \quad \cos\left(\frac{x}{2}\right) = \sqrt{\frac{1 + \cos x}{2}}.$$

From these we can derive

$$(\tan x)^2 = \frac{1 - \cos 2x}{1 + \cos 2x} \quad \text{and} \quad \tan\left(\frac{x}{2}\right) = \frac{1 - \cos x}{\sin x} = \frac{\sin x}{1 + \cos x}.$$

EXAMPLE 8
Find the exact value of $\sin\left(\frac{3\pi}{8}\right)$.

Solution   If $x = \frac{3\pi}{4}$ in the formula for $\sin\left(\frac{x}{2}\right)$, then $\frac{x}{2} = \frac{3\pi}{8}$, and

$$\sin\left(\frac{3\pi}{8}\right) = \sqrt{\frac{1 - \cos\left(\frac{3\pi}{4}\right)}{2}}$$

$$= \sqrt{\frac{1 - \left(-\frac{\sqrt{2}}{2}\right)}{2}} = \sqrt{\frac{2 + \sqrt{2}}{4}}$$

$$= \frac{\sqrt{2 + \sqrt{2}}}{2}.$$

$\square$

In the next example the half-angle formulas are used to reduce powers of trigonometric functions to forms that are easier to work with.

EXAMPLE 9
Rewrite the expression $(\cos x)^4$ so that it involves the sum or difference of only constants and sine and cosine functions to the first power.

Solution   Use the half angle-formula for the cosine to write

$$(\cos x)^4 = ((\cos x)^2)^2$$

$$= \left(\frac{1 + \cos 2x}{2}\right)^2$$

$$= \frac{1}{4}\left(1 + 2\cos 2x + (\cos 2x)^2\right).$$

Now use the half-angle formula again on $(\cos 2x)^2$.

$$(\cos 2x)^2 = \frac{1 + \cos 4x}{2}$$

$$(\cos x)^4 = \frac{1}{4}\left(1 + 2\cos 2x + (\cos 2x)^2\right)$$

$$= \frac{1}{4}\left(1 + 2\cos 2x + \frac{1 + \cos 4x}{2}\right)$$

$$= \frac{1}{4} + \frac{\cos 2x}{2} + \frac{1 + \cos 4x}{8}$$

$$= \frac{3}{8} + \frac{\cos 2x}{2} + \frac{\cos 4x}{8}$$

□

Product-to-Sum and Sum-to-Product Identities

The product-to-sum identities are:

- $\sin x \cos y = \frac{1}{2}[\sin(x + y) + \sin(x - y)]$

- $\cos x \sin y = \frac{1}{2}[\sin(x + y) - \sin(x - y)]$

- $\cos x \cos y = \frac{1}{2}[\cos(x + y) + \cos(x - y)]$

- $\sin x \sin y = \frac{1}{2}[\cos(x - y) - \cos(x + y)]$

EXAMPLE 10

Write the product $\cos 3t \sin 5t$ as a sum or difference.

Solution Using the second identity and the fact that $\sin(-x) = -\sin x$ gives

$$\cos 3t \sin 5t = \frac{1}{2}[\sin(3t + 5t) - \sin(3t - 5t)]$$

$$= \frac{1}{2}[\sin(8t) - \sin(-2t)]$$

$$= \frac{1}{2}\sin(8t) + \frac{1}{2}\sin(2t).$$

□

There are four sum-to-product formulas that can be used to write sums of sines and cosines as products.

- $\sin x + \sin y = 2\sin\frac{x+y}{2}\cos\frac{x-y}{2}$

- $\sin x - \sin y = 2\cos\frac{x+y}{2}\sin\frac{x-y}{2}$

- $\cos x + \cos y = 2\cos\frac{x+y}{2}\cos\frac{x-y}{2}$

- $\cos x - \cos y = -2\sin\frac{x+y}{2}\sin\frac{x-y}{2}$

EXAMPLE 11

Rewrite $\cos 5t + \cos 2t$ as a product.

Solution Using the third identity

$$\cos 5t + \cos 2t = 2 \cos \frac{5t + 2t}{2} \cos \frac{5t - 2t}{2}$$

$$= 2 \cos \frac{7t}{2} \cos \frac{3t}{2}.$$

$\square$

Solving Trigonometric Equations

EXAMPLE 12

Find all values of x in the interval $[0, 2\pi]$ that satisfy the equation $\sin 2x = \sqrt{3} \sin x$.

Solution First use the double-angle formula to write $\sin 2x = 2 \sin x \cos x$. Then

$$\sin 2x = \sqrt{3} \sin x$$

$$2 \sin x \cos x = \sqrt{3} \sin x$$

$$2 \sin x \cos x - \sqrt{3} \sin x = 0$$

$$\sin x (2 \cos x - \sqrt{3}) = 0$$

$$\sin x = 0, \quad \cos x = \frac{\sqrt{3}}{2}$$

$$x = 0, \pi, 2\pi, \quad x = \frac{\pi}{6}, \frac{11\pi}{6}.$$

$\square$

EXAMPLE 13

Find all values of x in the interval $[0, 2\pi]$ that satisfy $\cos 2x \cos 3x = \sin 2x \sin 3x$.

Solution If we bring the expression on the right side of the equation to the left side, we recognize the new expression as fitting the sum formula for cosines.

$$\cos 2x \cos 3x - \sin 2x \sin 3x = 0$$

$$\cos(2x + 3x) = 0$$

$$\cos(5x) = 0$$

$$5x = \frac{\pi}{2}, \frac{3\pi}{2}, \frac{5\pi}{2}, \frac{7\pi}{2}, \frac{9\pi}{2}, \frac{11\pi}{2}, \frac{13\pi}{2}, \frac{15\pi}{2}, \frac{17\pi}{2}, \frac{19\pi}{2}$$

$$x = \frac{\pi}{10}, \frac{3\pi}{10}, \frac{\pi}{2}, \frac{7\pi}{10}, \frac{9\pi}{10}, \frac{11\pi}{10}, \frac{13\pi}{10}, \frac{3\pi}{2}, \frac{17\pi}{10}, \frac{19\pi}{10}$$

Keep in mind that, $0 \leq x \leq 2\pi$ if and only if $0 \leq 5x \leq 10\pi$.

$\square$

Exercise Set 4.7 (Page 250)

1. $\sin\left(\frac{\pi}{2} - \frac{5\pi}{3}\right) = \sin\frac{\pi}{2}\cos\frac{5\pi}{3} - \cos\frac{\pi}{2}\sin\frac{5\pi}{3} = 1\left(\frac{1}{2}\right) - 0\left(-\frac{\sqrt{3}}{2}\right) = \frac{1}{2}$

3. $\cos\left(\frac{7\pi}{12}\right) = \cos\left(\frac{\pi}{3} + \frac{\pi}{4}\right) = \cos\frac{\pi}{3}\cos\frac{\pi}{4} - \sin\frac{\pi}{3}\sin\frac{\pi}{4} = \frac{1}{2}\left(\frac{\sqrt{2}}{2}\right) - \frac{\sqrt{3}}{2}\left(\frac{\sqrt{2}}{2}\right) = \frac{\sqrt{2}}{4}\left(1 - \sqrt{3}\right)$

5. $\tan\left(\frac{\pi}{12}\right) = \tan\left(\frac{\pi}{3} - \frac{\pi}{4}\right) = \frac{\tan\left(\frac{\pi}{3}\right) - \tan\left(\frac{\pi}{4}\right)}{1 + \tan\left(\frac{\pi}{3}\right)\tan\left(\frac{\pi}{4}\right)} = \frac{\sqrt{3}-1}{1+\sqrt{3}(1)} = \frac{\sqrt{3}-1}{\sqrt{3}+1}$

7. Since $\left(\sin\frac{7\pi}{12}\right)^2 = \frac{1}{2}\left(1 - \cos\frac{7\pi}{6}\right) = \frac{1}{2}\left(1 - \left(-\frac{\sqrt{3}}{2}\right)\right) = \frac{1}{2}\left(1 + \frac{\sqrt{3}}{2}\right)$ and since $\frac{\pi}{2} < \frac{7\pi}{12} < \pi$, the sine is positive, so

$$\sin\frac{7\pi}{12} = \sqrt{\frac{1}{2}\left(1 + \frac{\sqrt{3}}{2}\right)} = \sqrt{\frac{2+\sqrt{3}}{4}} = \frac{\sqrt{2+\sqrt{3}}}{2}.$$

9. Since $\left(\cos\frac{5\pi}{8}\right)^2 = \frac{1}{2}\left(1 + \cos\frac{5\pi}{4}\right) = \frac{1}{2}\left(1 - \frac{\sqrt{2}}{2}\right)$ and since $\frac{\pi}{2} < \frac{5\pi}{8} < \pi$, the cosine is negative, so

$$\cos\frac{5\pi}{8} = -\sqrt{\frac{1}{2}\left(1 - \frac{\sqrt{2}}{2}\right)} = -\sqrt{\frac{2-\sqrt{2}}{4}} = -\frac{\sqrt{2-\sqrt{2}}}{2}.$$

11. Since $\left(\sin\frac{13\pi}{12}\right)^2 = \frac{1}{2}\left(1 - \cos\frac{13\pi}{6}\right) = \frac{1}{2}\left(1 - \frac{\sqrt{3}}{2}\right)$ and since $\pi < \frac{13\pi}{12} < \frac{3\pi}{2}$, the sine is negative, so

$$\sin\frac{13\pi}{12} = -\sqrt{\frac{1}{2}\left(1 - \frac{\sqrt{3}}{2}\right)} = -\sqrt{\frac{2-\sqrt{3}}{4}} = -\frac{\sqrt{2-\sqrt{3}}}{2}.$$

13. We are given $\cos t = \frac{4}{5}, 0 < t < \frac{\pi}{2}$. Since $0 < t < \frac{\pi}{2}$, $\sin t > 0$.

  a. $\cos 2t = 2(\cos t)^2 - 1 = 2\left(\frac{4}{5}\right)^2 - 1 = 2\left(\frac{16}{25}\right) - 1 = \frac{7}{25}$

  b. To use the formula $\sin 2t = 2\sin t\cos t$, first find $\sin t$. That is,

$$(\cos t)^2 + (\sin t)^2 = 1 \text{ implies } (\sin t)^2 = 1 - \left(\frac{4}{5}\right)^2 = \frac{9}{25} \text{ so } \sin t = \frac{3}{5}.$$

Then $\sin 2t = 2 \sin t \cos t = 2 \left(\frac{3}{5} \right) \left(\frac{4}{5} \right) = \frac{24}{25}$.

c. $\left(\cos \frac{t}{2} \right)^2 = \frac{1}{2}(1 + \cos t) = \frac{1}{2} \left(1 + \frac{4}{5} \right) = \frac{9}{10}$ so $\cos \frac{t}{2} = \pm \sqrt{\frac{9}{10}}$. Since $0 < t < \frac{\pi}{2}$ implies $0 < \frac{t}{2} < \frac{\pi}{4}$, we have $\cos \frac{t}{2} > 0$ so $\cos \frac{t}{2} = \sqrt{\frac{9}{10}} = \frac{3}{\sqrt{10}} = \frac{3\sqrt{10}}{10}$.

d. $\left(\sin \frac{t}{2} \right)^2 = \frac{1}{2}(1 - \cos t) = \frac{1}{2} \left(1 - \frac{4}{5} \right) = \frac{1}{10}$ so $\sin \frac{t}{2} = \pm \sqrt{\frac{1}{10}}$. Since $0 < t < \frac{\pi}{2}$ implies $0 < \frac{t}{2} < \frac{\pi}{4}$, we have $\sin \frac{t}{2} > 0$ so $\sin \frac{t}{2} = \sqrt{\frac{1}{10}} = \frac{1}{\sqrt{10}} = \frac{\sqrt{10}}{10}$.

15. We are given $\tan t = \frac{5}{12}$, $\sin t < 0$. Since $\tan t > 0$ and $\sin t < 0$ this implies $\cos t < 0$. First determine $\sin t$ and $\cos t$. From the identity,

$$(\tan t)^2 + 1 = (\sec t)^2 = \frac{1}{(\cos t)^2} \text{ implies } \left(\frac{5}{12} \right)^2 + 1 = \frac{1}{(\cos t)^2},$$

and $(\cos t)^2 = \frac{144}{169}$ so $\cos t = -\frac{12}{13}$. Then

$$(\sin t)^2 = 1 - (\cos t)^2 = 1 - \frac{144}{169} = \frac{25}{169} \text{ so } \sin t = -\sqrt{\frac{25}{169}} = -\frac{5}{13}.$$

a.
$$\cos 2t = 2(\cos t)^2 - 1 = 2 \left(\frac{144}{169} \right) - 1 = \frac{119}{169}$$

b.
$$\sin 2t = 2 \sin t \cos t = 2 \left(-\frac{5}{13} \right) \left(-\frac{12}{13} \right) = \frac{120}{169}$$

c.
$$\left(\cos \frac{t}{2} \right)^2 = \frac{1}{2}(1 + \cos t) = \frac{1}{2} \left(1 - \frac{12}{13} \right) = \frac{1}{26} \text{ so } \cos \frac{t}{2} = \pm \sqrt{\frac{1}{26}}$$

Since $\cos t < 0$ and $\sin t < 0$, t is in quadrant III, so $\pi < t < \frac{3\pi}{2}$ implies $\frac{\pi}{2} < \frac{t}{2} < \frac{3\pi}{4}$, and we have

$$\cos \frac{t}{2} < 0 \text{ so } \cos \frac{t}{2} = -\frac{1}{\sqrt{26}} = -\frac{\sqrt{26}}{26}.$$

d.
$$\left(\sin \frac{t}{2} \right)^2 = \frac{1}{2}(1 - \cos t) = \frac{1}{2} \left(1 + \frac{12}{13} \right) = \frac{25}{26} \text{ so } \sin \frac{t}{2} = \pm \sqrt{\frac{25}{26}}$$

Since $\pi < t < \frac{3\pi}{2}$ implies $\frac{\pi}{2} < \frac{t}{2} < \frac{3\pi}{4}$, and we have

$$\sin \frac{t}{2} > 0 \text{ so } \sin \frac{t}{2} = \sqrt{\frac{25}{26}} = \frac{5\sqrt{26}}{26}.$$

17. $\sin\left(t + \frac{3\pi}{2}\right) = \sin t \cos\frac{3\pi}{2} + \cos t \sin\frac{3\pi}{2} = (\sin t)(0) + (\cos t)(-1) = -\cos t$

19. $\sin\left(t + \frac{\pi}{2}\right) = \sin t \cos\frac{\pi}{2} + \cos t \sin\frac{\pi}{2} = (\sin t)(0) + (\cos t)(1) = \cos t$

21. $\sin\left(\frac{3\pi}{2} - t\right) = \sin\frac{3\pi}{2}\cos t - \cos\frac{3\pi}{2}\sin t = (-1)(\cos t) - (0)(\sin t) = -\cos t$

23. $(\cos 2x)^2 = \frac{1 + \cos 4x}{2} = \frac{1}{2} + \frac{1}{2}\cos 4x$

25.

$$(\cos x)^4 = ((\cos x)^2)^2 = \left(\frac{1 + \cos 2x}{2}\right)^2 = \frac{1}{4}\left(1 + 2\cos 2x + (\cos 2x)^2\right)$$

$$= \frac{1}{4}\left(1 + 2\cos 2x + \frac{1 + \cos 4x}{2}\right) = \frac{3}{8} + \frac{1}{2}\cos 2x + \frac{1}{8}\cos 4x$$

27. $\sin 6t \cos 5t = \frac{1}{2}(\sin(6t + 5t) + \sin(6t - 5t)) = \frac{1}{2}(\sin 11t + \sin t)$

29. $\cos 2t \cos 3t = \frac{1}{2}(\cos(2t + 3t) + \cos(2t - 3t)) = \frac{1}{2}(\cos 5t + \cos(-t)) = \frac{1}{2}(\cos 5t + \cos t)$

31. $(1 - (\cos x)^2)(\sec x)^2 = (\sin x)^2 \frac{1}{(\cos x)^2} = (\tan x)^2$

33. $\tan x - \cot x = \frac{\sin x}{\cos x} - \frac{\cos x}{\sin x} = \frac{(\sin x)^2 - (\cos x)^2}{\cos x \sin x} = -\frac{\cos 2x}{\cos x \sin x} = -\frac{2\cos 2x}{\sin 2x} = -2\cot 2x$

35. $\sin x \sin 2x + \cos x \cos 2x = \cos(2x - x) = \cos x$

37. $\sec x - \cos x = \frac{1}{\cos x} - \cos x = \frac{1 - (\cos x)^2}{\cos x} = \frac{(\sin x)^2}{\cos x} = \sin x \frac{\sin x}{\cos x} = \sin x \tan x$

39. $\sin 2x = \sin x$ so $\sin 2x - \sin x = 2\sin x \cos x - \sin x = \sin x(2\cos x - 1) = 0$ implies $\sin x = 0$ or $\cos x = \frac{1}{2}$ so $x = 0, \pi, 2\pi, \frac{\pi}{3}, \frac{5\pi}{3}$

41. $2(\sin x)^2 + \cos x - 1 = 2(1 - (\cos x)^2) + \cos x - 1 = -2(\cos x)^2 + \cos x + 1 = 2(\cos x)^2 - \cos x - 1 = (2\cos x + 1)(\cos x - 1) = 0$ implies $\cos x = -\frac{1}{2}$ or $\cos x = 1$ so
$x = \frac{2\pi}{3}, \frac{4\pi}{3}, 0, 2\pi$

43. We have $\tan x + \cot x = \frac{2}{\sin 2x}$ implies $\frac{\sin x}{\cos x} + \frac{\cos x}{\sin x} = \frac{(\cos x)^2 + (\sin x)^2}{\cos x \sin x} = \frac{2}{\sin 2x}$ so
$\sin 2x = 2\sin x \cos x$. So the equation is an identity and hence holds for all applicable x.

45. $2 \sin \frac{x+y}{2} \cos \frac{x-y}{2} = 2 \cdot \frac{1}{2} \left(\sin \left(\frac{x+y}{2} - \frac{x-y}{2} \right) + \sin \left(\frac{x+y}{2} + \frac{x-y}{2} \right) \right) = \sin y + \sin x$

47. $2 \cos \frac{x+y}{2} \cos \frac{x-y}{2} = 2 \cdot \frac{1}{2} \left(\cos \left(\frac{x+y}{2} + \frac{x-y}{2} \right) + \cos \left(\frac{x+y}{2} - \frac{x-y}{2} \right) \right) = \cos x + \cos y$

49. It is an identity since the graphs of the two functions coincide.

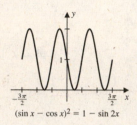

$(\sin x - \cos x)^2 = 1 - \sin 2x$

51. It is an identity.

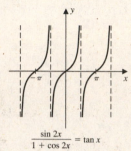

$\frac{\sin 2x}{1 + \cos 2x} = \tan x$

53. It is not an identity since the graphs of the two functions do not coincide.

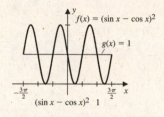

$(\sin x - \cos x)^2 \quad 1$

4.8   Inverse Trigonometric Functions

Only one-to-one functions have inverses. The trigonometric functions are not one-to-one, which is easily seen by the horizontal line test. For each of the trigonometric functions, there are horizontal lines that intersect the graph in many points, not just one. To define the inverse trigonometric functions, the domains are restricted so that the ranges remain the same but the functions become one-to-one.

Summary of the Inverse Trigonometric Functions

The table summarizes the definitions of all the inverse trigonometric functions.

Definition	Domain	Range
$\arcsin x = y \Leftrightarrow \sin y = x$	x in $[-1, 1]$	y in $\left[-\frac{\pi}{2}, \frac{\pi}{2}\right]$
$\arccos x = y \Leftrightarrow \cos y = x$	x in $[-1, 1]$	y in $[0, \pi]$
$\arctan x = y \Leftrightarrow \tan y = x$	x in $(-\infty, \infty)$	y in $\left(-\frac{\pi}{2}, \frac{\pi}{2}\right)$
$\text{arccot } x = y \Leftrightarrow \cot y = x$	x in $(-\infty, \infty)$	y in $(0, \pi)$
$\text{arcsec } x = y \Leftrightarrow \sec y = x$	x in $(-\infty, -1] \cup [1, \infty)$	y in $\left[0, \frac{\pi}{2}\right) \cup \left(\frac{\pi}{2}, \pi\right]$
$\text{arccsc } x = y \Leftrightarrow \csc y = x$	x in $(-\infty, -1] \cup [1, \infty)$	y in $\left[-\frac{\pi}{2}, 0\right) \cup \left(0, \frac{\pi}{2}\right]$

EXAMPLE 1

Find the exact value.

(a) $\arccos\left(\cos\left(\frac{\pi}{6}\right)\right)$ (b) $\sin\left(\arctan\left(\sqrt{3}\right)\right)$ (c) $\arctan\left(\tan\left(\frac{5\pi}{6}\right)\right)$

Solution (a) The restricted domain of the cosine function is $[0, \pi]$, which contains $\frac{\pi}{6}$ so

$$\arccos\left(\cos\left(\frac{\pi}{6}\right)\right) = \frac{\pi}{6}.$$

(b) The domain of arctan is all real numbers and since

$$\tan\left(\frac{\pi}{3}\right) = \frac{\sin\left(\frac{\pi}{3}\right)}{\cos\left(\frac{\pi}{3}\right)} = \frac{\frac{\sqrt{3}}{2}}{\frac{1}{2}} = \sqrt{3},$$

the result is

$$\sin\left(\arctan\left(\sqrt{3}\right)\right) = \sin\left(\frac{\pi}{3}\right) = \frac{\sqrt{3}}{2}.$$

(c) Since the restricted domain of tan is $\left(-\frac{\pi}{2}, \frac{\pi}{2}\right)$, which does not contain $\frac{5\pi}{6}$, the answer is *not* $\frac{5\pi}{6}$. First compute the tangent and then find a value in the domain of the inverse tangent whose tangent agrees. The reference angle for $\frac{5\pi}{6}$ is $\frac{\pi}{6}$, and $\frac{5\pi}{6}$ is in the second quadrant. In this quadrant, the sine is positive and the cosine is negative, so

$$\tan\left(\frac{5\pi}{6}\right) = -\tan\left(\frac{\pi}{6}\right) = -\frac{\sin\left(\frac{\pi}{6}\right)}{\cos\left(\frac{\pi}{6}\right)} = -\frac{\frac{1}{2}}{\frac{\sqrt{3}}{2}} = -\frac{1}{\sqrt{3}} = -\frac{\sqrt{3}}{3}.$$

Since $\tan\left(-\frac{\pi}{6}\right)$ is also $-\frac{\sqrt{3}}{3}$,

$$\arctan\left(\tan\left(\frac{5\pi}{6}\right)\right) = -\frac{\pi}{6}.$$

$\square$

EXAMPLE 2

Find the exact value.
(a) $\sin\left(\arccos\left(\frac{4}{5}\right)\right)$ (b) $\sin\left(\arcsin\left(\frac{4}{5}\right) + \arccos\left(\frac{3}{5}\right)\right)$

Solution (a) Let t satisfy

$$t = \arccos\left(\frac{4}{5}\right) \Rightarrow \cos t = \frac{4}{5}.$$

By the Pythagorean Identity,

$$(\cos t)^2 + (\sin t)^2 = 1$$

$$\left(\frac{4}{5}\right)^2 + (\sin t)^2 = 1$$

$$(\sin t)^2 = 1 - \frac{16}{25} = \frac{9}{25}$$

$$\sin t = \pm\sqrt{\frac{9}{25}} = \pm\frac{3}{5}.$$

Since $t = \arccos\left(\frac{4}{5}\right)$, t is in the interval $[0, \pi]$, where the sine is always nonnegative, and

$$\sin t = \sin\left(\arccos\left(\frac{4}{5}\right)\right) = \frac{3}{5}.$$

(b) To simplify the expression, first apply the sum formula for sine, $\sin(a + b) = \sin a \cos b + \cos a \sin b$.

$$\sin\left(\arcsin\left(\frac{4}{5}\right) + \arccos\left(\frac{3}{5}\right)\right)$$

$$= \sin\left(\arcsin\left(\frac{4}{5}\right)\right) \cos\left(\arccos\left(\frac{3}{5}\right)\right) + \cos\left(\arcsin\left(\frac{4}{5}\right)\right) \sin\left(\arccos\left(\frac{3}{5}\right)\right)$$

$$= \left(\frac{4}{5}\right)\left(\frac{3}{5}\right) + \cos\left(\arcsin\left(\frac{4}{5}\right)\right) \sin\left(\arccos\left(\frac{3}{5}\right)\right)$$

The figure shows a triangle with angle θ satisfying $\sin\theta = \frac{4}{5}$, that is, $\theta = \arcsin\left(\frac{4}{5}\right)$.

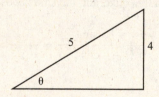

By the Pythagorean Theorem, the missing side has length

$$\sqrt{5^2 - 4^2} = \sqrt{25 - 16} = \sqrt{9} = 3,$$

so $\cos\theta = \frac{3}{5}$ or $\theta = \arccos\left(\frac{3}{5}\right)$. Then

$$\frac{3}{5} = \cos\theta = \cos\left(\arcsin\left(\frac{4}{5}\right)\right),$$

and

$$\frac{4}{5} = \sin\theta = \sin\left(\arccos\left(\frac{3}{5}\right)\right).$$

Finally,

$$\sin\left(\arcsin\left(\frac{4}{5}\right) + \arccos\left(\frac{3}{5}\right)\right) = \frac{4}{5} \cdot \frac{3}{5} + \frac{3}{5} \cdot \frac{4}{5}$$
$$= \frac{24}{25}.$$

$\square$

EXAMPLE 3

Solve the equation on the given interval, express the solution for x in terms of inverse functions, and use a calculator to approximate the solutions.

(a) $\sin 2x - \cos x = 0$, $\left[0, \frac{\pi}{2}\right]$ (b) $(\tan x)^2 - 3\tan x - 4 = 0$, $\left(-\frac{\pi}{2}, \frac{\pi}{2}\right)$

Solution (a) Use the double-angle formula for sine, $\sin 2x = 2\sin x \cos x$.

$$\sin 2x - \cos x = 0$$
$$2\sin x \cos x - \cos x = 0$$
$$\cos x (2\sin x - 1) = 0$$
$$\cos x = 0, \quad \sin x = \frac{1}{2}$$
$$x = \arccos(0) = \frac{\pi}{2} \approx 1.571, \quad x = \arcsin\left(\frac{1}{2}\right) = \frac{\pi}{6} \approx 0.524$$

(b) Let $u = \tan x$.

$$0 = (\tan x)^2 - 3\tan x - 4 = u^2 - 3u - 4$$
$$= (u - 4)(u + 1)$$
$$= (\tan x - 4)(\tan x + 1)$$
$$\tan x = 4, \quad \tan x = -1$$
$$x = \arctan(4) \approx 1.326, \quad x = \arctan(-1) = -\frac{\pi}{4} \approx -0.785$$

$\square$

EXAMPLE 4

A lighthouse is located on an island 2 miles from the nearest point on a straight shoreline. If the light from the lighthouse is moving along the shoreline, express the angle θ formed by the beam of light and the shoreline in terms of the distance x in the figure.

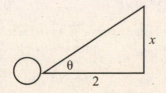

Solution The parameters θ, x, and 2 can be related by a tangent function, since the sides of length x and 2 are the opposite and adjacent sides of the angle θ. So

$$\tan\theta = \frac{x}{2},$$

and

$$\theta = \arctan\left(\frac{x}{2}\right).$$

$\square$

Exercise Set 4.8 (Page 259)

1. $y = \arccos\left(\frac{\sqrt{3}}{2}\right)$ implies $\cos y = \frac{\sqrt{3}}{2}$ with $0 \le y \le \pi$, so $y = \arccos\left(\frac{\sqrt{3}}{2}\right) = \frac{\pi}{6}$.

3. $y = \arcsin(1)$ implies $\sin y = 1$ with $-\frac{\pi}{2} \le y \le \frac{\pi}{2}$, so $y = \arcsin(1) = \frac{\pi}{2}$.

5. $y = \arccos\left(-\frac{\sqrt{2}}{2}\right)$ implies $\cos y = -\frac{\sqrt{2}}{2}$ with $0 \le y \le \pi$, so $y = \arccos\left(-\frac{\sqrt{2}}{2}\right) = \frac{3\pi}{4}$.

7. $y = \arctan\left(\sqrt{3}\right)$ implies $\tan y = \sqrt{3}$ with $-\frac{\pi}{2} < y < \frac{\pi}{2}$, so $y = \arctan\left(\sqrt{3}\right) = \frac{\pi}{3}$.

9. $y = \arctan(-1)$ implies $\tan y = -1$ with $-\frac{\pi}{2} < y < \frac{\pi}{2}$, so $y = \arctan(-1) = -\frac{\pi}{4}$.

11. $y = \operatorname{arcsec}\left(\sqrt{2}\right)$ implies $\sec y = \sqrt{2}$ implies $\cos y = \frac{1}{\sqrt{2}} = \frac{\sqrt{2}}{2}$ with $0 \le y < \frac{\pi}{2}$ or $\frac{\pi}{2} < y \le \pi$, so
$y = \operatorname{arcsec}\left(\sqrt{2}\right) = \frac{\pi}{4}$.

13. $y = \arccos(2)$ implies $\cos y = 2$, which can never occur since $-1 \le \cos y \le 1$.

15. Since $\cos(\arccos x) = x$ for x in $[-1, 1]$, we have $\cos\left(\arccos\left(\frac{1}{2}\right)\right) = \frac{1}{2}$.

17. The $\arccos\left(\frac{\sqrt{2}}{2}\right)$ is the angle in $[0, \pi]$ whose cosine is $\frac{\sqrt{2}}{2}$, so $\arccos\left(\frac{\sqrt{2}}{2}\right) = \frac{\pi}{4}$ and
$\sin\left(\arccos\left(\frac{\sqrt{2}}{2}\right)\right) = \sin\left(\frac{\pi}{4}\right) = \frac{\sqrt{2}}{2}$.

19. The arcsin $\left(\frac{\sqrt{3}}{2}\right)$ is the angle in $\left[-\frac{\pi}{2}, \frac{\pi}{2}\right]$ whose sine is $\frac{\sqrt{3}}{2}$, so arcsin $\left(\frac{\sqrt{3}}{2}\right) = \frac{\pi}{3}$ and

$\tan\left(\arcsin\left(\frac{\sqrt{3}}{2}\right)\right) = \tan\left(\frac{\pi}{3}\right) = \sqrt{3}$.

21. Since arcsin $(\sin x) = x$ for x in $\left[-\frac{\pi}{2}, \frac{\pi}{2}\right]$, we have arcsin $\left(\sin\left(-\frac{\pi}{4}\right)\right) = -\frac{\pi}{4}$.

23. The arccos $(\cos x) = x$ for x in $[0, \pi]$, but since $-\frac{\pi}{4}$ does not lie in this interval, the identity can not be applied. However,

$$\cos\left(-\frac{\pi}{4}\right) = \cos\left(\frac{\pi}{4}\right) \text{ implies } \arccos\left(\cos\left(-\frac{\pi}{4}\right)\right) = \arccos\left(\cos\left(\frac{\pi}{4}\right)\right) = \frac{\pi}{4}.$$

25. Since arctan $(\tan x) = x$ for x in $\left(-\frac{\pi}{2}, \frac{\pi}{2}\right)$, and $\frac{\pi}{3}$ is in this interval, arctan $\left(\tan\left(\frac{\pi}{3}\right)\right) = \frac{\pi}{3}$.

27. The arctan $(\tan x) = x$ for x in $\left(-\frac{\pi}{2}, \frac{\pi}{2}\right)$, but since $\frac{7\pi}{6}$ does not lie in this interval, the identity can not be applied. However,

$$\tan\left(\frac{7\pi}{6}\right) = \tan\left(\frac{\pi}{6}\right) \text{ implies } \arctan\left(\tan\left(\frac{7\pi}{6}\right)\right) = \arctan\left(\tan\left(\frac{\pi}{6}\right)\right) = \frac{\pi}{6}.$$

29. Let $t = \arcsin\left(\frac{4}{5}\right)$ so $\sin t = \frac{4}{5}$. The triangle in the figure has $\sin t = \frac{4}{5}$ which implies the side adjacent t is $\sqrt{25 - 16} = 3$ so

$$\cos t = \cos\left(\arcsin\left(\frac{4}{5}\right)\right) = \frac{3}{5}.$$

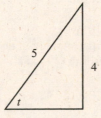

31. Let $t = \arccos\left(\frac{4}{5}\right)$ so $\cos t = \frac{4}{5}$. The triangle in the figure has $\cos t = \frac{4}{5}$ which implies the side opposite t is $\sqrt{25 - 16} = 3$ so

$$\tan t = \tan\left(\arccos\left(\frac{4}{5}\right)\right) = \frac{3}{4}.$$

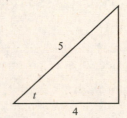

33. Let $t = \arctan 4$ so $\tan t = 4$. The triangle in the figure has $\tan t = 4$ which implies the hypotenuse is $\sqrt{16 + 1} = \sqrt{17}$ so

$$\cos t = \cos(\arctan 4) = \frac{1}{\sqrt{17}} = \frac{\sqrt{17}}{17}.$$

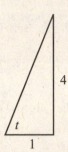

35. Apply the formula $\cos(a + b) = \cos a \cos b - \sin a \sin b$, where $a = \arcsin(3/5)$ and $b = \arccos(4/5)$. Then

$$\cos\left(\arcsin\left(\frac{3}{5}\right) + \arccos\left(\frac{4}{5}\right)\right) = \cos\left(\arcsin\left(\frac{3}{5}\right)\right)\cos\left(\arccos\left(\frac{4}{5}\right)\right)$$

$$- \sin\left(\arcsin\left(\frac{3}{5}\right)\right)\sin\left(\arccos\left(\frac{4}{5}\right)\right)$$

$$= \frac{4}{5} \cdot \frac{4}{5} - \frac{3}{5} \cdot \frac{3}{5} = \frac{7}{25}.$$

37. Let $y = \arcsin(-x)$. Then

$$-x = \sin y \text{ implies } x = -\sin y = \sin(-y) \text{ so } -y = \arcsin x \text{ and } y = -\arcsin x.$$

39. Let $y = \arccos x$. Then $\cos y = x$. A right triangle with one angle y, adjacent side x, and hypotenuse 1 has opposite side $\sqrt{1 - x^2}$ so $\cos y = x$, and $\sin y = \sqrt{1 - x^2}$. So $y = \arcsin\left(\sqrt{1 - x^2}\right)$.

41. Let $y = \arcsin x$. Then $\sin y = x$. A right triangle with one angle y, opposite side x, and hypotenuse 1 has adjacent side $\sqrt{1 - x^2}$ so $\sin y = x$, and $\tan y = \frac{x}{\sqrt{1-x^2}}$. So $\tan(\arcsin x) = \frac{x}{\sqrt{1-x^2}}$.

43. a. We have

$$(\tan x)^2 - \tan x - 2 = (\tan x + 1)(\tan x - 2) = 0,$$

which implies $\tan x = -1$ or $\tan x = 2$, so $x = \arctan(-1)$, $x = \arctan(2)$.

b. We have $x = \arctan(-1) = -\frac{\pi}{4} \approx -0.785$ or $x = \arctan(2) \approx 1.107$.

45. The illustration indicates that $\tan \theta = \frac{x}{4}$.

47. Approximately 11,242 miles.

. . . .

4.9   Applications of Trigonometric Functions

Law of Cosines

To *solve* a triangle means to find the lengths of all sides and all angles. The *Law of Cosines* can be used to solve a triangle when we know two sides and the angle between them or when we know all three sides.

Law of Cosines:

$$a^2 = b^2 + c^2 - 2bc \cos \alpha$$
$$b^2 = a^2 + c^2 - 2ac \cos \beta$$
$$c^2 = a^2 + b^2 - 2ab \cos \gamma$$

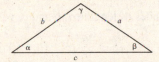

EXAMPLE 1

Let the angles of a triangle be α, β, and γ with opposite sides of lengths a, b, and c, respectively. If $\alpha = 37°$, $b = 16$, and $c = 24$, use the Law of Cosines to solve the triangle, rounding all answers to one decimal place.

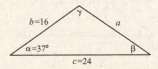

Solution   The missing parts are β, γ, and a. Since we are given the two sides b and c and the angle α between them, the side a can be found.

$$a^2 = b^2 + c^2 - 2bc \cos \alpha$$
$$a^2 = 16^2 + 24^2 - 2(16)(24) \cos 37°$$
$$= 256 + 576 - 768 \cos 37°$$
$$a = \sqrt{832 - 768 \cos 37°} \approx 14.8$$

Now that we have an approximation to side a, the Law of Cosines is used to find either angle β or angle γ. We elect to find β.

$$b^2 = a^2 + c^2 - 2ac\cos\beta$$
$$16^2 = (14.8)^2 + 24^2 - 2(14.8)(24)\cos\beta$$
$$256 = 219.04 + 576 - 710.4\cos\beta$$
$$\cos\beta = \frac{256 - 795.04}{-710.4} = \frac{539.04}{710.4}$$
$$\beta = \arccos\left(\frac{539.04}{710.4}\right) \approx 0.709$$

The angle β in degrees is

$$\beta \approx 0.709\left(\frac{180}{\pi}\right) \approx 40.6°.$$

Since the sum of the angles of a triangle is 180°, we have,

$$\gamma \approx 180 - 37 - 40.6 = 102.4°.$$

□

Law of Sines

The *Law of Sines* applied to a triangle with sides of lengths a, b and c, with corresponding opposite angles α, β and γ, states that

> **Law of Sines:**

$$\frac{\sin\alpha}{a} = \frac{\sin\beta}{b} = \frac{\sin\gamma}{c}.$$

The Law of Sines can be used to solve a triangle when you know a side and the angle opposite it.

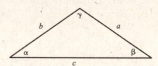

EXAMPLE 2

Suppose that $\alpha = 50°$, $\beta = 46°$, and $a = 25$. Use the Law of Sines to solve the triangle, rounding all answers to one decimal place.

Solution The missing parts are γ, b and c. Since the sum of the angles of a triangle is 180°, and we are given two angles, α and β, we can find γ from

$$\gamma = 180 - 50 - 46 = 84°.$$

To find b, the Law of Sines gives

$$\frac{\sin\beta}{b} = \frac{\sin\alpha}{a}$$
$$b = \frac{a\sin\beta}{\sin\alpha} = \frac{25\sin 46°}{\sin 50°} \approx 23.5.$$

To find c we again use the Law of Sines.

$$\frac{\sin \gamma}{c} = \frac{\sin \alpha}{a}$$

$$c = \frac{a \sin \gamma}{\sin \alpha} = \frac{25 \sin 84°}{\sin 50°} \approx 32.5.$$

□

EXAMPLE 3

Suppose that $\alpha = 150°$, $b = 3$ and $c = 5$. Use the Law of Cosines and the Law of Sines to solve the triangle.

Solution The situation is shown in the figure.

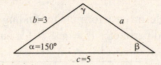

Since two sides and the angle α between them are given, the Law of Cosines can be used to find the side opposite the angle α.

$$a^2 = b^2 + c^2 - 2bc \cos \alpha$$

$$a^2 = 3^2 + 5^2 - 2(3)(5) \cos 150°$$

$$a^2 = 9 + 25 - 30 \left(-\frac{\sqrt{3}}{2} \right)$$

$$a^2 = 34 + 15\sqrt{3}$$

$$a = \sqrt{34 + 15\sqrt{3}} \approx 7.75$$

Now we use the Law of Sines:

$$\frac{\sin \beta}{b} = \frac{\sin \alpha}{a}$$

$$\sin \beta = \frac{b \sin \alpha}{a} = \frac{3 \sin 150°}{7.75}$$

$$= \frac{3 \left(\frac{1}{2} \right)}{7.75} \approx 0.1936,$$

and

$$\beta = \arcsin \left(\frac{3}{15.5} \right) \approx 11.16°.$$

Since the sum of the angles of a triangle is 180°, we have

$$\gamma = 180 - \alpha - \beta \approx 180 - 150 - 11.16 = 18.84°.$$

□

EXAMPLE 4

Show there is no triangle satisfying the conditions $a = 3$, $b = 8$, and $\alpha = 26.2°$.

Solution If there was such a triangle, then the angle β could be found by the Law of Sines.

$$\frac{\sin \beta}{b} = \frac{\sin \alpha}{a}$$

$$\sin \beta = \frac{b \sin \alpha}{a} = \frac{8 \sin 26.2°}{3} \approx 1.18$$

Since the sine of an angle is never greater than one, we can conclude that no such triangle exists.   □

Heron's Formula

Given the lengths of the three sides of a triangle, an easy formula for finding the area of the triangle is *Heron's formula*. A triangle with sides of lengths a, b and c and perimeter $P = a + b + c$ has area

$$A = \frac{1}{4}\sqrt{P(P - 2a)(P - 2b)(P - 2c)}.$$

EXAMPLE 5

Find the area of a triangle with sides of lengths 12 ft, 18 ft and 24 ft.

Solution To use Heron's formula, first compute the perimeter of the triangle.

$$P = a + b + c$$
$$= 12 + 18 + 24$$
$$= 54$$

Then the area is

$$A = \frac{1}{4}\sqrt{54(54 - 24)(54 - 36)(54 - 48)}$$
$$= \frac{1}{4}\sqrt{54(30)(18)(6)}$$
$$= \frac{1}{4}\sqrt{174960} = \frac{1}{4}108\sqrt{15} = 27\sqrt{15}$$
$$\approx 104.57 \text{ ft}^2.$$

□

EXAMPLE 6

The lengths of the sides of a triangular parcel of land are approximately 200 ft, 400 ft, and 500 ft, and the land is valued at \$20,000.00 per acre. What is the value of the parcel of land?

Solution The area of the parcel can be found using Heron's formula. Since the perimeter is

$$P = 200 + 400 + 500 = 1100,$$

the area is

$$A = \frac{1}{4}\sqrt{1100(1100 - 400)(1100 - 800)(1100 - 1000)}$$
$$= \frac{1}{4}\sqrt{1100(700)(300)(100)} = \frac{1}{4}\sqrt{23100000000}$$
$$\approx 37,997 \text{ ft}^2.$$

Since one acre is 43,560 square feet, the number of acres in the parcel is approximately

$$\frac{37997}{43560} = 0.87 \text{ acres,}$$

and the value of the parcel of land is approximately

$$\left(\frac{37997}{43560}\right) 20000 \approx \$17,450.00.$$

$\square$

Exercise Set 4.9 (Page 271)

1. Since $\alpha = 55°$, $b = 12$, $c = 20$, we have

$$a^2 = b^2 + c^2 - 2bc \cos \alpha = 12^2 + 20^2 - 2(12)(20) \cos 55° \text{ so } a = \sqrt{544 - 480 \cos 55°} \approx 16.4,$$

and

$$b^2 = a^2 + c^2 - 2ac \cos \beta \text{ implies } \cos \beta = \frac{12^2 - (16.4)^2 - 20^2}{-2(16.4)(20)}.$$

So

$$\beta = \arccos\left(\frac{12^2 - (16.4)^2 - 20^2}{-2(16.4)(20)}\right) \approx 0.64309 = 0.64309\left(\frac{180}{\pi}\right)° \approx 36.8°,$$

and $\gamma \approx 180 - 55 - 36.8 = 88.2°$.

3. Since $\beta = 30°$, $a = 25$, $c = 32$ we have

$$b = \sqrt{25^2 + 32^2 - 2(25)(32) \cos 30°} \approx 16.2,$$

and

$$a^2 = b^2 + c^2 - 2bc \cos \alpha \text{ implies } \cos \alpha = \frac{25^2 - (16.2)^2 - (32)^2}{-2(16.2)(32)}.$$

So

$$\alpha = \arccos\left(\frac{25^2 - (16.2)^2 - (32)^2}{-2(16.2)(32)}\right) \approx 0.87895 = 0.87895\left(\frac{180}{\pi}\right)^{\circ} \approx 50.4^{\circ},$$

and $\gamma \approx 180 - 30 - 50.4 = 99.6^{\circ}$.

5. Since $\alpha = 40^{\circ}$, $\beta = 87^{\circ}$, and $c = 115$, we have $\gamma = 180 - 40 - 87 = 53^{\circ}$, and

$$\frac{\sin\alpha}{a} = \frac{\sin\gamma}{c} \text{ so } a = \frac{115\sin 40^{\circ}}{\sin 53^{\circ}} \approx 92.6; \quad \frac{\sin\beta}{b} = \frac{\sin\gamma}{c} \text{ so } b = \frac{115\sin 87^{\circ}}{\sin 53^{\circ}} \approx 143.8.$$

7. Since $\beta = 100^{\circ}$, $\gamma = 30^{\circ}$, and $c = 20$, we have $\alpha = 180 - 100 - 30 = 50^{\circ}$, and

$$\frac{\sin\alpha}{a} = \frac{\sin\gamma}{c} \text{ so } a = \frac{20\sin 50^{\circ}}{\sin 30^{\circ}} \approx 30.6; \quad \frac{\sin\beta}{b} = \frac{\sin\gamma}{c} \text{ so } b = \frac{20\sin 100^{\circ}}{\sin 30^{\circ}} \approx 39.4.$$

9. Since $\alpha = 130^{\circ}$, $b = 5$, and $c = 7$, using the Law of Cosines we have

$$a^2 = b^2 + c^2 - 2bc\cos\alpha = 5^2 + 7^2 - 2(5)(7)\cos 130^{\circ} \text{ so}$$
$$a = \sqrt{74 - 70\cos 130^{\circ}} \approx 11.$$

Then by the Law of Sines,

$$\frac{\sin\alpha}{a} = \frac{\sin\beta}{b} \text{ so } \beta = \arcsin\left(\frac{5\sin 130^{\circ}}{11}\right) \approx 20.4^{\circ},$$

and $\gamma \approx 180 - 130 - 20.4 = 29.6^{\circ}$.

11. Since $a = 8$, $b = 9$, and $c = 13$, by the Law of Cosines we have

$$\cos\alpha = \frac{8^2 - (9)^2 - (13)^2}{-2(9)(13)} \approx 0.795,$$

so

$$\alpha = \arccos(0.795) \approx 0.65 = 0.65\left(\frac{180}{\pi}\right)^{\circ} \approx 37.4^{\circ}.$$

By the Law of Cosines,

$$\gamma = \arccos\left(\frac{13^2 - 9^2 - 8^2}{-2(8)(9)}\right) \approx 1.738\left(\frac{180}{\pi}\right) \approx 99.50^{\circ},$$

so $\beta \approx 180 - 37.4 - 99.5 = 43.1^{\circ}$.

13. In order for a triangle to satisfy the conditions $a = 3, b = 10, \alpha = 25.4°$, by the Law of Sines,

$$\frac{\sin \alpha}{a} = \frac{\sin \beta}{b} \text{ so } \sin \beta = \frac{b \sin \alpha}{a} = \frac{10 \sin 25.4°}{3} \approx 1.4.$$

Since the sine of an angle is never greater than 1, no such triangle can exist.

15. Since $a = 125, b = 150$, and $\alpha = 55°$, by the Law of Sines,

$$\frac{\sin \alpha}{a} = \frac{\sin \beta}{b} \text{ so } \sin \beta = \frac{b \sin \alpha}{a} = \frac{150 \sin 55°}{125}.$$

There are two angles between $0°$ and $180°$ that satisfy this condition. One such angle is

$$\beta = \arcsin\left(\frac{150 \sin 55°}{125}\right) \approx 1.386 = 1.386\left(\frac{180}{\pi}\right)^\circ \approx 79.4°.$$

A second angle is $\beta_1 \approx 180 - 79.4 = 100.6°$.

Using β :

$$\gamma = 180 - 55 - 79.4 = 45.6° \text{ and } \frac{\sin \alpha}{a} = \frac{\sin \gamma}{c} \text{ so } c = \frac{125 \sin 45.6°}{\sin 55°} \approx 109.$$

Using β_1 :

$$\gamma = 180 - 55 - 100.6 = 24.4° \text{ and } \frac{\sin \alpha}{a} = \frac{\sin \gamma}{c} \text{ so } c = \frac{125 \sin 24.4°}{\sin 55°} \approx 63.$$

17. By Heron's Formula, the area of the triangle with sides $a = 12, b = 14, c = 16$, is

$$A = \frac{1}{4}\sqrt{P(P - 2a)(P - 2b)(P - 2c)},$$

where

$$P = a + b + c = 12 + 14 + 16 = 42.$$

So

$$A = \frac{1}{4}\sqrt{42(42 - 24)(42 - 28)(42 - 32)} = \frac{1}{4}\sqrt{42(18)(14)(10)}$$

$$= \frac{1}{4}84\sqrt{15} = 21\sqrt{15} \approx 81.3 \text{ centimeters}^2.$$

19. Since point B is directly northwest of point C, the angle made by the line segment connecting C to B and the horizontal is $45°$. Then angle ACB is $180 - 45 = 135°$. By the Law of Cosines

$$\overline{AB}^2 = 2^2 + 3^2 - 2(2)(3)\cos 135° \text{ so } \overline{AB} = \sqrt{13 - 12\cos 135°} \approx 4.6 \text{ miles.}$$

a. The approximate cost of construction directly between points A and B is

$$125000\sqrt{13 - 12\cos 135°} \approx \$579,400.00.$$

The approximate cost of construction from A to C, then C to B is $(3+2)(100000) = \$500,000.00$. So the engineers should select the route that avoids the swamp.

b. Let P denote the cost per mile for construction through C. If the total cost of construction from A to B to C is to equal the cost directly from A to B, then

$$5P = 579400 \text{ so } P = \frac{579400}{5} = 115881.$$

So if the cost per mile through C is approximately $\$115,881$, then the cost of either alternative is about the same.

21. The angle at point B is $180 - 105 - 42 = 33$, so by the Law of Sines

$$\frac{\sin 33°}{45} = \frac{\sin 42°}{\overline{AB}} \text{ so } \overline{AB} = \frac{45 \sin 42°}{\sin 33°} \approx 55.3 \text{ feet.}$$

23. At 12:00 am, two hours after the ships have left port, the first ship has traveled 40 miles and the second ship 50 miles. The angle between the two ships is $62 + 75 = 137°$. If d denotes the distance between the ships, by the Law of Cosines,

$$d^2 = 40^2 + 50^2 - 2(40)(50)\cos 137°,$$

and

$$d = \sqrt{40^2 + 50^2 - 2(40)(50)\cos 137°} \approx 83.8 \text{ miles.}$$

25. To use the Law of Sines to find the distance d from the fire to the tower B, first find the missing angle, which is $180 - 50 - 63 = 67°$. Then

$$\frac{\sin 67°}{10} = \frac{\sin 63°}{d} \text{ so } d = \frac{10 \sin 63°}{\sin 67°} \approx 9.7 \text{ miles.}$$

Review Exercises for Chapter 4 (Page 273)

1. a. $P(t) = \left(\frac{1}{2}, \frac{\sqrt{3}}{2}\right)$ **b.** $\frac{\pi}{3}$ **c.** $\cos \frac{\pi}{3} = \frac{1}{2}$, $\sin \frac{\pi}{3} = \frac{\sqrt{3}}{2}$, $\tan \frac{\pi}{3} = \sqrt{3}$,
$\cot \frac{\pi}{3} = \frac{1}{\sqrt{3}} = \frac{\sqrt{3}}{3}$, $\sec \frac{\pi}{3} = 2$, $\csc \frac{\pi}{3} = \frac{2}{\sqrt{3}} = \frac{2\sqrt{3}}{3}$

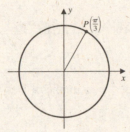

3. a. $P(t) = \left(-\frac{\sqrt{2}}{2}, -\frac{\sqrt{2}}{2}\right)$ **b.** $\frac{\pi}{4}$ **c.** $\cos \frac{5\pi}{4} = -\frac{\sqrt{2}}{2}$, $\sin \frac{5\pi}{4} = -\frac{\sqrt{2}}{2}$, $\tan \frac{5\pi}{4} = 1$,
$\cot \frac{5\pi}{4} = 1$, $\sec \frac{5\pi}{4} = -\frac{2}{\sqrt{2}} = -\sqrt{2}$, $\csc \frac{5\pi}{4} = -\frac{2}{\sqrt{2}} = -\sqrt{2}$

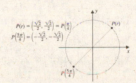

5. a. $P(t) = \left(-\frac{\sqrt{3}}{2}, \frac{1}{2}\right)$ **b.** $\frac{\pi}{6}$ **c.** $\cos\left(-\frac{19\pi}{6}\right) = -\frac{\sqrt{3}}{2}$, $\sin\left(-\frac{19\pi}{6}\right) = \frac{1}{2}$, $\tan\left(-\frac{19\pi}{6}\right) = -\frac{\sqrt{3}}{3}$,
$\cot\left(-\frac{19\pi}{6}\right) = -\sqrt{3}$, $\sec\left(-\frac{19\pi}{6}\right) = -\frac{2\sqrt{3}}{3}$, $\csc\left(-\frac{19\pi}{6}\right) = 2$

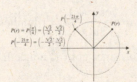

7. If $\cos t = \frac{3}{5}$ and $\frac{3\pi}{2} < t < 2\pi$, then $\sin t < 0$, and

$$(\cos t)^2 + (\sin t)^2 = 1 \text{ implies } (\sin t)^2 = 1 - \frac{9}{25} = \frac{16}{25} \text{ so } \sin t = \pm\sqrt{\frac{16}{25}} = \pm\frac{4}{5},$$

so $\sin t = -\frac{4}{5}$. Then

$$\tan t = -\frac{4}{3}, \quad \cot t = -\frac{3}{4}, \quad \sec t = \frac{5}{3}, \quad \csc t = -\frac{5}{4}.$$

9. If $\tan t = \frac{1}{4}$ and $0 < t < \frac{\pi}{2}$, then $\cos t > 0$ and $\sin t > 0$, and

$$(\tan t)^2 + 1 = (\sec t)^2 \text{ implies } (\sec t)^2 = 1 + \frac{1}{16} = \frac{17}{16} \text{ so } \sec t = \frac{1}{\cos t} = \sqrt{\frac{17}{16}} = \frac{\sqrt{17}}{4}.$$

Then

$$\cos t = \frac{4}{\sqrt{17}} = \frac{4\sqrt{17}}{17}, \quad \frac{1}{4} = \tan t = \frac{\sin t}{\cos t} \text{ so } \sin t = \frac{1}{4}\frac{4\sqrt{17}}{17} = \frac{\sqrt{17}}{17},$$

and

$$\cot t = 4, \quad \csc t = \frac{17}{\sqrt{17}} = \sqrt{17}.$$

11. If $0 \leq x \leq \pi$, then $0 \leq \frac{x}{3} \leq \frac{\pi}{3}$ and $\cos \frac{x}{3} = \frac{1}{2}$ implies $\frac{x}{3} = \frac{\pi}{3}$ so $x = \pi$.

13. Since

$$2(\sin x)^2 - 3\sin x + 1 = (2\sin x - 1)(\sin x - 1) = 0 \text{ implies } \sin x = \frac{1}{2}, \sin x = 1 \text{ so } x = \frac{\pi}{6}, x = \frac{5\pi}{6}, x = \frac{\pi}{2}.$$

15. Since

$$(\tan x)^3 - 4\tan x = \tan x[(\tan x)^2 - 4] = \tan x(\tan x - 2)(\tan x + 2) = 0,$$

we have

$$\tan x = 0, \tan x = 2, \tan x = -2,$$

so

$$x = 0, \pi, x = \arctan 2 \approx 1.1, x = \arctan(-2) = \pi - \arctan(2) \approx 2.03.$$

Note $x = \arctan(-2)$ is not in $[0, \pi]$.

17. We have

$$\cot x - \csc x = \frac{\cos x}{\sin x} - \frac{1}{\sin x} = \frac{\cos x - 1}{\sin x} = 1 \text{ implies } \cos x - 1 = \sin x \text{ so } \cos x - \sin x = 1,$$

so $x = 0, \frac{3\pi}{2}$. But there are no solutions since $\cot x$ is not defined at $x = 0$ and $x = \frac{3\pi}{2}$ is not in the interval $[0, \pi]$.

19. The function $f(x) = (\sin x)^3$ is odd since

$$f(-x) = (\sin(-x))^3 = (-\sin x)^3 = -(\sin x)^3 = -f(x).$$

21. The function $f(x) = x^2(\cos x)^2$ is even since

$$f(-x) = (-x)^2(\cos(-x))^2 = x^2(\cos x)^2 = f(x).$$

23. Since (a) and (c) have period $\frac{2\pi}{2} = \pi$ and (b) and (d) have period $\frac{2\pi}{1/2} = 4\pi$, we have (a) or (c) is either (i) or (ii) and (b) or (d) is either (iii) or (iv).

a. (ii) since $\sin 2 \left(0 + \frac{\pi}{2}\right) = \sin \pi = 0$.

b. (iv) since the curve has the form of a sine wave shifted to the left $\frac{\pi}{2}$ units.

c. (i).   **d.** (iii).

25. $y = 3 \sin \frac{x}{3}$

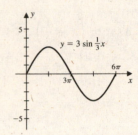

27. $y = -3 \cos 2x$

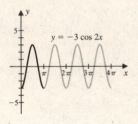

29. $y = \cos(3x - \pi) = \cos 3 \left(x - \frac{\pi}{3}\right)$

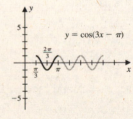

31. $y = \cot(x + \pi/6)$

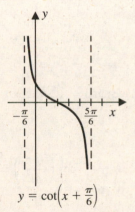

33. $y = \sec 3\pi x$

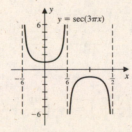

35. a. The curve has an equation of the form $y = A \sin(Bx)$, where the amplitude is 1 and the period is $\frac{\pi}{2}$. So

$$A = 1, \frac{2\pi}{B} = \frac{\pi}{2} \text{ so } B = 4 \text{ and } y = \sin 4x.$$

b. The curve has the form $y = A \cos B(x + C)$, with

$$A = 1, \frac{2\pi}{B} = \frac{\pi}{2} \text{ so } B = 4, C = -\frac{\pi}{8} \text{ so } y = \cos 4\left(x - \frac{\pi}{8}\right).$$

37. a. The curve has an equation of the form $y = A \sin B(x + C)$, where the amplitude is 3, the period is 1, and the curve is shifted to the left $\frac{1}{4}$ unit. So

$$A = 3, \frac{2\pi}{B} = 1 \text{ so } B = 2\pi, C = \frac{1}{4} \text{ and } y = 3 \sin 2\pi \left(x + \frac{1}{4}\right).$$

b. The curve has the form $y = A \cos Bx$, with

$$A = 3, B = 2\pi, \text{ so } y = 3 \cos 2\pi x.$$

39. For $\alpha = 60°, \overline{AC} = 13$ we have

a.

$$\sin 60° = \frac{\overline{BC}}{13} \text{ so } \overline{BC} = 13 \sin 60° = \frac{13\sqrt{3}}{2};$$
$$\cos 60° = \frac{\overline{AB}}{13} \text{ so } \overline{AB} = 13 \cos 60° = \frac{13}{2};$$
$$\gamma = 90 - 60 = 30°.$$

b. We have $\sin \alpha = \sqrt{3}/2, \cos \alpha = 1/2, \tan \alpha = \sqrt{3}, \cot \alpha = \sqrt{3}/3, \sec \alpha = 2,$ and $\csc \alpha = 2\sqrt{3}/3.$

41.

$$\cos\left(\frac{11\pi}{12}\right) = \cos\left(\frac{8\pi}{12} + \frac{3\pi}{12}\right) = \cos\left(\frac{2\pi}{3} + \frac{\pi}{4}\right)$$

$$= \cos\frac{2\pi}{3} \cos\frac{\pi}{4} - \sin\frac{2\pi}{3} \sin\frac{\pi}{4} = \left(-\frac{1}{2}\right)\frac{\sqrt{2}}{2} - \left(\frac{\sqrt{3}}{2}\right)\frac{\sqrt{2}}{2}$$

$$= -\frac{\sqrt{2}}{4}\left(1 + \sqrt{3}\right)$$

43.

$$\cos\left(-\frac{13\pi}{12}\right) = \cos\left(\frac{13\pi}{12}\right) = \cos\left(\frac{9\pi}{12} + \frac{4\pi}{12}\right) = \cos\left(\frac{3\pi}{4} + \frac{\pi}{3}\right)$$

$$= \cos\frac{3\pi}{4} \cos\frac{\pi}{3} - \sin\frac{3\pi}{4} \sin\frac{\pi}{3} = \left(-\frac{\sqrt{2}}{2}\right)\frac{1}{2} - \left(\frac{\sqrt{2}}{2}\right)\frac{\sqrt{3}}{2}$$

$$= -\frac{\sqrt{2}}{4}\left(\sqrt{3}+1\right)$$

45. $\sin\frac{5\pi}{8} = \sqrt{\frac{1-\cos\frac{5\pi}{4}}{2}} = \sqrt{\frac{1+\sqrt{2}/2}{2}} = \sqrt{\frac{2+\sqrt{2}}{4}} = \frac{\sqrt{2+\sqrt{2}}}{2}$

47. $\tan\frac{7\pi}{12} = \frac{\sin\frac{7\pi}{12}}{\cos\frac{7\pi}{12}} = \frac{\sqrt{\frac{1-\cos\frac{7\pi}{6}}{2}}}{-\sqrt{\frac{1+\cos\frac{7\pi}{6}}{2}}} = \frac{\sqrt{\frac{1+\sqrt{3}/2}{2}}}{-\sqrt{\frac{1-\sqrt{3}/2}{2}}} = -\frac{\sqrt{2+\sqrt{3}}}{\sqrt{2-\sqrt{3}}} = -\sqrt{7+4\sqrt{3}} = -2-\sqrt{3}$

49. $\sin\left(t-\frac{\pi}{2}\right) = \sin t\cos\frac{\pi}{2} - \cos t\sin\frac{\pi}{2} = (\sin t)(0) - (\cos t)(1) = -\cos t$

51. $\sin\left(\frac{3\pi}{2}-t\right) = \sin\frac{3\pi}{2}\cos t - \cos\frac{3\pi}{2}\sin t = (-1)(\cos t) - (0)(\sin t) = -\cos t$

53. $(\cos 5x)^2 = \frac{1+\cos 10x}{2} = \frac{1}{2} + \frac{1}{2}\cos 10x$

55. $(\sin x)^4 = ((\sin x)^2)^2 = \left(\frac{1-\cos 2x}{2}\right)^2 = \frac{1}{4}\left(1 - 2\cos 2x + (\cos 2x)^2\right)$

$= \frac{1}{4}\left(1 - 2\cos 2x + \frac{1+\cos 4x}{2}\right) = \frac{3}{8} - \frac{1}{2}\cos 2x + \frac{1}{8}\cos 4x$

57. $\sin 3t\cos 4t = \frac{1}{2}[\sin(3t+4t) + \sin(3t-4t)] = \frac{1}{2}[\sin 7t - \sin t]$

59. $\cos 2t\cos 4t = \frac{1}{2}[\cos(2t+4t) + \cos(2t-4t)] = \frac{1}{2}[\cos 6t + \cos 2t]$

61. $\sin 2t + \sin 6t = 2\sin\frac{2t+6t}{2}\cos\frac{2t-6t}{2} = 2\sin 4t\cos 2t$

63. $\cos 4t + \cos 2t = 2\cos\frac{4t+2t}{2}\cos\frac{4t-2t}{2} = 2\cos 3t\cos t$

65. $(\cos x)^4 - (\sin x)^4 = ((\cos x)^2 - (\sin x)^2)((\cos x)^2 + (\sin x)^2) = (\cos x)^2 - (\sin x)^2 = \cos 2x$

67.

$$\frac{\sin x}{1-\cos x} = \frac{\sin x}{1-\cos x}\cdot\frac{1+\cos x}{1+\cos x} = \frac{\sin x(1+\cos x)}{1-(\cos x)^2}$$

$$= \frac{\sin x(1+\cos x)}{(\sin x)^2} = \frac{1+\cos x}{\sin x} = \cot x + \csc x$$

69. $\sin\left(\arctan\sqrt{3}\right) = \sin\left(\frac{\pi}{3}\right) = \frac{\sqrt{3}}{2}$

71. Since $\sin(\arcsin x) = x$ for x in $[-1, 1]$, we have

$$\sin\left(\arcsin\left(\frac{3}{5}\right) - \arcsin\left(\frac{5}{13}\right)\right) = \sin\left(\arcsin\left(\frac{3}{5}\right)\right)\cos\left(\arcsin\left(\frac{5}{13}\right)\right)$$
$$- \cos\left(\arcsin\left(\frac{3}{5}\right)\right)\sin\left(\arcsin\left(\frac{5}{13}\right)\right)$$
$$= \frac{3}{5}\cos\left(\arcsin\left(\frac{5}{13}\right)\right) - \cos\left(\arcsin\left(\frac{3}{5}\right)\right)\frac{5}{13}$$
$$= \frac{3}{5}\left(\frac{12}{13}\right) - \frac{4}{5}\left(\frac{5}{13}\right) = \frac{16}{65}.$$

73. Since $\alpha = 25°$, $b = 12$, $c = 20$, and $a^2 = b^2 + c^2 - 2bc\cos\alpha$, we have

$$a = \sqrt{12^2 + 20^2 - 2(12)(20)\cos 25°} \approx 10.4.$$

Also,

$$\frac{\sin\beta}{b} = \frac{\sin\alpha}{a}\beta = \arcsin\left(\frac{12\sin 25°}{10.4}\right) \approx 0.50938 = 0.50938\left(\frac{180}{\pi}\right)° \approx 29.2°,$$

and $\gamma = 180 - 25 - 29.2 = 125.8°$.

75. Since $a = 6$, $b = 8$, and $c = 10$, we have $a^2 + b^2 = c^2$ and $\gamma = 90°$. Also,

$$\cos\alpha = \frac{6^2 - 8^2 - 10^2}{-2(8)(10)} = \frac{4}{5},$$

so

$$\alpha = \arccos\left(\frac{4}{5}\right) \approx 0.6435 = 0.6435\left(\frac{180}{\pi}\right)° \approx 36.9°,$$

and $\beta = 90° - \alpha \approx 53.1°$.

77. Since $\beta = 76°$, $\gamma = 50°$, and $b = 10.5$, we have $\alpha = 180 - 76 - 50 = 54°$, and

$$\frac{\sin\alpha}{a} = \frac{\sin\beta}{b} \text{ so } a = \frac{10.5\sin 54°}{\sin 76°} \approx 8.8; \quad \frac{\sin\gamma}{c} = \frac{\sin\beta}{b} \text{ so } c = \frac{10.5\sin 50°}{\sin 76°} \approx 8.3.$$

79. The points of intersection of the curves $y = \sin x$ and $y = x^2$ occur at approximately $x = 0.88$, and $x = 0$.

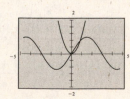

81. The period of $f(x) = 4\cos(125x)$ is $\frac{2\pi}{125} \approx 0.05$, and since the amplitude is 4, a reasonable viewing rectangle is $\left[-\frac{2\pi}{125}, \frac{2\pi}{125}\right] \times [-4, 4]$.

83. Since the area of a trapezoid is the product of its height and the average of its bases, we have

$$A = \frac{1}{2}[b + (b + 2b\cos\theta)]b\sin\theta = b^2(1 + \cos\theta)\sin\theta.$$

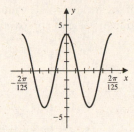

85. If the plane is traveling at 380 miles/hour, then after two and one half hours the plane is 950 mi from the airport. Its bearing from the airport is 150° measured clockwise from north, or 60° measured clockwise from east. Then the plane is

$$950\cos 60° = 950\left(\frac{1}{2}\right) = 475$$

miles east of the airport and

$$950\sin 60° = 950\frac{\sqrt{3}}{2} \approx 822.7$$

miles south of the airport.

Chapter 4 Exercises for Calculus (Page 275)

1. The graph indicates that the minimum occurs at $x = -2$ with a minimum value of approximately -1.1. The maximum occurs at $x = 3$ with a maximum value of approximately 2.9.

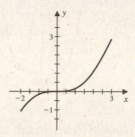

3. **a.** $\sqrt{a^2 - u^2} = \sqrt{a^2 - (a\sin t)^2} = \sqrt{a^2(1 - (\sin t)^2)} = a\sqrt{(\cos t)^2} = a\cos t$

    b. $\sqrt{u^2 + a^2} = \sqrt{(a\tan t)^2 + a^2} = \sqrt{a^2((\tan t)^2 + 1)} = a\sqrt{(\sec t)^2} = a\sec t$

    c. $\sqrt{u^2 - a^2} = \sqrt{(a\sec t)^2 - a^2} = \sqrt{a^2((\sec t)^2 - 1)} = a\sqrt{(\tan t)^2} = a\tan t$

    d. $\dfrac{\sqrt{u^2 - a^2}}{u} = \dfrac{\sqrt{(a\sec t)^2 - a^2}}{a\sec t} = \dfrac{a\tan t}{a\sec t} = \sin t$

5. Let x be the length of the line segment connecting the two vertices that lie approximately east-west of one another. By the Law of Cosines,

    $$x^2 = 10^2 + 7^2 - 2(10)(7)\cos 95° \text{ so } x = \sqrt{10^2 + 7^2 - 2(10)(7)\cos 95°} \approx 13.$$

    The perimeter of the upper triangle is $P_u = 10 + 7 + 13 = 30$ so by Heron's Formula, the area of the upper triangle is

    $$A_u = \frac{1}{4}\sqrt{30(30 - 20)(30 - 14)(30 - 26)} = (\frac{1}{4})80\sqrt{3} = 20\sqrt{3} \approx 34.64.$$

    For the lower triangle, we have $P_l = 9 + 12 + 13 = 34$. By Heron's Formula, the area of the lower triangle is

    $$A_l = \frac{1}{4}\sqrt{34(34 - 18)(34 - 24)(34 - 26)} = (\frac{1}{4})16\sqrt{170} \approx 52.15.$$

    The area of the quadrilateral is then $A_u + A_l \approx 34.64 + 52.15 = 86.79 \text{ feet}^2$.

7. If α is the angle of the line of sight from ship A to the plane, β is the angle from ship B to the plane, and d is the distance from A to B, then from Exercise 6,

    $$33000 = \frac{d}{\cot\alpha - \cot\beta} = \frac{d}{\cot 32° - \cot 47°} \text{ so } d = 33000(\cot 32° - \cot 47°) \approx 22038.$$

    So the two ships are approximately 22000 feet apart, rounded to the nearest 100 feet.

9. Let d be the distance between A and C and let α be the angle ACB. Then $\alpha = 180 - 80 - 59 = 41°$. By the Law of Sines

$$\frac{\sin 41°}{300} = \frac{\sin 59°}{d} \text{ so } d = \frac{300 \sin 59°}{\sin 41°} \approx 392 \text{ feet.}$$

11. a. Let b denote the base of the inscribed triangle and h the height. Since

$$\sin \frac{\pi}{n} = \frac{b/2}{1} = \frac{b}{2} \text{ so } b = 2 \sin \frac{\pi}{n},$$

and since

$$\cos \frac{\pi}{n} = \frac{h}{1} \text{ so } h = \cos \frac{\pi}{n},$$

the area of each triangle is

$$A = \frac{1}{2} \left(2 \sin \frac{\pi}{n} \right) \left(\cos \frac{\pi}{n} \right) = \frac{1}{2} \left(2 \sin \frac{\pi}{n} \cos \frac{\pi}{n} \right) = \frac{1}{2} \sin \frac{2\pi}{n}.$$

b. Since the area of a circle of radius 1 is known to be π we would expect $\frac{n}{2} \sin \frac{2\pi}{n}$ to better approximate the area as n increases. So as n approaches ∞ we expect $\frac{n}{2} \sin \frac{2\pi}{n}$ will approach π.

- -

Chapter 4 Chapter Test (Page 277)

1. False. The number of degrees in $\frac{11\pi}{4}$ radians is 495°.

2. False. The number of degrees in $-5\pi/6$ radians is $-150°$.

3. True

4. True.

5. True.

6. False. The reference angle for 480° is 60°.

7. False. If $\cos\theta > 0$ and $\tan\theta < 0$, then θ is in quadrant IV.

8. False. If $\sin\theta > 0$ and the $\sec\theta > 0$, then θ is in quadrant I.

9. True.

10. True.

11. False. If $\tan\theta > 0$, then θ is in quadrant I or in quadrant III.

12. False. If $\csc\theta < 0$, then θ is in quadrant III or in quadrant IV.

13. False. If $\theta = \frac{5\pi}{6}$, then $\sin\theta = \frac{1}{2}$.

14. False. If $\theta = -11\pi/4$, then $\sin\theta = -\sqrt{2}/2$.

15. True.

16. True.

17. True.

18. False. If $\theta = -5\pi/3$, then $\cot \theta = \sqrt{3}/3$.

19. False. If $\theta = \pi$, then $\sec \theta = -1$.

20. True.

21. True.

22. False. The missing angle of the triangle is $45°$.

23. False. The side a of the triangle has length 3.

24. True.

25. True.

26. True.

27. False. The side a of the triangle has length $4 \tan 60°$.

28. False. The hypotenuse of the triangle has length $x = 4 \csc 30°$.

29. True.

30. False. $\cos \theta = \frac{\sqrt{5}}{3}$

31. False. $\tan \theta = \frac{2\sqrt{5}}{5}$

32. True.

33. False. If $\cos \theta = \frac{4}{5}$ and θ lies in quadrant IV, then $\sin \theta = -\frac{3}{5}$.

34. False. If $\tan \theta = -\frac{\sqrt{3}}{2}$ and θ lies in quadrant II, then

$$\cos \theta = -\frac{2\sqrt{7}}{7}.$$

35. True.

36. True.

37. False. For θ in the interval $[0, 2\pi]$, we have 2θ in $[0, 4\pi]$. So θ satisfies $\cos 2\theta = \frac{1}{2}$ when

$$\theta = \frac{\pi}{6}, \frac{5\pi}{6}, \frac{7\pi}{6}, \frac{11\pi}{6}.$$

38. True.

39. True.

40. True.

41. True.

42. False. The curve is given by $y = 1 + 2 \cos \left(2x - \frac{\pi}{2} \right)$.

43. True.

44. True.

45. True.

46. True.

47. False. The graph of $y = 2 \cos \frac{1}{2} x$ is obtained from the graph of $y = \cos x$ through a horizontal stretching by a factor of 2 and a vertical stretching by a factor of 2.

48. False. The graph of $y = \frac{1}{2} \sin 2x$ is obtained from the graph of $y = \sin x$ by a horizontal compression by a factor of 2 and a vertical compression by a factor of 2.

49. True.

50. True.

51. True.

52. False. The graph of

$$y = \sin \left(2x + \frac{\pi}{2} \right) = \sin 2 \left(x + \frac{\pi}{4} \right)$$

is obtained from the graph of $y = \sin x$ by a horizontal compression by a factor of 2, followed by a shift to the left $\frac{\pi}{4}$ units.

53. False. The graph of $y = 1 + \sin(2x - 3)$ is obtained from the graph of $y = \sin x$ from a horizontal compression by a factor of 2, followed by a shift to the right $3/2$ units and upward 1 unit.

54. False. The graph of $y = -1 + 2 \cos \left(2x - \frac{\pi}{3} \right)$ is obtained from the graph of $y = \cos x$ from a horizontal compression by a factor of 2 and a vertical stretching by a factor of 2, followed by a shift to the right $\frac{\pi}{6}$ units and downward 1 unit.

55. True.

56. False.   The only values of θ in the interval $[0, 2\pi]$ that satisfy $2\sin 2\theta = 4\sin\theta$ are $\theta = 0, \pi, 2\pi$.

57. True.

58. False. $\cos\left(\dfrac{\pi}{12}\right) = \dfrac{\sqrt{2}}{4}\left(\sqrt{3}+1\right)$

59. True.

60. False.   For all θ, $\sin\left(\theta - \dfrac{\pi}{2}\right) = -\cos\theta$.

61. True.

62. True.

63. False. $\arccos\left(\dfrac{\sqrt{3}}{2}\right) = \dfrac{\pi}{6}$.

64. False.   Since the values of $\arcsin x$ are in the interval $[-\frac{\pi}{2}, \frac{\pi}{2}]$, we have $\arcsin(-\frac{1}{2}) = -\frac{\pi}{6}$.

65. True.

66. False.   If $\theta = \arctan\left(\frac{1}{3}\right)$, then $\sin\theta = \frac{\sqrt{10}}{10}$.

67. True.

68. True.

69. True.

70. True.

CHAPTER 5: Exponential and Logarithm Functions

5.1   Introduction

The exponential functions extend the notion of exponent to include all real numbers. The *natural exponential* function $f(x) = e^x$ and the *natural logarithm* function $g(x) = \ln x$ are perhaps the most important function-inverse pair in science and mathematics.

5.2   The Natural Exponential Function

For $a > 0$, the *exponential function with base a* is defined as $f(x) = a^x$, for all real numbers x. The value of the base a determines the general shape of the graph of the exponential function.

$f(x) = a^x, a > 1$

- domain: $(-\infty, \infty)$
- range: $(0, \infty)$
- graph increasing: for all x
- graph decreasing: for no x
- horizontal asymptote: $y = 0$

$$f(x) \to 0 \quad \text{as} \quad x \to -\infty$$
$$f(x) \to \infty \quad \text{as} \quad x \to \infty$$

- y-intercept: $(0, 1)$
- x-intercept: none

$f(x) = a^x, 0 < a < 1$

- domain: $(-\infty, \infty)$
- range: $(0, \infty)$
- graph increasing: for no x
- graph decreasing: for all x
- horizontal asymptote: $y = 0$

$$f(x) \to \infty \quad \text{as} \quad x \to -\infty$$
$$f(x) \to 0 \quad \text{as} \quad x \to \infty$$

- y-intercept: $(0, 1)$
- x-intercept: none

Note that when $a = 1$, the function $f(x) = a^x = 1$ is a constant function.

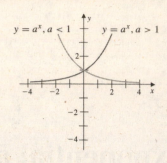

Graphs of Exponential Functions

EXAMPLE 1

Use the graph of $y = 2^x$ shown in the figure to sketch the graph of the function.

(a) $y = 2^{(x-1)} + 1$ (b) $y = -2^{(x+1)} - 2$

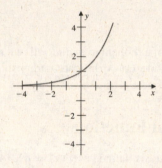

Solution The shifting and scaling properties allow us to sketch reasonable graphs quickly once we know the general shape of the basic curve.

(a) The basic function is $y = 2^x$, and since the argument is changed to $(x - 1)$, the graph of $y = 2^{(x-1)}$ is obtained by shifting the basic curve to the right 1 unit. To obtain the graph of $y = 2^{(x-1)} + 1$ from the graph of $y = 2^x$, first shift the basic curve to the right 1 unit, and then shift the resulting curve upward 1 unit.

- y-intercept: Set $x = 0$.

$$y = 2^{-1} + 1$$
$$= \frac{1}{2} + 1 = \frac{3}{2}$$

- Horizontal asymptote: The line $y = 1$, since the horizontal asymptote of $y = 2^x$ is the line $y = 0$, and the graph is shifted upward 1 unit.

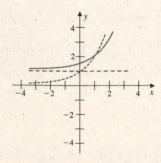

(b) The graph of $y = -2^x$ is the reflection of the graph of $y = 2^x$ about the x-axis. Changing the argument from x to $x + 1$ shifts the reflected graph to the left 1 unit. Finally, shifting this graph downward 2 units gives the graph of $y = -2^{(x+1)} - 2$.

    • y-intercept: Set $x = 0$.

$$y = -2^1 - 2$$
$$= -4$$

    • Horizontal asymptote: The line $y = -2$, since $y = -2^{(x+1)}$ was shifted downward 2 units.
     Also $y = 2^x$ approaches the horizontal line $y = 0$ from above. The graph of $y = -2^{(x+1)} - 2$ is obtained from $y = 2^x$ by first reflecting it about the x-axis, so it approaches the line $y = -2$ from below.

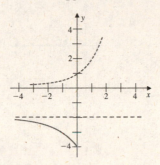

□

EXAMPLE 2

Use the graph of $y = \left(\frac{1}{2}\right)^x$ shown in the figure to sketch the graph of $y = 3 \cdot 2^{(1-x)} + 1$.

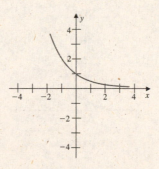

Solution First note that

$$2^{-x} = \frac{1}{2^x} = \left(\frac{1}{2}\right)^x.$$

The equation can now be rewritten using a positive exponent.

$$y = 3 \cdot 2^{(1-x)} + 1 = 3 \cdot 2^{-(x-1)} + 1 = 3 \cdot \left(\frac{1}{2}\right)^{x-1} + 1$$

The strategy is:

- Start with the graph of $y = \left(\frac{1}{2}\right)^x$.
- Vertically stretch the graph by a factor of 3.
- Shift the resulting graph to the right 1 unit.
- Shift the resulting graph upward 1 unit.
- y-intercept: Set $x = 0$.

$$y = 3 \cdot 2^1 + 1 = 7$$

- Horizontal asymptote: $y = 1$

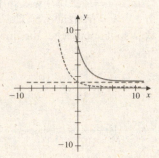

□

The Natural Exponential Function

As $n \to \infty$, the real numbers $\left(1 + \frac{1}{n}\right)^n$ approach an irrational number denoted $e \approx 2.71828$. The function

$$f(x) = e^x$$

is called the *natural exponential function*.

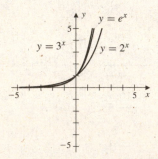

The graph shows the relationship between the graphs of $y = e^x$, $y = 2^x$, and $y = 3^x$.

EXAMPLE 3

Sketch the graph of the function.

(a) $f(x) = 1 - e^{-(x-2)}$ (b) $f(x) = e^{-|x-1|}$

Solution (a)

- Start with the graph of $y = e^x$.

- Graph of $y = e^{-x}$: Reflect the graph of $y = e^x$ about the y−axis.
- Graph of $y = -e^{-x}$: Reflect the graph of $y = e^{-x}$ about the x−axis.
- Graph of $y = -e^{-(x-2)}$: Shift the graph of $y = -e^{-x}$ to the right 2 units.
- Graph of $y = 1 - e^{-(x-2)}$: Shift the graph of $y = -e^{-(x-2)}$ upward 1 unit.

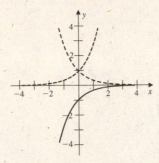 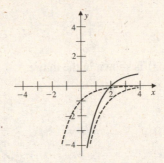

(b) Replacing x with $|x|$ changes the graph of $y = e^{-x}$ to one that is symmetric with respect to the y-axis, since

$$e^{-|x|} = e^{-|-x|}.$$

For example, if $x > 0$, the points with first coordinate x and first coordinate $-x$ have the same y-coordinates. Now shift the graph of $y = e^{-|x|}$ to the right 1 unit.

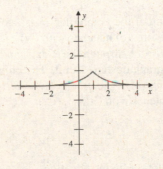

$\square$

EXAMPLE 4

Match the equation with the curve.

(a) $y = e^{x-1} - 2$ (b) $y = -e^{x+1} + 1$ (c) $y = 3e^x$ (d) $y = 2e^{-x}$

(i) 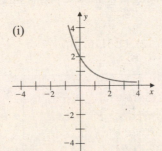   (ii)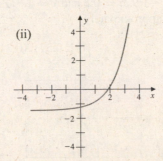

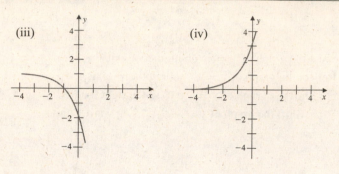

Solution The graph of $y = e^x$ is shown in the figure.

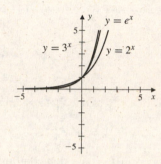

The graph of $y = e^{-x}$ is the reflection of the graph of $y = e^x$ through the y-axis, and $y = 2e^{-x}$ has the same shape so (d) matches with (i).

The graph of $y = 3e^x$ is a vertical stretching of the graph of $y = e^x$, so it has a similar shape. However, it passes through $(0, 3)$ and rises more quickly. Therefore, (c) matches with (iv).

The graph of $y = -e^x$ is the reflection of $y = e^x$ through the x-axis, and $y = -e^{x+1}$ requires an additional shift to the left 1 unit. The addition of 1 shifts the graph upward 1 unit. The only curve that requires a reflection of $y = e^x$ through the x-axis is (iii), so (b) matches with (iii). The graph of $y = e^{x-1} - 2$ is obtained by shifting $y = e^x$ to the right 1 unit and downward 2 units, so (a) matches with (ii).

$\square$

EXAMPLE 5

Use a graphing device to approximate all solutions to the equation $x^2 e^x = 2x^2 + 3x - 1$.

Solution To approximate the solutions, plot $y = x^2 e^x$ and $y = 2x^2 + 3x - 1$ together on the same coordinate axes, and determine the points of intersection. When using a graphing device, it is essential to have a viewing rectangle that shows the important features. A viewing rectangle of $[-5, 5] \times [-5, 10]$ shows three points of intersection. A viewing rectangle of $[-5, 5] \times [-5, 5]$ shows only two of those points. From the initial plot, the points of intersection occur at approximately $x = -1.9$, $x = 0.3$ and $x = 1.3$. Zooming in near these points we get better accuracy with the points of intersection being approximately

$$(-1.91, 0.57), \quad (0.31, 0.12), \quad (1.31, 6.36).$$

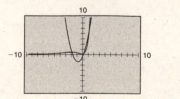 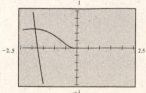

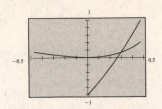
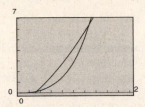

□

EXAMPLE 6

Use a graphing device to compare the rates of growth of $f(x) = e^x$ and $g(x) = x^{15}$ by graphing the functions together in several appropriate viewing rectangles. Approximate the solutions to $e^x = x^{15}$.

Solution We want to determine whether, for large x,

$$x^{15} > e^x \quad \text{or} \quad e^x > x^{15}.$$

The graph of $y = x^{15}$ and $y = e^x$ that is eventually above the other, will be the dominant function.
Since both functions are nonnegative for all x, there is no need to show much of the negative y-axis. The first viewing rectangle is

$$[-5, 5] \times [-5, 5].$$

In this view, it appears that $g(x) = x^{15}$ eventually grows faster than $f(x) = e^x$. We should not jump to a quick conclusion. For a second viewing rectangle, select

$$[0, 50] \times [0, 10^{10}].$$

It still appears that $g(x) = x^{15}$ grows faster than $f(x) = e^x$. One last viewing rectangle reveals another story. Try

$$[0, 100] \times [0, 10^{27}].$$

This time the function $f(x) = e^x$ eventually overtakes $g(x) = x^{15}$, and in fact, it eventually grows much faster. The figures show points of intersection occurring when

$$x \approx 1.1, \quad \text{and} \quad x \approx 61.8,$$

which are the approximate solutions to the equation $e^x = x^{15}$.

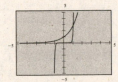
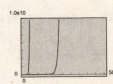
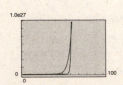

□

Solving Equations Involving e^x

EXAMPLE 7

Solve the equation.
(a) $2x^3 e^x + x^4 e^x = 0$ (b) $x^2 e^x - 2x e^x = 3e^x$

Solution (a) Since e^x is common to both terms, it can be factored from these terms. So

$$e^x(2x^3 + x^4) = 0.$$

Since $e^x > 0$ for all x, both sides of the equation can be divided by e^x, and the last equation will be 0 only when

$$2x^3 + x^4 = 0$$
$$x^3(2 + x) = 0$$
$$x = 0, \ x = -2.$$

(b) First rewrite the equation so that one side is zero.

$$x^2 e^x - 2x e^x = 3e^x$$
$$x^2 e^x - 2x e^x - 3e^x = 0$$
$$e^x(x^2 - 2x - 3) = 0$$

Since $e^x > 0$, each side of the last equation can be divided by e^x and

$$x^2 - 2x - 3 = 0$$
$$(x - 3)(x + 1) = 0$$
$$x = 3, \ x = -1.$$

$\square$

Compound Interest

If an initial amount of A_0 dollars is invested at an interest rate i compounded n times a year, the investment after t years has a value

$$A_n(t) = A_0 \left(1 + \frac{i}{n}\right)^{nt} \text{ dollars.}$$

If the interest is compounded *continuously,* then the amount after t years is

$$A_c(t) = A_0 e^{it} \text{ dollars.}$$

EXAMPLE 8

Suppose $5,000.00 is invested at 10% interest, and the interest rate remains fixed for 8 years. Determine the value of the investment if the interest is compounded annually, semiannually, quarterly, monthly, weekly, daily, hourly and continuously.

Solution We have

Annually: $A_1(8) = 5000(1 + 0.1)^8 \approx \$10,717.94$

Semiannually: $A_2(8) = 5000 \left(1 + \frac{0.1}{2}\right)^{16} \approx \$10,914.37$

Quarterly: $A_4(8) = 5000 \left(1 + \frac{0.1}{4}\right)^{32} \approx \$11,018.79$

Monthly: $A_{12}(8) = 5000 \left(1 + \frac{0.1}{12}\right)^{96} \approx \$11,090.88$

Weekly: $A_{52}(8) = 5000 \left(1 + \frac{0.1}{52}\right)^{416} \approx \$11,119.16$

Daily: $A_{365}(8) = 5000 \left(1 + \frac{0.1}{365}\right)^{2920} \approx \$11,126.49$

Hourly: There are 24 hours in a day and 365 days in the year, so the interest is compounded $24 \cdot 365 = 8760$ times.

$$A_{8760}(8) = 5000 \left(1 + \frac{0.1}{8760}\right)^{70080} \approx \$11,127.65$$

Continuously: $A_c(8) = 5000e^{0.8} \approx \$11,127.71$ □

EXAMPLE 9

What initial investment of 8% compounded semiannually for 7 years will accumulate to $25,000.00?

Solution The value of the investment after t years is

$$A_2(t) = A_0 \left(1 + \frac{0.08}{2}\right)^{2t}.$$

And after 7 years the value is

$$A_2(7) = A_0 \left(1 + \frac{0.08}{2}\right)^{14}.$$

The initial investment that will yield $25,000.00 can be found from the equation

$$25000 = A_2(7) = A_0 \left(1 + \frac{0.08}{2}\right)^{14}$$

$$25000 = A_0(1.04)^{14}$$

$$A_0 = \frac{25000}{(1.04)^{14}} \approx \$14,437.$$

□

Exercise Set 5.2 (Page 292)

1. $f(x) = 2^x + 1$

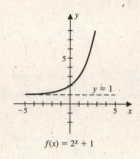

$f(x) = 2^x + 1$

3. $f(x) = -4^x$

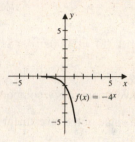

$f(x) = -4^x$

5. $f(x) = 3 \cdot \left(\frac{1}{4}\right)^{x-1} + 2$

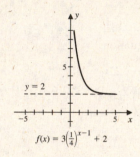

$f(x) = 3\left(\frac{1}{4}\right)^{x-1} + 2$

7. $f(x) = -e^{x-1}$

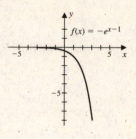

$f(x) = -e^{x-1}$

9. $f(x) = 2 - e^{-(x-3)}$

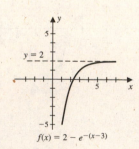

$f(x) = 2 - e^{-(x-3)}$

11. $f(x) = e^{2x}$

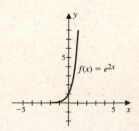

$f(x) = e^{2x}$

13. $f(x) = -e^{|x|}$

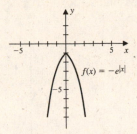

15. The graph shows that $e^x > 1$ when $x > 0$.

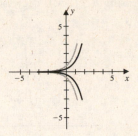

17. $\left(\frac{1}{4}\right)^x = \frac{1}{4^x} \geq 2$ implies $4^x \leq \frac{1}{2}$, so $x \leq -\frac{1}{2}$

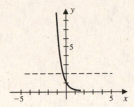

19. a. ii   **b.** iii   **c.** iv   **d.** i

21. The points of intersection of $y = e^{x-2}$ and $y = x$ are $x \approx 0.2$, $x \approx 3.2$.

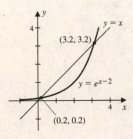

23. The points of intersection of $y = e^{-x}$ and $y = (x-2)^2$ are $x \approx 1.5$, $x \approx 2.3$, $x = -3.3$.

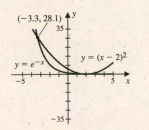
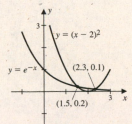

25. a. $f(x) = xe^x$

  b. Increasing: $(-1, \infty)$; decreasing: $(-\infty, -1)$.

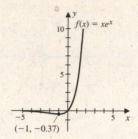

27. a. $f(x) = e^{-x^2 - x}$

  b. Increasing: $(-\infty, -0.5)$; decreasing: $(-0.5, \infty)$.

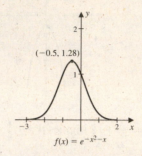

29. a. $f(x) = e^x - e^{-x}$

  b. Increasing for all x.

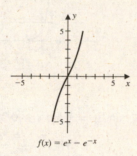

31. Let $3^x = e^{kx}$. If $x = 1$, then $3 = e^k$. The graphs of $y = e^x$ and $y = 3$ intersect when $x \approx 1.1$, so $k \approx 1.1$.

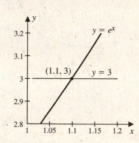

33. a. The function $f(x) = 2^x$ eventually grows much faster than $g(x) = x^5$. In Figure (iii) we see $f(x) = 2^x$ crosses $g(x) = x^5$ and remains above $y = g(x)$ from that point on.

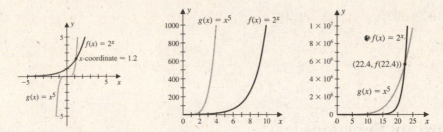

b. The points of intersection of the two graphs occur when $x \approx 1.2$ and $y \approx 2^{1.2}$, and when $x \approx 22.4$ and $y \approx 2^{22.4}$. These values of x are approximate solutions to $2^x = x^5$.

35. The value of the CD at 6.5% interest compounded n times per year and which matures in 5 years is
$A_n(5) = 5000 \left(1 + \frac{0.065}{n}\right)^{5n}$.

Interest Compounded	Value of CD
(a) annually: $n = 1$	$6850.43
(b) monthly: $n = 12$	$6914.09
(c) daily: $n = 365$	$6919.95
(d) continuously: $A_c(5) = 5000e^{(0.065)5}$	$6920.15

37. If $10,000.00 is invested and the interest is compounded quarterly, after 5 years and an interest rate of

a. 8%, the value of the investment is

$$A_4(5) = 10000 \left(1 + \frac{0.08}{4}\right)^{20} = \$14,859.47;$$

b. 6.5%, the value of the investment is

$$A_4(5) = 10000 \left(1 + \frac{0.065}{4}\right)^{20} = \$13,804.20;$$

c. 6%, the value of the investment is

$$A_4(5) = 10000 \left(1 + \frac{0.06}{4}\right)^{20} = \$13,468.55;$$

d. 5.5%, the value of the investment is

$$A_4(5) = 10000 \left(1 + \frac{0.055}{4}\right)^{20} = \$13,140.66.$$

39. Solve $10000 = A_0 \left(1 + \frac{0.08}{2}\right)^{5(2)}$ for A_0, which gives $A_0 = \frac{10000}{(1+0.04)^{10}} \approx \$6,756.00$.

41. $A(t) = 1000(1.02)^t$, so $A(7) = 11,487$.

··

5.3   Logarithm Functions

For $a \neq 1$ the inverse function of the exponential function to the base a, $f(x) = a^x$, is called the *logarithm function to the base a,* written $g(x) = \log_a x$.

For each x in $(0, \infty)$,

$$y = \log_a x \text{ if and only if } x = a^y.$$

The important properties that result from the inverse relation between the exponential and logarithm functions are:

- For each x in $(0, \infty)$, $\log_a a^y = y$.

- For each real number y, $a^{\log_a x} = x$.

Recall that the domain of the exponential function is the set of all real numbers, so the range of the logarithm function is also the set of all real numbers. The range of the exponential function is $(0, \infty)$, so the domain of the logarithm function is $(0, \infty)$.

The inverse to the natural exponential function $f(x) = e^x$ is the *natural logarithm function.* The natural logarithm function is the logarithm to the base e, written $y = \log_e x = \ln x$.

- For each x in $(0, \infty)$,

$$y = \ln x \text{ if and only if } x = e^y.$$

- For each x in $(0, \infty)$, $\ln e^y = y$.

- For each real number y, $e^{\ln x} = x$.

Evaluation of Logarithms

EXAMPLE 1
Evaluate the expression.

(a) $\log_2 64$ (b) $\log_4 16$ (c) $\log_{1/3} 3$ (d) $\log_3 \frac{1}{27}$ (e) $e^{\ln 6}$ (f) $\ln e^{\sqrt{2}}$

Solution   (a) Using the inverse relation between the logarithm function and the exponential function gives

$$x = \log_2 64$$
$$2^x = 64$$
$$x = 6.$$

(b) We have

$$x = \log_4 16$$
$$4^x = 16$$
$$x = 2.$$

(c) We have

$$x = \log_{1/3} 3$$
$$\left(\frac{1}{3}\right)^x = 3$$
$$\frac{1}{3^x} = 3$$
$$3^x = \frac{1}{3}$$
$$x = -1.$$

(d) Since $3^{-3} = \frac{1}{3^3} = \frac{1}{27}$, we have

$$x = \log_3 \frac{1}{27}$$
$$3^x = \frac{1}{27}$$
$$x = -3,$$

(e) The exponential and logarithm functions are inverses of one another, and therefore each undoes the process of the other. That is, if a value is input to ln, and the resulting number then used as input to e, the final output is the original value. So, for example,
$$e^{\ln 6} = 6.$$

(f) The inverse relation between the exponential and logarithm functions goes both ways, so, for example,
$$\ln e^{\sqrt{2}} = \sqrt{2}.$$

$\square$

EXAMPLE 2
Solve the equation.
(a) $\log_3(2x - 5) = 2$ (b) $\log_2(3x^2 + 10x) = 3$ (c) $e^{2x-1} = 2$

Solution   (a) We have

$$\log_3(2x - 5) = 2$$
$$2x - 5 = 3^2$$
$$2x = 14$$

$$x = 7.$$

(b) We have

$$\log_2(3x^2 + 10x) = 3$$
$$3x^2 + 10x = 2^3$$
$$3x^2 + 10x = 8.$$

To solve the quadratic, rewrite the expression with one side zero and factor.

$$3x^2 + 10x = 8$$
$$3x^2 + 10x - 8 = 0$$
$$(3x - 2)(x + 4) = 0$$
$$3x - 2 = 0 \quad \text{or} \quad x + 4 = 0$$
$$x = \frac{2}{3} \quad \text{or} \quad x = -4$$

(c) The inverse relationship between the exponential and logarithm functions is the key here. If we take the natural logarithm of the left side the result is the input to the natural exponential function, $2x - 1$. So as not to change the equation, we take the natural logarithm of both sides and simplify.

$$e^{2x-1} = 2$$
$$\ln e^{2x-1} = \ln 2$$
$$2x - 1 = \ln 2$$
$$2x = 1 + \ln 2$$
$$x = \frac{1 + \ln 2}{2}$$

$\square$

Arithmetic Properties of Logarithms

There are three important properties of the logarithm functions. For emphasis we also list the properties for the natural logarithm function.

- $\log_a (x_1 x_2) = \log_a x_1 + \log_a x_2$

- $\log_a \left(\frac{x_1}{x_2} \right) = \log_a x_1 - \log_a x_2$

- $\log_a x_1^r = r \log_a x_1$

- $\ln (x_1 x_2) = \ln x_1 + \ln x_2$

- $\ln \left(\frac{x_1}{x_2} \right) = \ln x_1 - \ln x_2$

- $\ln x_1^r = r \ln x_1$

EXAMPLE 3
Use the properties of logarithms to simplify the expression so that the result does not contain logarithms of products, quotients, or powers.

(a) $\ln x(x^2 + 1)$ (b) $\ln \dfrac{1}{x^2}$ (c) $\ln \dfrac{x\sqrt[5]{x^2}}{(x+2)^2}$ (d) $\log_5 \sqrt{\dfrac{x^3}{4x^2 - 1}}$

Solution (a) The expression in the natural logarithm function is the product of the two terms x and $(x^2 + 1)$, so the product rule for logarithms can be used to give

$$\ln x(x^2 + 1) = \ln x + \ln(x^2 + 1).$$

This is the final answer! The expression $\ln(x^2 + 1)$ cannot be further simplified.
(b) Because $e^0 = 1$, the inverse relation gives $\ln 1 = 0$, and

$$\ln \frac{1}{x^2} = \ln 1 - \ln x^2 = -2 \ln x.$$

This can also be recognized using the power rule.

$$\ln \frac{1}{x^2} = \ln x^{-2} = -2 \ln x$$

(c) This problem may look complicated, but just be careful to apply the properties in a proper sequence. The first property to apply is the quotient rule, since the entire expression is one quotient. Once the quotient rule is applied, the two resulting pieces will be treated separately and the properties applied to them individually. To simplify the steps, first rewrite

$$\sqrt[5]{x^2} = \left(x^2\right)^{1/5} = x^{\frac{2}{5}}.$$

Then the arithmetic properties give

$$\ln \frac{x \cdot x^{2/5}}{(x + 2)^2} = \ln x^{7/5} - \ln(x + 2)^2 = \frac{7}{5} \ln x - 2 \ln(x + 2).$$

(d) We have

$$\log_5 \sqrt{\frac{x^3}{4x^2 - 1}} = \log_5 \left(\frac{x^3}{4x^2 - 1}\right)^{1/2}$$

$$= \frac{1}{2} \log_5 \frac{x^3}{4x^2 - 1}$$

$$= \frac{1}{2} \left[\log_5 x^3 - \log_5(4x^2 - 1)\right]$$

$$= \frac{1}{2} \log_5 x^3 - \frac{1}{2} \log_5(4x^2 - 1)$$

$$= \frac{3}{2} \log_5 x - \frac{1}{2} \log_5(4x^2 - 1).$$

It may appear that the simplification is done, but some further simplification is possible by recognizing that

$$(4x^2 - 1) = 4\left(x^2 - \frac{1}{4}\right)$$

$$= 4\left(x - \frac{1}{2}\right)\left(x + \frac{1}{2}\right)$$

Now apply the product rule, first on $4\left[\left(x - \frac{1}{2}\right)\left(x + \frac{1}{2}\right)\right]$, and then on $\left[\left(x - \frac{1}{2}\right)\left(x + \frac{1}{2}\right)\right]$.

$$
\begin{aligned}
\log_5 \sqrt{\frac{x^3}{4x^2 - 1}} &= \frac{3}{2}\log_5 x - \frac{1}{2}\log_5(4x^2 - 1) \\
&= \frac{3}{2}\log_5 x - \frac{1}{2}\log_5 4\left(x - \frac{1}{2}\right)\left(x + \frac{1}{2}\right) \\
&= \frac{3}{2}\log_5 x - \frac{1}{2}\log_5 4 - \frac{1}{2}\log_5\left(x - \frac{1}{2}\right)\left(x + \frac{1}{2}\right) \\
&= \frac{3}{2}\log_5 x - \frac{1}{2}\log_5 4 - \frac{1}{2}\log_5\left(x - \frac{1}{2}\right) - \frac{1}{2}\log_5\left(x + \frac{1}{2}\right)
\end{aligned}
$$

$\square$

EXAMPLE 4

Rewrite the expression as a single logarithm.
(a) $2\ln x + 3\ln(x + 1)$ (b) $4\ln x - \frac{1}{2}\ln(x + 1)$

Solution (a) First use the property

$$r\ln a = \ln a^r$$

to write the expression as

$$2\ln x + 3\ln(x + 1) = \ln x^2 + \ln(x + 1)^3.$$

Now use the sum property

$$\ln a + \ln b = \ln ab$$

to get

$$2\ln x + 3\ln(x + 1) = \ln x^2 + \ln(x + 1)^3 = \ln x^2(x + 1)^3.$$

(b) We have

$$4\ln x - \frac{1}{2}\ln(x + 1) = \ln x^4 - \ln(x + 1)^{1/2} = \ln\frac{x^4}{(x + 1)^{1/2}}.$$

$\square$

EXAMPLE 5

Solve the equation for x.
(a) $\ln x + \ln(x + 1) = \ln 2$ (b) $2\ln(x + 3) - \ln x = \ln 12$ (c) $e^{3x} = 3^{2x-1}$

Solution (a) Using the properties of logarithms to combine the ln expressions into one expression allows us to solve the equation.

$$
\begin{aligned}
\ln x + \ln(x + 1) &= \ln 2 \\
\ln x(x + 1) &= \ln 2 \\
x(x + 1) &= 2 \\
x^2 + x - 2 &= 0
\end{aligned}
$$

$$(x + 2)(x - 1) = 0$$
$$x = -2, \; x = 1$$

Since $\ln(x + 1)$ is not defined at $x = -2$, the only solution is $x = 1$. So $x = -2$ is an extraneous solution. Note that $\ln x(x + 1) = \ln 2$ gives $x(x + 1) = 2$ for several reasons. The fact that the function $\ln$ is one-to-one, means $\ln a = \ln b$ if and only if $a = b$. Alternatively, if each side is raised to an exponent of e, then $e^{\ln x(x+1)} = e^{\ln 2}$, which by the inverse relation gives $x(x + 1) = 2$.

(b) We have

$$2\ln(x + 3) - \ln x = \ln 12$$
$$\ln(x + 3)^2 - \ln x = \ln 12$$
$$\ln \frac{(x + 3)^2}{x} = \ln 12$$
$$\frac{(x + 3)^2}{x} = 12$$
$$x^2 + 6x + 9 = 12x$$
$$x^2 - 6x + 9 = 0$$
$$(x - 3)^2 = 0$$
$$x = 3.$$

(c) Taking the natural logarithm of both sides gives

$$e^{3x} = 3^{2x-1}$$
$$\ln e^{3x} = \ln 3^{2x-1}$$
$$3x = (2x - 1)\ln 3$$
$$3x = 2x \ln 3 - \ln 3$$
$$3x - 2x \ln 3 = -\ln 3$$
$$x(3 - 2\ln 3) = -\ln 3$$
$$x = \frac{-\ln 3}{3 - 2\ln 3} = \frac{\ln 3}{2\ln 3 - 3} = \frac{\ln 3}{\ln 3^2 - 3} = \frac{\ln 3}{\ln 9 - 3}.$$

$\square$

Graphs of Logarithm Functions

There are several important properties to remember when sketching graphs of logarithms. The most common graphs are for $a > 1$, which have the following properties.

- Domain: $(0, \infty)$,

  The graph of $y = \log_a x$ is on the right side of the y-axis.

- Range: $(-\infty, \infty)$

The logarithm grows arbitrarily large since,

$$\log_a x \to \infty \quad \text{as} \quad x \to \infty.$$

Although the logarithm grows arbitrarily large as x grows large, the growth is very slow. On the other hand exponential growth is very rapid.

- Vertical Asymptote: $x = 0$

The logarithm approaches the line $x = 0$ from the right since,

$$\log_a x \to -\infty \quad \text{as} \quad x \to 0^+.$$

Notice that the logarithm function is not defined at $x = 0$, and since the domain is only positive real numbers, x can only approach 0 from the right side. This is the reason for the $+$ exponent on the 0.

- x-intercept: $(1, 0)$,

The graph crosses the x-axis when $x = 1$, since

$$\log_a x = 0 \text{ if and only if } a^0 = x, \text{ so } x = 1.$$

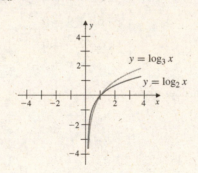

EXAMPLE 6

Sketch the graph of the function.
(a) $y = 2 - \ln(x - 1)$ (b) $y = \ln(-x + 1)$

Solution (a) To sketch the graph we will use the following steps.

- Use $y = \ln x$ as the basic graph.
- Plot $y = -\ln x$.
- Plot $y = -\ln(x - 1)$.
- Plot $y = 2 - \ln(x - 1)$.

The graph of $y = -\ln x$ is obtained by reflecting the basic graph of $y = \ln x$ about the x-axis. So

$$-\ln x \to \infty \quad \text{as} \quad x \to 0^+$$
$$-\ln x \to -\infty \quad \text{as} \quad x \to \infty.$$

Now shift $y = -\ln x$ to the right 1 unit to obtain the graph of $y = -\ln(x - 1)$. Finally, shift the graph of $y = -\ln(x - 1)$ upward 2 units to obtain the graph of $y = 2 - \ln(x - 1)$.

- Vertical asymptote: $x = 1$

● x-intercept:

$$2 - \ln(x-1) = 0$$
$$-\ln(x-1) = -2$$
$$\ln(x-1) = 2$$
$$e^{\ln(x-1)} = e^2$$
$$x - 1 = e^2$$
$$x = e^2 + 1 \approx 8.4$$

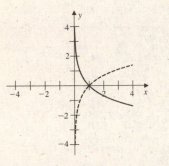   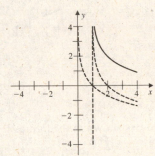

(b) The graph of $y = \ln(-x)$ is the reflection of the graph of $y = \ln x$ about the y-axis. To obtain $y = \ln(-x+1) = \ln(-(x-1))$, shift the graph of $y = \ln(-x)$ to the right 1 unit.

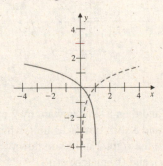

□

EXAMPLE 7

Match the equation with the curve.

(a) $y = -\ln x + 2$ (b) $y = \ln(x-2) - 1$

(c) $y = \ln(-x) + 1$ (d) $y = 2\ln(x-2) - 1$

(i)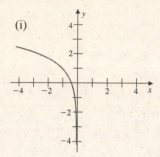

(ii)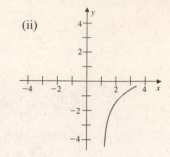

(iii)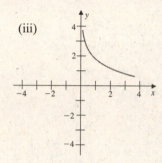

(iv)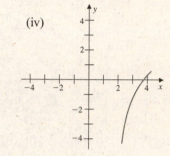

Solution The graph of $y = \ln x$ is shown in the figure.

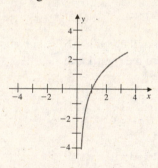

(a) A reflection of $y = \ln x$ about the x-axis and then a shift upward by 2 units, which is (iii).

(c) A reflection of $y = \ln x$ about the y-axis and a shift upward 1 unit, which is (i).

Both (b) and (d) involve a shift of $y = \ln x$ to the right 2 units and downward 1 unit. Since $y = 2\ln(x - 2) - 1$ also involves a vertical stretching, (d) matches (iv), and (b) matches (ii).   □

EXAMPLE 8

Determine the length of time it takes an initial investment to triple in value if it earns 10% compounded continuously.

Solution The value of the investment after t years is given by the exponential formula

$$A_c(t) = A_0 e^{0.1t},$$

where A_0 is some initial investment and 0.1 represents the 10% interest rate. To determine the length of time it takes the initial investment to triple solve for time t in the equation

$$3A_0 = A_0 e^{0.1t}$$
$$3 = e^{0.1t}$$

$$\ln 3 = \ln e^{0.1t}$$

$$\ln 3 = 0.1t$$

$$t = \frac{\ln 3}{0.1} = \frac{\ln 3}{\frac{1}{10}} = 10 \ln 3 \text{ years} \approx 11 \text{ years.}$$

$\square$

Exercise Set 5.3 (Page 303)

1. $x = \log_4 4^3$ implies $4^x = 4^3$ so $x = 3$

3. $x = \log_4 64$ implies $4^x = 64$ so $x = 3$

5. $x = \log_4 2$ implies $4^x = 2$ so $x = \frac{1}{2}$

7. $x = \log_{10} 0.001$ implies $10^x = 0.001$ or $10^x = \frac{1}{1000}$ so $x = -3$

9. $x = \log_2 \frac{1}{8}$ implies $2^x = \frac{1}{8}$ so $x = -3$

11. Since $e^{\ln x} = x$ for $x > 0$ we have $e^{\ln 5} = 5$.

13. Since $\ln e^x = x$ for all x we have $\ln e^{\frac{1}{3}} = \frac{1}{3}$.

15. Since $e^{\ln x} = x$ for $x > 0$ we have $e^{2\ln \pi} = e^{\ln \pi^2} = \pi^2$.

17. $\ln x(x+1) = \ln x + \ln(x+1)$

19. $\log_3 \frac{x^4}{x+1} = \log_3 x^4 - \log_3(x+1) = 4\log_3 x - \log_3(x+1)$

21. $\ln \frac{2x^3}{(x+4)^2} = \ln 2x^3 - \ln(x+4)^2 = \ln 2 + \ln x^3 - 2\ln(x+4) = \ln 2 + 3\ln x - 2\ln(x+4)$

23. We have

$$\log_3 \frac{(3x+2)^{3/2}(x-1)^3}{x\sqrt{x+1}} = \log_3\left[(3x+2)^{3/2}(x-1)^3\right] - \log_3\left(x\sqrt{x+1}\right)$$

$$= \log_3(3x+2)^{3/2} + \log_3(x-1)^3 - \log_3 x - \log_3(x+1)^{1/2}$$

$$= \frac{3}{2}\log_3(3x+2) + 3\log_3(x-1) - \log_3 x - \frac{1}{2}\log_3(x+1).$$

25. We have

$$\ln\sqrt{x\sqrt{x+1}} = \ln\left(x(x+1)^{1/2}\right)^{1/2} = \frac{1}{2}\ln\left(x(x+1)^{1/2}\right)$$

$$= \frac{1}{2}\left[\ln x + \ln(x+1)^{1/2}\right] = \frac{1}{2}\ln x + \frac{1}{4}\ln(x+1).$$

27. $\ln x + 2\ln(x+1) = \ln x + \ln(x+1)^2 = \ln\left(x(x+1)^2\right)$

29. $\frac{1}{2}\ln x - 2\ln(x-1) = \ln x^{1/2} - \ln(x-1)^2 = \ln\frac{\sqrt{x}}{(x-1)^2}$

31. $\ln(x-1) + \frac{1}{2}\ln x - 2\ln x = \ln(x-1)\sqrt{x} - \ln x^2 = \ln\frac{(x-1)\sqrt{x}}{x^2} = \ln\frac{x-1}{x^{3/2}}$

33. $\log_3 x = 4$ implies $3^4 = x$ so $x = 81$

35. $\log_2(3x-4) = 3$ implies $2^3 = 3x - 4$ or $3x = 12$ so $x = 4$

37. $\log_x 4 = 2$ implies $x^2 = 4$ or $x = \pm 2$ so $x = 2$

39. $\ln(2-x) = 4$ implies $e^{\ln(2-x)} = e^4$ or $2 - x = e^4$ so $x = 2 - e^4$

41. $\ln 2 + \ln(x+1) = \ln(4x-7)$ implies $\ln 2(x+1) = \ln(4x-7)$
so $2x + 2 = 4x - 7$ implies $2x = 9$ and $x = \frac{9}{2}$

43. $2\ln x = \ln(4x+6) - \ln 2$ implies $\ln x^2 = \ln\frac{4x+6}{2}$
so $x^2 = 2x + 3$ implies $x^2 - 2x - 3 = (x-3)(x+1) = 0$ and $x = 3, x = -1$
Hence $x = 3$, since $\ln x$ and $\ln(4x-6)$ are not defined at $x = -1$, so $x = -1$ is not a solution.

45. $\ln(2x - 1) - \ln(x - 1) = \ln 5$ implies $\ln \frac{2x-1}{x-1} = \ln 5$

so $\frac{2x-1}{x-1} = 5$ implies $2x - 1 = 5x - 5$ so $3x = 4$ and $x = \frac{4}{3}$

47. $\log_3(2x^2 + 17x) = 2$ implies $2x^2 + 17x = 9$ or $0 = 2x^2 + 17x - 9 = (2x - 1)(x + 9)$

so $2x - 1 = 0, x = -9$ implies $x = \frac{1}{2}, x = -9$

49. $4^x = 3$ implies $\ln 4^x = \ln 3$ so $x \ln 4 = \ln 3$ and $x = \frac{\ln 3}{\ln 4} = \log_4 3$

51. $e^{2x} = 3^{x-4}$ implies $\ln e^{2x} = \ln 3^{x-4}$ or $2x \ln e = (x - 4) \ln 3$

so $2x = (\ln 3)x - 4 \ln 3$ implies $(\ln 3 - 2)x = 4 \ln 3$ and $x = \frac{4 \ln 3}{\ln 3 - 2}$

53. $2 \cdot 3^{-x} = 2^{3x}$ implies that $\ln(2 \cdot 3^{-x}) = \ln 2^{3x}$ or $\ln 2 - x \ln 3 = 3x \ln 2$

so $\ln 2 = 3x \ln 2 + x \ln 3 = x(3 \ln 2 + \ln 3)$ and $x = \frac{\ln 2}{3 \ln 2 + \ln 3}$

55. $y = \log_2(x - 3)$

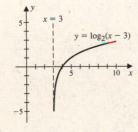

57. $y = 2 - \log_2(x - 1)$

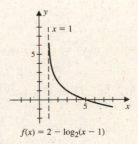

59. $y = 2\ln(x + 1) - 3$

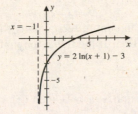

61. $y = \ln(-x)$

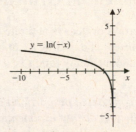

63. $y = |\ln x|$

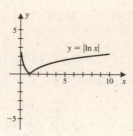

65. a. iii   **b.** iv   **c.** i   **d.** ii

67. $y = \ln(4 - x^2)$

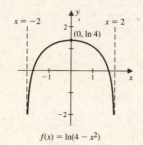

69. $y = (\ln x)/x$

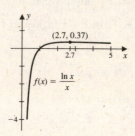

71. The function $g(x) = \sqrt[n]{x}$ grows more rapidly than $f(x) = a + \ln x$ for all $n > 0$.

73. About $\ln 2$, $\ln 1.02 = 35$ years.

. . . .

5.4 Exponential Growth and Decay

If a quantity grows or decays at a rate that is directly proportional to the amount of the quantity that is present, then the quantity present at any time t can be modeled by an exponential function. If the initial amount of the quantity is Q_0, then the amount at any time t is

$$Q(t) = Q_0 e^{kt},$$

where k is the *constant of proportionality* that depends on the specific situation. If $k > 0$, then *Q grows exponentially*, and if $k < 0$, then *Q decays exponentially*.

EXAMPLE 1
A bacteria culture starts with 300 bacteria and 6 hours later has 3000 bacteria.
(a) Find an expression for the number of bacteria after t hours.

(b) Find the number of bacteria that will be present after 8 hours.

(c) When will the population reach 20000?

(d) How long does it take the population to double in size?

Solution (a) Use the information given to find the specific values for Q_0 and k. Since the initial amount of bacteria is 300,

$$Q_0 = 300.$$

To find k, use the fact that after 6 hours, that is, when $t = 6$, there are 3000 bacteria present. Then $Q(6) = 3000$, so

$$3000 = Q(6) = 300e^{k(6)}$$

$$e^{6k} = \frac{3000}{300} = 10$$

$$\ln e^{6k} = \ln 10$$

$$6k = \ln 10$$

$$k = \frac{1}{6}\ln 10.$$

The number of bacteria after t hours is

$$Q(t) = 300e^{\frac{1}{6}\ln 10 \, t}.$$

(b) After $t = 8$ hours the number of bacteria present is

$$Q(6) = 300e^{\frac{1}{6}\ln 10 \, t} \approx 6463.$$

Notice that the expression for $Q(6)$ can be rewritten in the form

$$300e^{\frac{1}{6}\ln 10 \, t(8)} = 300e^{(\ln 10)\frac{8}{6}} = 300\left(e^{\ln 10}\right)^{\frac{4}{3}}$$

$$= 300(10)^{\frac{4}{3}}.$$

(c) Find the value for time t so that $Q(t) = 20000$.

$$Q(t) = 300e^{\frac{1}{6}\ln 10 \, t} = 20000$$

$$e^{\frac{1}{6}\ln 10 \, t} = \frac{20000}{300} = \frac{200}{3}$$

$$\ln e^{\frac{1}{6}\ln 10 \, t} = \ln\frac{200}{3}$$

$$\frac{1}{6}\ln 10 \, t = \ln\frac{200}{3}$$

$$t = \frac{6\ln\frac{200}{3}}{\ln 10} \approx 11 \text{ hours}$$

(d) Find the time t so that $Q(t) = 600$.

$$Q(t) = 300e^{\frac{1}{6}\ln 10 \, t} = 600$$

$$e^{\frac{1}{6}\ln 10t} = 2$$

$$\frac{1}{6}\ln 10t = \ln 2$$

$$t = \frac{6\ln 2}{\ln 10} \approx 1.8 \text{ hours}$$

□

EXAMPLE 2

A radioactive substance has an initial mass of 50 mg and a half-life of approximately 25 years.
(a) Find an expression for the mass after t hours.
(b) How much will remain after 100 years?
(c) When will the mass decay to 10 mg?

Solution (a) The *half-life* of a radioactive substance is the amount of time it takes for one half of the substance to decay. To find the constant of proportionality, use the fact that the half-life is 25 years, so that at time $t = 25$ years the mass of the substance will be 25 mg.

$$Q(t) = 50e^{kt}$$

$$25 = Q(25) = 50e^{25k}$$

$$e^{25k} = \frac{25}{50} = \frac{1}{2}$$

$$25k = \ln\frac{1}{2} = \ln 1 - \ln 2$$

$$25k = -\ln 2$$

$$k = -\frac{\ln 2}{25}$$

The quantity at time t is

$$Q(t) = 50e^{-\frac{1}{25}\ln 2\, t}.$$

Note that we used the arithmetic property of logarithms,

$$\ln\left(\frac{x}{y}\right) = \ln x - \ln y$$

and the fact that $\ln 1 = 0$. Note also that $\ln 2 > 0$, so the proportionality constant is less than 0, and the exponential function represents exponential decay.
(b)

$$Q(100) = 50e^{-\frac{1}{25}\ln 2(100)} = 50e^{-4\ln 2}$$

$$\approx 3.125 \text{ mg}$$

(c) Solving for t in the equation

$$10 = Q(t) = 50e^{-\frac{1}{25}\ln 2\, t}$$

$$e^{-\frac{1}{25}\ln 2t} = \frac{1}{5}$$

$$-\frac{1}{25}\ln 2t = \ln \frac{1}{5} = \ln 1 - \ln 5$$

$$-\frac{1}{25}\ln 2t = -\ln 5$$

$$t = \frac{25\ln 5}{\ln 2} \approx 58.1 \text{ years}$$

The computer generated graph in the figure shows the decaying mass function.

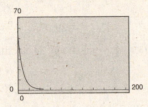

□

EXAMPLE 3
Find the half-life of a radioactive substance that decays 2% in 10 years.

Solution The quantity of the radioactive substance present at time t is

$$Q(t) = Q_0 e^{kt}.$$

If the substance decays 2% in 10 years, then after 10 years there still remains 98% of the original amount. This gives the equation

$$0.98Q_0 = Q(10) = Q_0 e^{10k},$$

from which the proportionality constant k can be found. The actual initial amount of the substance is not needed.

$$0.98Q_0 = Q(10) = Q_0 e^{10k}$$

$$e^{10k} = 0.98$$

$$10k = \ln(0.98)$$

$$k = \frac{1}{10}\ln 0.98$$

$$Q(t) = Q_0 e^{\frac{1}{10}\ln 0.98\, t}$$

To find the half-life, find t so that

$$\frac{1}{2}Q_0 = Q_0 e^{\frac{1}{10}\ln 0.98\, t}$$

$$e^{\frac{1}{10}\ln 0.98\, t} = \frac{1}{2}$$

$$\frac{1}{10}\ln 0.98\, t = -\ln 2$$

$$t = -\frac{10\ln 2}{\ln 0.98} \approx 343.1 \text{ years.}$$

Notice the number of years is positive, since $\ln 0.98 < 0$.   □

EXAMPLE 4

The parents of a newly born child put \$20,000 into an account with the hope that the amount will grow to \$150,000 when the child starts college in 18 years. What rate of continuously compounded interest is necessary for this goal to be met?

Solution If an initial amount of A_0 dollars is invested at an interest rate i compounded continuously, the investment after t years has a value

$$A_c(t) = A_0 e^{it}.$$

If the parents start with $A_0 = 20000$ and after 18 years want an accumulation of $A_c(18) = 150000$, the required interest rate can be found from the equation

$$150000 = 20000 e^{18i}$$
$$7.5 = e^{18i}$$
$$\ln 7.5 = \ln e^{18i}$$
$$\ln 7.5 = 18i$$
$$i = \frac{\ln 7.5}{18} \approx 0.11 \text{ or } 11\%.$$

□

Note: the "rule of 70" states that an investment that pays $i\%$ interest compounded continuously will double in $\frac{70}{i}$ years.

Exercise Set 5.4 (Page 310)

1. We have $Q(t) = Q_0 e^{kt}$ with $Q_0 = 2000$.

   a. Since the bacteria doubles every 3 hours,

   $$2(2000) = Q(3) = 2000 e^{3k} \text{ implies } e^{3k} = 2 \text{ so } 3k = \ln 2 \text{ and } k = \frac{\ln 2}{3},$$

   and $Q(t) = 2000 e^{\frac{\ln 2}{3} t}$.

   b. $Q(6) = 2000 e^{\frac{\ln 2}{3} 6} = 8000$

   c. $2000 e^{\frac{\ln 2}{3} t} = 22000$ implies $e^{\frac{\ln 2}{3} t} = 11$ so $\frac{\ln 2}{3} t = \ln 11$ and $t = \frac{3\ln 11}{\ln 2} \approx 10.4$ hours

3. We have $Q(t) = Q_0 e^{kt}$ with $Q_0 = 500$.

    a. Since after 5 hours there are 4000 bacteria,

$$4000 = Q(5) = 500e^{5k} \text{ implies } e^{5k} = 8 \text{ or } \ln e^{5k} = \ln 8 \text{ so } 5k = \ln 8 \text{ and } k = \frac{\ln 8}{5},$$

and $Q(t) = 500e^{\frac{\ln 8}{5}t}$.

    b. $Q(6) = 500e^{\frac{\ln 8}{5}6} \approx 6063$

    c. $500e^{\frac{\ln 8}{5}t} = 15000$ implies $e^{\frac{\ln 8}{5}t} = 30$ so $\frac{\ln 8}{5}t = \ln 30$ and $t = \frac{5\ln 30}{\ln 8} \approx 8.2$ hours

    d. $500e^{\frac{\ln 8}{5}t} = 1000$ implies $e^{\frac{\ln 8}{5}t} = 2$ so $\frac{\ln 8}{5}t = \ln 2$ and $t = \frac{5\ln 2}{\ln 8} \approx 1.7$ hours

5. We have $Q(t) = Q_0 e^{kt}$ with $Q_0 = 64$.

    a. If the half life is 578 hours,

$$32 = Q(578) = 64e^{578k} \text{ implies } e^{578k} = \frac{1}{2} \text{ so } 578k = \ln\frac{1}{2} = -\ln 2 \text{ and } k = -\frac{\ln 2}{578},$$

and $Q(t) = 64e^{-\frac{\ln 2}{578}t}$.

    b. $Q(75) = 64e^{-\frac{\ln 2}{578}75} \approx 58.5$ milligrams

    c. $64e^{-\frac{\ln 2}{578}t} = 12$ implies $e^{-\frac{\ln 2}{578}t} = \frac{3}{16}$ so $-\frac{\ln 2}{578}t = \ln\frac{3}{16} = \ln 3 - \ln 16$ and $t = -\frac{578[\ln 3 - \ln 16]}{\ln 2} \approx 1395.9$ hours

7. We have $Q(t) = Q_0 e^{kt}$. If the culture doubles in size in 2 hours, then

$$2Q_0 = Q(2) = Q_0 e^{2k} \text{ implies } e^{2k} = 2 \text{ so } 2k = \ln 2 \text{ and } k = \frac{\ln 2}{2},$$

and $Q(t) = Q_0 e^{\frac{\ln 2}{2}t}$. The culture will triple in size when,

$$Q_0 e^{\frac{\ln 2}{2}t} = 3Q_0 \text{ implies } e^{\frac{\ln 2}{2}t} = 3 \text{ so } \frac{\ln 2}{2}t = \ln 3 \text{ and } t = \frac{2\ln 3}{\ln 2} \approx 3.2 \text{ hours.}$$

9. We have $Q(t) = 220e^{kt}$. If the mass decays to 200 grams in 4 years, then

$$Q(4) = 220e^{4k} = 200 \text{ implies } e^{4k} = \frac{200}{220} = \frac{10}{11} \text{ and } 4k = \ln\frac{10}{11} = \ln 10 - \ln 11.$$

So $k = \frac{\ln 10 - \ln 11}{4}$ and $Q(t) = 220e^{\frac{\ln 10 - \ln 11}{4}t}$.

The half life is the time required for half the substance to decay, so

$$110 = 220e^{\frac{\ln 10 - \ln 11}{4}t} \text{ implies } e^{\frac{\ln 10 - \ln 11}{4}t} = \frac{1}{2} \text{ and } \frac{\ln 10 - \ln 11}{4}t = \ln\frac{1}{2} = -\ln 2.$$

So $t = -\frac{4\ln 2}{\ln 10 - \ln 11} \approx 29.09$ years.

11. a. Let $Q(t)$ represent the population t years after 1950. Since the initial population is the 1950 statistic, $Q(0) = 2555$. Since 1960 is ten years after the initial date of 1950,

$$3040 = Q(10) = 2555e^{10k} \text{ implies } e^{10k} = \frac{3040}{2555} \text{ and } 10k = \ln\frac{3040}{2555}.$$

So $k = \frac{1}{10}\ln\frac{3040}{2555}$ and $Q(t) = 2555e^{\frac{1}{10}\ln\frac{3040}{2555}t}$.

The population in 2050, which is 100 years after the initial year of 1950, is

$$Q(100) = 2555e^{\frac{1}{10}\left(\ln\frac{3040}{2555}\right)100} \approx 14528 \text{ million.}$$

b. Since

$$Q(0) = 5275 \text{ implies } 6079 = Q(10) = 5275e^{10k} \text{ we have } e^{10k} = \frac{6079}{5275} \text{ and } k = \frac{1}{10}\ln\frac{6079}{5275},$$

and

$$Q(t) = 5275e^{\frac{1}{10}\ln\frac{6079}{5275}t}.$$

The population in 2050, which is 60 years after the initial year of 1990, is

$$Q(60) = 5275e^{\frac{1}{10}\left(\ln\frac{6079}{5275}\right)60} \approx 12356 \text{ million.}$$

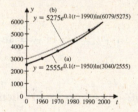

13. We have $A(t) = 10000e^{0.08t}$.

 a. $20000 = 10000e^{0.08t}$ implies $2 = e^{0.08t}$ so $0.08t = \ln 2$ and $t = \frac{\ln 2}{0.08} = 8.7$ years

 b. $30000 = 10000e^{0.08t}$ implies $3 = e^{0.08t}$ so $0.08t = \ln 3$ and $t = \frac{\ln 3}{0.08} = 13.7$ years

15. We have $A(t) = 10000e^{it}$. Then
$25000 = A(5) = 10000e^{5i}$ implies $e^{5i} = \frac{5}{2}$ so $5i = \ln\frac{5}{2}$ and $i = \frac{1}{5}\ln\frac{5}{2} \approx 0.18$ or 18%.

17. We have $T(t) = T_m + (T_0 - T_m)e^{kt}$, where $T_0 = -3°C$ and $T_m = 20°C$, so $T(t) = 20 - 23e^{kt}$. Since one minute later the temperature reads $5°C$,

$$5 = T(1) = 20 - 23e^{k} \text{ implies } e^{k} = \frac{15}{23} \text{ so } k = \ln\frac{15}{23},$$

and $T(t) = 20 - 23e^{\ln\frac{15}{23}t}$. The thermometer will read $19.5°C$, when

$$20 - 23e^{\ln\frac{15}{23}t} = 19.5 \text{ implies } e^{\ln\frac{15}{23}t} = \frac{0.5}{23} = \frac{1}{46} \text{ and } \ln\frac{15}{23}t = \ln\frac{1}{46} = -\ln 46.$$

So $t = -\frac{\ln 46}{\ln\frac{15}{23}} \approx 9$ minutes.

19. By Newton's Law of Cooling the temperature of the body at time t is

$$T(t) = T_m + (T_0 - T_m)e^{kt},$$

where $T_m = 62°F$, the constant temperature of the lake, and $T_0 = 67°C$, the temperature when the body was found at 11:00 a.m. So $T(t) = 62 + 5e^{kt}$. At noon, or 1 hour after the body was found the temperature was $66°F$, so

$$66 = T(1) = 62 + 5e^k \text{ implies } 5e^k = 4 \text{ so } e^k = \frac{4}{5} \text{ and } k = \ln\frac{4}{5},$$

so $T(t) = 62 + 5e^{\ln\frac{4}{5}t}$. The victim died at the time t when the temperature of the body was $98.6°$, so

$$98.6 = 62 + 5e^{\ln\frac{4}{5}t} \text{ implies } e^{\ln\frac{4}{5}t} = \frac{36.6}{5} \text{ so } \ln\frac{4}{5}t = \ln\frac{36.6}{5},$$

and

$$t = \frac{\ln\frac{36.6}{5}}{\ln\frac{4}{5}} \approx -8.9 \text{ hours.}$$

So the death occurred about 8.9 hours before 11:00 a.m., at about 2:06 a.m.

21. The amount of the original carbon 14 left after t years is $A(t) = A(0)e^{-t\ln 2/5730}$. Solving for t when $A(t) = 0.3397A(0)$ gives

$$t = -\frac{\ln 0.3397}{\ln 2} \cdot 5730 \approx 8925 \text{ years.}$$

Hence the oysters were alive in about 6920 B.C.E.

··

Review Exercises for Chapter 5 (Page 311)

1. a. iii   **b.** i   **c.** iv   **d.** ii

3. $f(x) = 2^{x-1} - 3$ **5.** $f(x) = e^{-x} - 2$

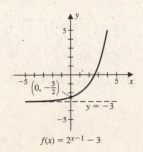

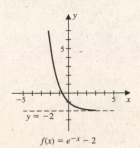

7. $y = 3e^{1-x} = 3e^{-(x-1)}$

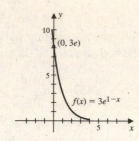

9. $f(x) = 2\ln x$

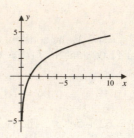

11. $f(x) = 3 - \log_2(x+1)$

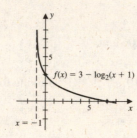

13. $f(x) = e^{-x^2+5x-6}$

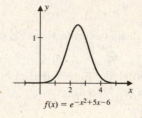

15. $x = \log_5 1$ implies $5^x = 1$ so $x = 0$

17. Since $2^{\log_2 x} = x$ for $x > 0$, we have $2^{\log_2 15} = 15$.

19. $x = \log_4 2$ implies $4^x = 2$ so $x = \frac{1}{2}$

21. Since $e^{\ln x} = x$ for $x > 0$, we have $e^{3\ln 4} = e^{\ln 4^3} = 64$.

23. $\ln \frac{3x^2}{\sqrt{x-1}} = \ln 3x^2 - \ln(x-1)^{1/2} = \ln 3 + \ln x^2 - \frac{1}{2}\ln(x-1) = \ln 3 + 2\ln x - \frac{1}{2}\ln(x-1)$

25. We have

$$
\begin{aligned}
\log_{10} \frac{\sqrt{x+1}\,\sqrt[3]{x-1}}{x(x+3)^{5/2}} &= \log_{10} \frac{(x+1)^{1/2}(x-1)^{1/3}}{x(x+3)^{5/2}} \\
&= \log_{10}\left((x+1)^{1/2}(x-1)^{1/3}\right) - \log_{10}\left(x(x+3)^{5/2}\right) \\
&= \log_{10}(x+1)^{1/2} + \log_{10}(x-1)^{1/3} - \log_{10}x - \log_{10}(x+3)^{5/2} \\
&= \frac{1}{2}\log_{10}(x+1) + \frac{1}{3}\log_{10}(x-1) - \log_{10}x - \frac{5}{2}\log_{10}(x+3)
\end{aligned}
$$

27. $\ln\left(x^{4/3}(x+1)^{1/3}(x-1)^2\right)$

29. We have

$$3\ln(x^3+2)+\ln 5-\frac{1}{2}\ln(x^5-1)=\ln(x^3+2)^3+\ln 5-\ln\sqrt{x^5-1}$$

$$=\ln\left(5(x^3+2)^3\right)-\ln\sqrt{x^5-1}=\ln\frac{5(x^3+2)^3}{\sqrt{x^5-1}}$$

31. $\ln(2x-3)=4$ implies that $e^{\ln(2x-3)}=e^4$. So $2x-3=e^4$ and $x=\frac{e^4+3}{2}$

33. We have $\ln(2x-1)+\ln(3x-2)=\ln 7$ which implies that $\ln\left((2x-1)(3x-2)\right)=\ln 7$. So $(2x-1)(3x-2)=7$ which implies that $6x^2-7x-5=(2x+1)(3x-5)=0$ and $x=-\frac{1}{2}$, That is, $x=\frac{5}{3}$. Hence $x=\frac{5}{3}$, since the natural logarithms are not defined at $x=-\frac{1}{2}$.

35. $3^x\cdot 5^{x-2}=3^{4x}$ which implies that $\ln\left(3^x5^{x-2}\right)=\ln 3^{4x}$, or that $\ln 3^x+\ln 5^{x-2}=4x\ln 3$. So $x\ln 3+(x-2)\ln 5=4x\ln 3$, and this implies that $x(\ln 3+\ln 5-4\ln 3)=2\ln 5$ and

$$x=\frac{2\ln 5}{\ln 3+\ln 5-4\ln 3}=\frac{2\ln 5}{\ln 5-3\ln 3}.$$

37. We have $2e^xx^2-e^xx=e^x$ which impies that $2e^xx^2-e^xx-e^x=e^x(2x^2-x-1)=2x^2-x-1=0$, since $e^x>0$ for all x. So $(2x+1)(x-1)=0$ and $x=-\frac{1}{2}$ or $x=1$.

39. For $e^{x^2}=x-2$ there are no solutions since the graphs never intersect.

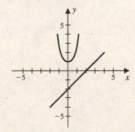

41. We have $e^x > x^4$ on $(-0.8, 1.4) \cup (8.6, \infty)$.

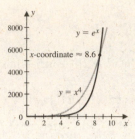 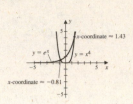

43. We have $e^{x-1} - 3 < x^5$ on $(-1.2, 14.3)$.

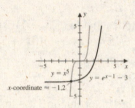 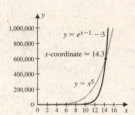

45. For $f(x) = x^2 e^{1-x^2}$, we have the following. Increasing: $(-\infty, -1) \cup (0, 1)$; decreasing: $(-1, 0) \cup (1, \infty)$; local maximums: $(-1, 1)$ and $(1, 1)$; local minimum: $(0, 0)$.

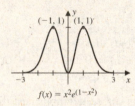

47. The value of an initial investment A_0 deposited at 6% compounded continuously is $A(t) = A_0 e^{0.06t}$. If the investment is to double,

$$2A_0 = A_0 e^{0.06t} \text{ implies } e^{0.06t} = 2 \text{ so } 0.06t = \ln 2 \text{ and } t = \frac{\ln 2}{0.06} \approx 11.6 \text{ years.}$$

49. We have $Q(t) = Q_0 e^{kt}$.

a. To find k use the initial information given. That is,

$$1000 = Q(1) = Q_0 e^{k(1)} \text{ implies } e^k = \frac{1000}{Q_0} \text{ so } k = \ln \frac{1000}{Q_0},$$

and

$$3000 = Q(4) = Q_0 e^{k(4)} \text{ implies } e^{4k} = \frac{3000}{Q_0} \text{ so } 4k = \ln \frac{3000}{Q_0}$$

and $4 \ln \frac{1000}{Q_0} = \ln \frac{3000}{Q_0}$. Hence

$$\left(\frac{1000}{Q_0}\right)^4 = \frac{3000}{Q_0} \text{ implies } Q_0^3 = \frac{10^{12}}{3000} = \frac{10^9}{3} \text{ so } Q_0 = \frac{10^3}{\sqrt[3]{3}}.$$

Then $k = \ln \sqrt[3]{3} = \frac{1}{3} \ln 3$, and $Q(t) = \frac{10^3}{\sqrt[3]{3}} e^{\frac{\ln 3}{3} t}$.

b. $Q(5) = \frac{10^3}{\sqrt[3]{3}} e^{\frac{\ln 3}{3} 5} \approx 4327$.

c. Since

$$20000 = \frac{10^3}{\sqrt[3]{3}} e^{\frac{\ln 3}{3} t} \text{ implies } e^{\frac{\ln 3}{3} t} = 20\sqrt[3]{3} \text{ so } \frac{\ln 3}{3} t = \ln 20\sqrt[3]{3},$$

and

$$t = \frac{3 \ln 20\sqrt[3]{3}}{\ln 3} \approx 9.2 \text{ hours.}$$

d. $3Q_0 = Q(t) = Q_0 e^{\frac{\ln 3}{3} t}$ implies $3 = e^{\frac{\ln 3}{3} t}$ so $\ln 3 = \frac{\ln 3}{3} t$ and $t = 3$ hours

51. We have $A_4(t) = A_0 \left(1 + \frac{0.1}{4}\right)^{4t}$, and the time it takes for the investment to double can be found from

$$2A_0 = A_0 \left(1 + \frac{0.1}{4}\right)^{4t} \text{ implies } 2 = \left(1 + \frac{0.1}{4}\right)^{4t} \text{ so } \ln 2 = 4t \ln\left(1 + \frac{0.1}{4}\right)$$

$$\text{and } t = \frac{\ln 2}{4 \ln\left(1 + \frac{0.1}{4}\right)} \approx 7 \text{ years.}$$

53. We have $A_c(t) = A_0 e^{0.09t}$, and the time it takes for the investment to double can be found from

$$2A_0 = A_0 e^{0.09t} \text{ implies } 2 = e^{0.09t} \text{ so } \ln 2 = 0.09t \text{ and } t = \frac{\ln 2}{0.09} \approx 7.7 \text{ years.}$$

Chapter 5 Exercises for Calculus (Page 314)

1. a. The graphs of $f(x) = 2 \ln x$ and $g(x) = e^{\frac{x}{2}}$ are reflections of one another through $y = x$, so $f = g^{-1}$.

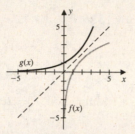

b. The graphs of $f(x) = \ln \frac{x}{2}$ and $g(x) = e^{2x}$ are not reflections of one another through $y = x$, so $f \neq g^{-1}$.

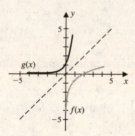

c. Neither $f(x) = \ln |x|$ nor $g(x) = e^{|x|}$ are $1 - 1$ functions, and hence can not have inverses.

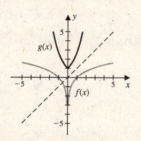

d. The graphs of $f(x) = -\ln x$ and $g(x) = e^{-x}$ are reflections of one another through $y = x$, so $f = g^{-1}$.

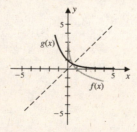

e. The graphs of $f(x) = 1 + \ln x$ and $g(x) = e^{x-1}$ are reflections of one another through $y = x$, so $f = g^{-1}$.

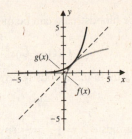

f. The graphs of $f(x) = 2 \ln x$ and $g(x) = \frac{1}{2} e^x$ are not reflections of one another through $y = x$, so $f \neq g^{-1}$.

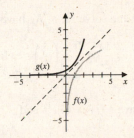

3. To order the functions according to how fast they grow as x approaches ∞, plot pairs of functions together for large x to place them in order. For example, to order $\ln x$, x^x and e^{3x}, first plot $y = \ln x$ and $y = x^x$ to see that $y = x^x$ grows faster. Then compare $y = e^{3x}$ and $y = x^x$ to see that $y = x^x$ grows faster. Finally compare $y = e^{3x}$ and $y = \ln x$ to see that $y = e^{3x}$ grows faster, so the ordering of these three functions is $\ln x$, e^{3x}, x^x. The complete ordering from smallest to largest as x approaches ∞ is

$$\frac{x^{10}}{e^x}, \quad \frac{1}{x^4}, \quad \ln x, \quad x^{1/20}, \quad x^{20}, \quad e^{3x}, \quad \frac{e^{6x}}{x^8}, \quad x^x.$$

5. We have $y = 3e^{x-2} = 3e^x e^{-2} = \frac{3}{e^2} e^x$, so the graph of $y = 3e^{x-2}$ is just a vertical scaling of the graph of $y = e^x$. Since $0 < \frac{3}{e^2} < 1$, it is a vertical compression.

7. We have $y = 3 + \ln 2x = 3 + \ln 2 + \ln x = (3 + \ln 2) + \ln x$, so the graph of $y = 3 + \ln 2x$ is a just a vertical translation of the graph of $y = \ln x$. Since $3 + \ln 2 > 0$, the shift is upward.

9. If the interest on an initial investment of A_0 dollars is compounded continuously at a fixed rate of $r\%$, then the value after t years is $A(t) = A_0 e^{\frac{r}{100}t}$. Then the time at which the investment doubles is given by,

$$2A_0 = A_0 e^{\frac{r}{100}t} \text{ implies } 2 = e^{\frac{r}{100}t} \text{ so } \frac{r}{100}t = \ln 2 \text{ and } t = \frac{100 \ln 2}{r} \approx \frac{70}{r}.$$

So $\frac{70}{r}$ is a reasonable estimate for the time it takes for the investment to double in value. For example, if the interest rate is 8.75%, then $\frac{70}{8.75} = 8.0$.

---

11. a. The concentration of the drug in the bloodstream can be modeled using exponential decay. So the concentration at time t has the form

$$C(t) = C_0 e^{kt}.$$

Since the initial concentration of the drug in the bloodstream is 20, we have $C_0 = 20$. If 3 hours later the concentration is 12,

$$12 = C(3) = 20e^{3k} \text{ implies } e^{3k} = \frac{12}{20} = \frac{3}{5} \text{ so } 3k = \ln\left(\frac{3}{5}\right) \text{ and } k = \frac{1}{3}\ln\left(\frac{3}{5}\right).$$

Then the concentration at time t is

$$C(t) = 20e^{\frac{1}{3}\ln\left(\frac{3}{5}\right)t}.$$

b. To find the half-life, set $C(t) = 10$ and solve for t. We have

$$10 = 20e^{\frac{1}{3}\ln\left(\frac{3}{5}\right)t} \text{ implies } e^{\frac{1}{3}\ln\left(\frac{3}{5}\right)t} = \frac{1}{2} \text{ so } \frac{1}{3}\ln\left(\frac{3}{5}\right)t = \ln\frac{1}{2} = -\ln 2 \text{ and } t = -\frac{3\ln 2}{\ln\left(\frac{3}{5}\right)} \approx 4.07 \text{ hours.}$$

c. Since the half-life is 5 hours, we have $C(t) = C_0 e^{kt}$ and

$$\frac{1}{2}C_0 = C(5) = C_0 e^{5k} \text{ implies } e^{5k} = \frac{1}{2} \text{ so } 5k = \ln\left(\frac{1}{2}\right) = -\ln 2 \text{ and } k = -\frac{\ln 2}{5}.$$

So

$$C(t) = C_0 e^{-\frac{\ln 2}{5}t}.$$

For a 25 kilogram dog, the amount $Q(t)$ of phenobarbital in the blood at time t is $25C(t)$. That is,

$$Q(t) = 25C_0 e^{-\frac{\ln 2}{5}t}.$$

When $t = 1$ hour, we want $Q(t) = (30)(25) = 750$. So

$$750 = Q(1) = 25C_0 e^{-\frac{\ln 2}{5}} \text{ so } C_0 = 30e^{\frac{\ln 2}{5}} \approx 34.46 \text{ mg/kg.}$$

Since the dog's weight is equivalent to 25 kilograms, the initial dose should be $(34.46)(25) = 861.5$ milligrams.

Chapter 5 Chapter Test (Page 315)

1. True.

2. False. When evaluated, $2 \log_a a^{1/2}$ is 1.

3. False. The solution to the equation $\log_2 x = 5$ is 32.

4. False. The solution to the equation $\log_3 x = 2$ is 9.

5. True.

6. True.

7. True.

8. False. The range of the function $f(x) = 1 + e^{x-2}$ is $(1, \infty)$.

9. True.

10. True.

11. False. The only solution to the equation $3^{2x+5} = 27$ is $x = -1$.

12. True.

13. True.

14. True.

15. True.

16. True.

17. False. The only solution to the equation $\ln x + \ln(x - 1) = \ln(4x + 6)$ is $x = 6$, since $x = -1$ is not in the domain.

18. False. The only solution to the equation $\ln(x + 6) - \ln(x + 1) = \ln(x - 2)$ is $x = 4$.

19. False. The graph of $y = e^x$ has a horizontal asymptote $y = 0$.

20. True.

21. False. The graph of $y = -1 + e^{x-2}$ is obtained by shifting the graph of $y = e^x$ to the right 2 units and downward 1 unit.

22. False. The graph of $y = e^{x+3}$ can be obtained by shifting the graph of $y = e^x$ to the left 1 unit and vertically stretching the result by a factor of e^2.

23. False. The graph of $y = \ln x$ has a vertical asymptote $x = 0$.

24. False. The graph of $y = -3 + \ln(x - 1)$ has a vertical asymptote $x = 1$.

25. False. The graph of $y = 2 + \ln(x - 1)$ is obtained by shifting the graph of $y = \ln x$ to the right 1 unit and upward 2 units.

26. False. The graph of $y = \ln(2x - 1)$ is obtained by shifting the graph of $y = \ln x$ to the right $\frac{1}{2}$ units and upward $\ln 2$ units.

27. True.

28. False. The red graph is given by $y = e^{x+1}$.

29. True.

30. False. The yellow graph is given by $y = 1 + 2e^{x-1}$.

31. False. The blue graph is given by $y = \ln(x - 1)$.

32. True.

33. False. The green graph is given by $y = -1 - \ln(x + 1)$.

34. False. The yellow graph is given by $y = 1 + 2\ln(x - 1)$.

35. True.

36. True.

37. False. If a bacteria culture starts with 1000 bacteria and after 3 hours there are 2500 bacteria, the number of bacteria at any time t is given by $Q(t) = 1000e^{(t \ln 2.5)/3}$.

38. False. If a radioactive substance decays 5% in 10 years, half of the initial mass will decay in $t = -\frac{10 \ln 2}{\ln(0.95)}$ years.

39. False. An initial investment deposited in an account returning 7% interest compounded continuously will double in approximately 10 years (rounded to the nearest year).

40. True.

CHAPTER 6: Conic Sections, Polar Coordinates, and Parametric Equations

6.1 Introduction

The graphs of the general *quadratic equation* in x and y,

$$Ax^2 + Bxy + Cy^2 + Dx + Ey + F = 0,$$

are called conic sections. The three basic figures are the parabola, ellipse and hyperbola, though certain special, degenerate curves can also occur. When $B = 0$ and $AC = 0$, the curve is a parabola, when $B = 0$ and $AC > 0$, the curve is an ellipse, and when $B = 0$ and $AC < 0$, the curve is a hyperbola. When $B \neq 0$, the curve is a rotated conic in the plane.

Polar coordinates and parametric equations provide two additional methods for describing curves in the plane. They allow for the visualization of a greater variety of curves.

6.2 Parabolas

The graph of the familiar equation of the form $y = ax^2 + bx + c$ is a *parabola*, with axis parallel to the y-axis. A more general geometric definition of a parabola is the set of points equidistant from a given point, called the *focus*, and a given line, called the *directrix*. When the directrix is one of the coordinate axes, the parabola is said to be in *standard form*. The *axis* of the parabola is the line through the focal point that is perpendicular to the directrix. The point of intersection of the axis and the parabola is the *vertex*. A useful tip to remember is that a parabola *never* crosses its directrix.

Standard Position Parabolas

Equation	Vertex	Focus	Directrix
$y = \frac{1}{4c}x^2$	$(0, 0)$	$(0, c)$	Horizontal: $y = -c$
$x = \frac{1}{4c}y^2$	$(0, 0)$	$(c, 0)$	Vertical: $x = -c$

When the vertex is shifted from $(0, 0)$ to the point (h, k), that is, h units in the horizontal direction and k units in the vertical direction, the parabola has an equation of the form

$$(y - k) = \frac{1}{4c}(x - h)^2 \text{ or } (x - h) = \frac{1}{4c}(y - k)^2.$$

333

EXAMPLE 1

Find the vertex, directrix and focus, and sketch the graph of the parabola.

(a) $y = -\frac{1}{8}x^2$ (b) $4y^2 = 9x$

Solution (a) The equation of the parabola is already in standard position, so we can read the value of c from the standard position.

$$y = -\frac{1}{8}x^2 = \frac{1}{4c}x^2$$
$$\frac{1}{4c} = -\frac{1}{8}$$
$$4c = -8$$
$$c = -2$$

- Vertex: $(0, 0)$
  The parabola is in standard position with vertex at the origin.
- Focus: $(0, c) = (0, -2)$
  The axis of the parabola is along the y-axis.
- Directrix: $y = -c = 2$
  Since the parabola never crosses its directrix, the parabola opens downward as shown in the figure.
- Maximum or Minimum Value: Since the curve is opening downward, the vertex $(0, 0)$ is a *maximum* point on the curve.
- Increasing: The curve increases on the interval $(-\infty, 0)$.
- Decreasing: The curve is decreasing on the interval $(0, \infty)$.

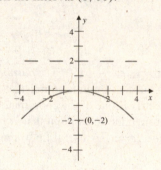

(b) A slight rewriting of the equation will put it in standard position.

$$4y^2 = 9x$$
$$x = \frac{4}{9}y^2$$
$$x = \frac{1}{9/4}y^2$$
$$4c = \frac{9}{4}$$
$$c = \frac{9}{16}$$

The parabola is in standard position with axis along the x-axis.

- Vertex: $(0, 0)$
- Focus: $(c, 0) = \left(\frac{9}{16}, 0\right)$
- Directrix: $x = -c = -\frac{9}{16}$

The parabola never crosses its directrix, so the parabola opens to the right as shown in the figure. Notice that the equation does not define a function. This is seen from the figure, which shows that the graph does not satisfy the vertical line test. For each $a > 0$, the vertical line $x = a$ crosses the curve in two points.

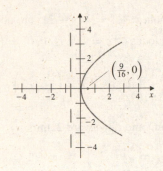

                                              □

EXAMPLE 2

Find the vertex, directrix and focus, and sketch the graph of the parabola.
(a) $y^2 + 4y + 6 - 2x = 0$ (b) $2x^2 + 8x - 3y + 11 = 0$

Solution (a) The first step is to group the x-terms and group the y-terms in order to rewrite the equation in standard form, with perhaps the vertex shifted. Since there are both y and y^2 terms present, completing the square on these terms is necessary. Completing the square on the y terms gives

$$
\begin{aligned}
y^2 + 4y &= y^2 + 4y + \left(\frac{4}{2}\right)^2 - \left(\frac{4}{2}\right)^2 \\
&= y^2 + 4y + 4 - 4 \\
&= (y + 2)^2 - 4.
\end{aligned}
$$

Then

$$
\begin{aligned}
y^2 + 4y + 6 - 2x &= 0 \\
y^2 + 4y &= 2x - 6 \\
(y + 2)^2 - 4 &= 2x - 6 \\
(y + 2)^2 &= 2x - 2 \\
(y + 2)^2 &= 2(x - 1) \\
x - 1 &= \frac{1}{2}(y + 2)^2.
\end{aligned}
$$

The vertex of the parabola is $(1, -2)$, and the graph of the parabola can be obtained from the graph of

$$x = \frac{1}{2}y^2,$$

which is a parabola in standard position. The vertex of this parabola is at the origin, with axis along the x-axis, and

$$4c = 2 \quad \text{so} \quad c = \frac{1}{2}.$$

The focus, vertex, and directrix of the parabola $x - 1 = \frac{1}{2}(y + 2)^2$ are obtained from shifting the focus, vertex, and directrix of the parabola $x = \frac{1}{2}y^2$. Since the focus of $x = \frac{1}{2}y^2$ is $(c, 0) = \left(\frac{1}{2}, 0\right)$, the focus of $x - 1 = \frac{1}{2}(y + 2)^2$ is

$$\left(\frac{1}{2} + 1, 0 - 2\right) = \left(\frac{3}{2}, -2\right),$$

that is, it is the point $\left(\frac{1}{2}, 0\right)$ shifted right 1 unit and downward 2 units.

Parabola	$x = \frac{1}{2}y^2$	$x - 1 = \frac{1}{2}(y + 2)^2$
c	$\frac{1}{2}$	$\frac{1}{2}$
Vertex	$(0, 0)$	$(1, -2)$
Focus	$(c, 0) = \left(\frac{1}{2}, 0\right)$	$\left(\frac{1}{2} + 1, 0 - 2\right) = \left(\frac{3}{2}, -2\right)$
Directrix	$x = -c = -\frac{1}{2}$	$x = -c + 1 = -\frac{1}{2} + 1 = \frac{1}{2}$

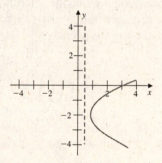

(b) To convert the equation to one that can be compared with a standard position equation complete the square on the x terms.

$$2x^2 + 8x - 3y + 11 = 0$$
$$2(x^2 + 4x) = 3y - 11$$
$$2(x^2 + 4x + 4 - 4) = 3y - 11$$
$$2(x + 2)^2 - 8 = 3y - 11$$
$$2(x + 2)^2 = 3y - 3$$
$$2(x + 2)^2 = 3(y - 1)$$

$$y - 1 = \frac{2}{3}(x + 2)^2$$

The parabola is obtained by shifting, to the left 2 units and upward 1 unit, the parabola $y = \frac{2}{3}x^2$ that is in standard position with axis along the y-axis. Since

$$\frac{1}{4c} = \frac{2}{3}, \quad 4c = \frac{3}{2}, \quad \text{and } c = \frac{3}{8}.$$

Parabola	$y = \frac{2}{3}x^2$	$y - 1 = \frac{2}{3}(x + 2)^2$
c	$\frac{3}{8}$	$\frac{3}{8}$
Vertex	$(0, 0)$	$(-2, 1)$
Focus	$(0, c) = \left(0, \frac{3}{8}\right)$	$\left(-2, 1 + \frac{3}{8}\right) = \left(-2, \frac{11}{8}\right)$
Directrix	$y = -c = -\frac{3}{8}$	$y = -c + 1 = -\frac{3}{8} + 1 = \frac{5}{8}$

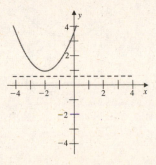

□

EXAMPLE 3

Determine the equation of the parabola that satisfies the given conditions.

(a) Focus at $(-1, 1)$ and directrix $x = 3$.

(b) Vertex at $(3, 4)$ and focus at $(3, 6)$.

Solution (a) The equation of a parabola requires the vertex and the value of c. The vertex of a parabola lies *midway* between the focus and the directrix on the line through the focus and perpendicular to the directrix. The horizontal distance between $(-1, 1)$ and the vertical line $x = 3$ is 4, so the vertex is 2 units to the right of the focus $(-1, 1)$, and hence is the point $(1, 1)$.

If the focus had been $(-2, 0)$ rather than $(-1, 1)$ and the directrix $x = 2$, then the conditions would describe a parabola in standard position with vertex at the origin and equation

$$x = \frac{1}{4(-2)}y^2 = -\frac{1}{8}y^2.$$

The parabola described by the conditions is obtained by a vertical shift upward 1 unit and a horizontal shift to the right 1 unit, of the standard position parabola. So its equation is

$$x - 1 = -\frac{1}{8}(y - 1)^2.$$

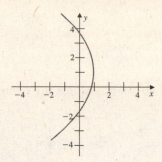

(b) Since the axis of the parabola is parallel to one of the coordinate axes and passes through the vertex (3, 4) and the focus (3, 6), the axis is vertical. The distance between the vertex and the focus is $6 - 4 = 2$ (both points are on the vertical line $x = 3$). Since the focus is above the vertex, the parabola opens upward, with $c = 2$. The directrix is 2 units below the vertex and has equation $y = 2$. The equation of the parabola is

$$y - 4 = \frac{1}{4(2)}(x - 3)^2 = \frac{1}{8}(x - 3)^2.$$

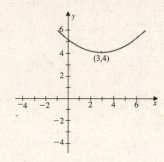

$\square$

EXAMPLE 4

Determine the equation of the parabola that satisfies the given conditions.
(a) Vertex: $(-1, 0)$; axis parallel to y-axis; passing through the point $(4, 10)$.
(b) Vertex: $(-1, 0)$; axis parallel to x-axis; passing through the point $(4, 10)$.

Solution (a) The parabola in standard position with axis the y-axis has equation

$$y = \frac{1}{4c}x^2,$$

and the parabola with vertex at $(-1, 0)$ is given by

$$y = \frac{1}{4c}(x + 1)^2.$$

This uses the first two pieces of information. To use the third, observe that if the curve passes through (4, 10), then

$$10 = \frac{1}{4c}(4 + 1)^2$$

$$10 = \frac{1}{4c} \cdot 25$$

$$\frac{2}{5} = \frac{1}{4c}$$

$$c = \frac{5}{8}.$$

The equation of the parabola is

$$y = \frac{1}{4\left(\frac{5}{8}\right)}(x + 1)^2 \qquad\qquad = \frac{2}{5}(x + 1)^2.$$

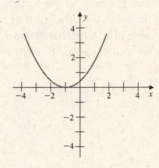

(b) The parabola in standard position with axis the x-axis has equation

$$x = \frac{1}{4c}y^2,$$

and the parabola with vertex at $(-1, 0)$ is given by

$$x + 1 = \frac{1}{4c}y^2.$$

The parabola passes through $(4, 10)$, so

$$4 + 1 = \frac{1}{4c}(10)^2$$
$$5 = \frac{1}{4c} \cdot 100 = \frac{25}{c}$$
$$c = 5.$$

The equation of the parabola is

$$x + 1 = \frac{1}{4(5)}y^2 = \frac{1}{20}y^2.$$

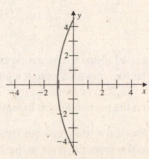

□

EXAMPLE 5

A flash light with a parabolic cross section has a depth of 1 inch and a cross section width of 2.5 inches. Where should the light source be placed to produce a parallel beam of light?

Solution Light rays emitted from the focus of a parabolic surface are reflected off the surface in parallel rays creating a concentrated beam of light. So the light source is placed at the focus of the parabolic reflector. The information given does not allow us to find the location of the focus directly, but we can find the general form for the equation of the parabola. This allows us to determine the value of c, which determines where to place the light.

The parabolic cross section of the light is shown in the figure. Since it is a parabola in standard position with axis along the x-axis, the equation has the form

$$x = \frac{1}{4c}y^2.$$

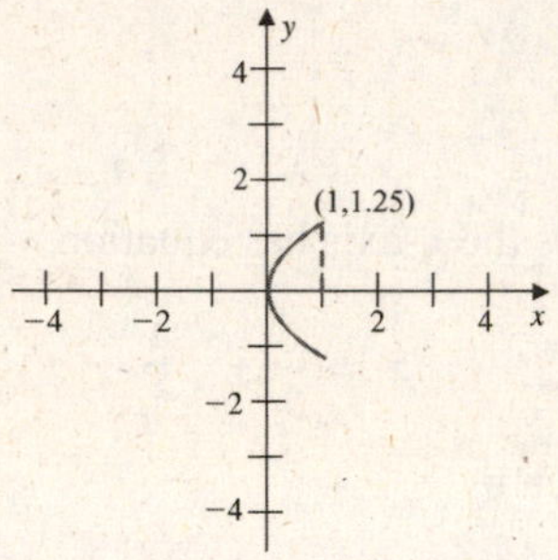

The parabola passes through the point $\left(1, \frac{1}{2}(2.5)\right) = (1, 1.25)$, so

$$1 = \frac{1}{4c}(1.25)^2 = \frac{1}{4c}\left(\frac{25}{16}\right)$$

$$\frac{64}{25} = \frac{1}{c}$$

$$c = \frac{25}{64}.$$

Therefore the focus of the parabola is $(25/64, 0)$. To produce a parallel beam of light, the light source is placed at the focus, which is,

$$\frac{25}{64} \approx 0.39 \text{ inches}$$

from the vertex of the light. □

EXAMPLE 6

A rock thrown horizontally from a bridge into a river follows a parabolic curve with vertex at the bridge and axis along a perpendicular from the river to the bridge. The rock passes through a point 50 feet from the perpendicular from the river to the bridge when it is a vertical distance of 20 feet from the top. How far is the bridge above the river if the rock lands a horizontal distance of 100 feet from the perpendicular from the river to the bridge?

Solution Set up an xy-coordinate system so that the x-axis runs out from the bridge above the path of the rock and the positive y-axis coincides with the perpendicular from the river to the bridge, as shown in the figure.

If we knew the equation of the parabolic path of the rock, then the solution is the value of x when $y = 0$. The information gives one point on the parabola and the standard position for the parabola, which is enough to find the equation.

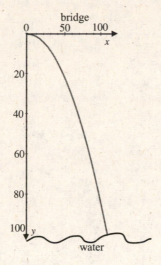

The path of the rock is a parabola in standard position with axis along the y-axis. The equation of the parabola is

$$y = \frac{1}{4c}x^2.$$

Since the rock is 20 feet below the bridge when it is horizontally 50 feet away from the edge, the parabola passes through the point $(50, 20)$. So

$$20 = \frac{1}{4c}(50)^2$$

$$4c = \frac{2500}{20} = 125,$$

and the equation of the parabola is

$$y = \frac{1}{125}x^2.$$

The rock hits the river at the point $(100, h)$, where h is the distance from the water to the bridge. Hence

$$h = \frac{1}{125}(100)^2 = \frac{10000}{125} = 80 \text{ feet.} \qquad \square$$

Exercise Set 6.2 (Page 325)

1. $y = 2x^2 = \frac{1}{4c}x^2$ implies $\frac{1}{4c} = 2$ so $c = \frac{1}{8}$
$V(0, 0)$; $F(0, 1/8)$; directrix, D: $y = -1/8$

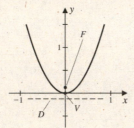

3. $9y = -16x^2$ implies $y = -\frac{16}{9}x^2$
so $\frac{1}{4c} = -\frac{16}{9}$ and $c = -\frac{9}{64}$
$V(0, 0)$; $F(0, -9/64)$; directrix, D: $y = 9/64$

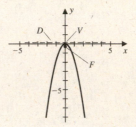

5. $y^2 = 2x$ implies $x = \frac{1}{2}y^2$ so $\frac{1}{4c} = \frac{1}{2}$
and $c = \frac{1}{2}$
$V(0, 0)$; $F(1/2, 0)$; directrix, D: $x = -1/2$

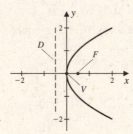

7. $9y^2 = -16x$ implies $x = -\frac{9}{16}y^2$
so $\frac{1}{4c} = -\frac{9}{16}$ and $c = -\frac{4}{9}$
$V(0, 0)$; $F(-4/9, 0)$; directrix, D: $x = 4/9$

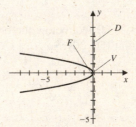

9. $x^2 - 6x + 9 = 2y$ implies $(x - 3)^2 = 2y$
or $y = \frac{1}{2}(x - 3)^2$ so $\frac{1}{4c} = \frac{1}{2}$ and $c = \frac{1}{2}$
$V(3, 0)$; $F(3, 1/2)$; directrix, D: $y = -1/2$

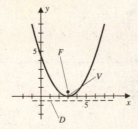

11. $x^2 - 4x - 2y + 2 = x^2 - 4x + 4 - 4 - 2y + 2 = 0$
implies $(x - 2)^2 = 2y + 2$
or $y + 1 = \frac{1}{2}(x - 2)^2$ so $\frac{1}{4c} = \frac{1}{2}$ and $c = \frac{1}{2}$
$V(2, -1)$; $F(2, -1/2)$; directrix, D: $y = -3/2$

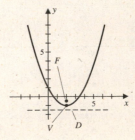

13. $y^2 - 8y + 12 = 2x$ implies $y^2 - 8y + 16 - 16 = 2x - 12$ or $(y - 4)^2 = 2x + 4$
implies $x + 2 = \frac{1}{2}(y - 4)^2$ so $\frac{1}{4c} = \frac{1}{2}$
and $c = \frac{1}{2}$
$V(-2, 4)$; $F(-3/2, 4)$; directrix, D: $x = -5/2$

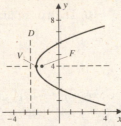

15. $2x^2 + 4x - 9y + 20 = 0$ implies $2(x^2 + 2x + 1 - 1) = 9y - 20$ or $2(x + 1)^2 = 9y - 18$
implies $y - 2 = \frac{2}{9}(x + 1)^2$ so $\frac{1}{4c} = \frac{2}{9}$
and $c = \frac{9}{8}$
$V(-1, 2)$; $F(-1, 25/8)$; directrix, D: $y = 7/8$

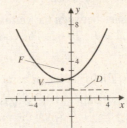

17. $3x^2 - 12x + 4y + 8 = 0$ implies $3(x^2 - 4x + 4 - 4) = -4y - 8$ or $3(x - 2)^2 = -4y + 4$
implies $y - 1 = -\frac{3}{4}(x - 2)^2$
so $\frac{1}{4c} = -\frac{3}{4}$ and $c = -\frac{1}{3}$
$V(2, 1)$; $F(2, 2/3)$; directrix, D: $y = 4/3$

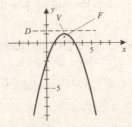

19. The parabola is in standard position with axis along the y-axis, so it has equation in the form $y = \frac{1}{4c}x^2$ and focus $(0, c)$, $c > 0$. Since the point $(2, c)$ lies on the curve, $c = \frac{1}{4c}(2)^2$ implies $c^2 = 1$ so $c = 1$. The equation is $y = \frac{1}{4}x^2$.

21. The parabola is in standard position with axis along the x-axis, so it has equation in the form $x = \frac{1}{4c}y^2$, and since the directrix is $x = -1$, the focus is $(1, 0)$. The equation is $x = \frac{1}{4(1)}y^2$ or $x = \frac{1}{4}y^2$.

23. Focus: $(-2, 2)$; Directrix: $y = -2$

Since the vertex lies midway between the focus and the directrix, and the distance from the focus to the directrix is 4, the vertex is $(-2, 0)$. The axis of the parabola is vertical with $c = 2$, so $y = \frac{1}{8}(x + 2)^2$.

25. Vertex: $(-2, 2)$; Directrix: $x = 2$

Since the vertex lies midway between the focus and the directrix, and the distance from the directrix to the vertex is 4, the focus is $(-6, 2)$. The axis of the parabola is horizontal with $c = -4$, so $x + 2 = \frac{1}{4(-4)}(y - 2)^2$ or $x + 2 = -\frac{1}{16}(y - 2)^2$.

27. Vertex: $(-2, 2)$; Focus: $(-2, 0)$

The distance between the vertex and the focus is 2, so the directrix is $y = 4$. The axis of the parabola is vertical and $c = -2$. The equation is $y - 2 = -\frac{1}{8}(x + 2)^2$.

29. Vertex: $(-2, 2)$; Focus: $(-4, 2)$

The distance between the vertex and the focus is 2, so the directrix is $x = 0$. The axis of the parabola is horizontal and $c = -2$. The equation is $x + 2 = -\frac{1}{8}(y - 2)^2$.

31. a. Vertex: $(0, 0)$; Point on parabola: $(4, 6)$

The axis is parallel to the y-axis, so since the vertex is at the origin, the axis is the y-axis, and the parabola has the form $y = \frac{1}{4c}x^2$. Since the parabola passes through the point $(4, 6)$, $6 = \frac{1}{4c}(4)^2$ implies $6 = \frac{4}{c}$ so $c = \frac{2}{3}$, and the equation of the parabola is $y = \frac{1}{4(2/3)}x^2 = \frac{3}{8}x^2$.

b. Vertex: $(1, 0)$; Point on parabola: $(5, 6)$

The axis is parallel to the y-axis, so since the vertex is at the point $(1, 0)$, the axis is the line $x = 1$, and the parabola has the form $y = \frac{1}{4c}(x - 1)^2$. Since the parabola passes through the point $(5, 6)$, $6 = \frac{1}{4c}(5 - 1)^2$ implies $6 = \frac{4}{c}$ so $c = \frac{2}{3}$, and the equation of the parabola is $y = \frac{1}{4(2/3)}(x - 1)^2 = \frac{3}{8}(x - 1)^2$.

c. Vertex: $(1, 2)$; Point on parabola: $(5, 8)$

The axis is parallel to the y-axis, so since the vertex is at the point $(1, 2)$, the axis is the line $x = 1$, and the parabola has the form $y - 2 = \frac{1}{4c}(x - 1)^2$. Since the parabola passes through the point $(5, 8)$, $8 - 2 = \frac{1}{4c}(5 - 1)^2$ implies $6 = \frac{4}{c}$ so $c = \frac{2}{3}$, and the equation of the parabola is $y - 2 = \frac{3}{8}(x - 1)^2$.

d. Vertex: $(0, 2)$; Point on parabola: $(4, 8)$

The axis is parallel to the y-axis, so since the vertex is at the point $(0, 2)$, the axis is the y-axis, and the parabola has the form $y - 2 = \frac{1}{4c}x^2$. Since the parabola passes through the point $(4, 8)$, $8 - 2 = \frac{1}{4c}(4)^2$ implies $6 = \frac{4}{c}$ so $c = \frac{2}{3}$, and the equation of the parabola is $y - 2 = \frac{3}{8}x^2$.

33. Since the axis of the parabola is the y-axis, the vertex lies on the y-axis and hence is of the form $(0, b)$. The equation of the parabola is $y - b = \frac{1}{4c}x^2$. Since the parabola passes through $(1, 2)$, $2 - b = \frac{1}{4c}(1)^2$ implies $c = \frac{1}{4(2-b)}$, and the equation is $y - b = \frac{1}{4(1/(4(2-b)))}x^2$ or $y - b = (2 - b)x^2, b \neq 2$.

35. To describe the light, use a parabola in standard position and axis along the x-axis, so the equation has the form $x = \frac{1}{4c}y^2, c > 0$. The information implies the point $(2, 2)$ lies on the parabola, so $2 = \frac{1}{4c}(2)^2$ implies $c = \frac{1}{2}$. So the focal point of the parabola is $\left(\frac{1}{2}, 0\right)$, and to produce a parallel beam of light the light source should be placed $\frac{1}{2}$ inch from the vertex of the light.

37. a. The parabolic path of the ball is a parabola with axis along the y-axis and vertex at $(0, 64)$. The equation of the path has the form $y - 64 = \frac{1}{4c}x^2$. The information implies the parabola passes through the point $(100, 64 - 16) = (100, 48)$. So

$$48 - 64 = \frac{1}{4c}(100)^2 \text{ implies } 4c = -\frac{10000}{16} = -625,$$

and

$$y - 64 = -\frac{1}{625}x^2.$$

To find where the ball hits the ground, find x when $y = 0$. That is,

$$-64 = -\frac{1}{625}x^2 \text{ implies } x^2 = 64(625) = 40000 \text{ so } x = \sqrt{40000} = 200.$$

The ball hits the ground 200 feet from the building.

b. Using the same analysis as in part (a), the equation of the parabolic path is

$$y - 1450 = -\frac{1}{625}x^2,$$

and if the distance above the ground is to be 0,

$$-1450 = -\frac{1}{625}x^2 \text{ so } x = \sqrt{(1450)(625)} \approx 952.$$

The ball hits the ground about 952 feet from the building.

6.3   Ellipses

An *ellipse* is a set of points in a plane for which the sum of the distances from two fixed points is a given constant. The two fixed points are called the *focal points*, the line passing through the focal points is called the *axis*, and the points of intersection of the axis and the ellipse are called the *vertices*. An ellipse centered at the origin and with axis along one of the coordinate axes is said to be in *standard position*.

Standard Position Ellipses

Equation: $a > b$	$\frac{x^2}{a^2} + \frac{y^2}{b^2} = 1$	$\frac{y^2}{a^2} + \frac{x^2}{b^2} = 1$
Axis	x-axis	y-axis
Vertices	$(-a, 0), (a, 0)$	$(0, -a), (0, a)$
Focal Points: $c^2 = a^2 - b^2$	$(-c, 0), (c, 0)$	$(0, -c), (0, c)$
Other Intercepts	$y : (0, -b), (0, b)$	$x : (-b, 0), (b, 0)$
Center	$(0, 0)$	$(0, 0)$
Eccentricity: $c = \sqrt{a^2 - b^2}$	$e = \frac{c}{a}$	$e = \frac{c}{a}$

The *center* of an ellipse in standard position is the origin, which is the midpoint of the line segment connecting the vertices. The axis of an ellipse is also called the *major axis*, and the line segment connecting the other intercepts and perpendicular to the axis is called the *minor axis*.

When the center is shifted to the point (h, k), the ellipse will have an equation of the form

$$\frac{(x-h)^2}{a^2} + \frac{(y-k)^2}{b^2} = 1 \quad \text{or} \quad \frac{(y-k)^2}{a^2} + \frac{(x-h)^2}{b^2} = 1.$$

EXAMPLE 1

Find the vertices and focal points and sketch the graph of the ellipse.

(a) $\frac{x^2}{16} + \frac{y^2}{9} = 1$ (b) $16x^2 + 9y^2 = 144$

Solution (a) The equation is in standard position, and since the denominator of the x^2 term is larger than the denominator of the y^2 term, the axis of the ellipse is on the x-axis, with center at the origin.

$$a^2 = 16, \ a = 4$$
$$b^2 = 9, \ b = 3$$
$$c = \sqrt{a^2 - b^2} = \sqrt{16 - 9} = \sqrt{7}$$

- Vertices: $(-a, 0) = (-4, 0)$, $(a, 0) = (4, 0)$
- Focal Points: $(-c, 0) = (-\sqrt{7}, 0)$, $(c, 0) = (\sqrt{7}, 0)$
- y-intercepts: $(0, -b) = (0, -3)$, $(0, b) = (0, 3)$

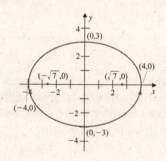

(b) First place the equation in standard position by dividing both sides of the equation by 144.

$$16x^2 + 9y^2 = 144$$
$$\frac{x^2}{9} + \frac{y^2}{16} = 1$$

Notice that the equation is similar to the equation in part (a), except that the roles of a and b are reversed. That is, the axis of the ellipse is the y-axis, although we still have $a = 4$, $b = 3$, and $c = \sqrt{7}$.

- Vertices: $(0, -a) = (0, -4)$, $(0, a) = (0, 4)$
- Focal Points: $(0, -c) = (0, -\sqrt{7})$, $(0, c) = (0, \sqrt{7})$
- x-intercepts: $(-b, 0) = (-3, 0)$, $(b, 0) = (3, 0)$

Notice that in both parts (a) and (b) the length of the major axis is 8, and the length of the minor axis is 6.

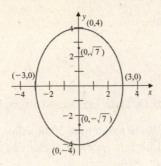

EXAMPLE 2

Find the vertices and focal points and sketch the graph of $4x^2 + 9y^2 + 8x - 36y + 4 = 0$.

Solution The key to recognizing the ellipse is to group the x terms and group the y terms and complete the square on each.

$$4x^2 + 8x + 9y^2 - 36y = -4$$

$$4(x^2 + 2x) + 9(y^2 - 4y) = -4$$

$$4(x^2 + 2x + 1 - 1) + 9(y^2 - 4y + 4 - 4) = -4$$

$$4(x + 1)^2 + 9(y - 2)^2 = -4 + 4 + 36$$

$$4(x + 1)^2 + 9(y - 2)^2 = 36$$

$$\frac{(x + 1)^2}{9} + \frac{(y - 2)^2}{4} = 1$$

The equation describes an ellipse with center $(-1, 2)$. The ellipse is a translation of the standard position ellipse.

$$\frac{x^2}{9} + \frac{y^2}{4} = 1,$$

which has major axis on the x-axis, with $a = 3$, $b = 2$, center $(0, 0)$, and vertices $(-3, 0)$ and $(3, 0)$. The focal points are $(-\sqrt{5}, 0)$ and $(\sqrt{5}, 0)$, since $c = \sqrt{a^2 - b^2} = \sqrt{9 - 4} = \sqrt{5}$.

The shifted ellipse satisfies:

- Major axis: On the horizontal line $y = 2$.
- Center: The point $(-1, 2)$ obtained by shifting the origin $(0, 0)$ to the left 1 unit and upward 2 units.
- Vertices: $(-a - 1, 2) = (-4, 2)$, $(a - 1, 2) = (2, 2)$
- Focal points: $(-c - 1, 2) = (-\sqrt{5} - 1, 2)$, $(c - 1, 2) = (\sqrt{5} - 1, 2)$
- Minor axis: On the vertical line $x = -1$.
- Minor axis intercepts: $(-1, -b + 2) = (-1, 0)$, $(-1, b + 2) = (-1, 4)$

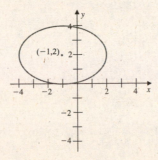

□

EXAMPLE 3

Find an equation of the ellipse that has foci at (3, 1) and (5, 1) and a vertex at (2, 1).

Solution Since the foci are on the line $y = 1$, the major axis of the ellipse is also along the line $y = 1$. We first find c, which is the distance from the midpoint of the line segment connecting the foci to a focus. This midpoint, which is also the center of the ellipse, occurs at

$$\left(\frac{5+3}{2}, \frac{1+1}{2}\right) = (4, 1),$$

so

$$c = 5 - 4 = 1.$$

Since one vertex is (2, 1), which is 1 unit horizontally to the left of the focus (3, 1), the other vertex is one unit to the right of the focus (5, 1) and is (6, 1). The length of the major axis, which is the distance between the vertices, is 4. So a, which is half the length of the major axis, is

$$a = \frac{1}{2}(6 - 2) = 2.$$

Now that we have both a and c, we can find b^2 and write the equation of the ellipse.

$$c^2 = a^2 - b^2$$
$$b^2 = a^2 - c^2$$
$$b^2 = 4 - 1 = 3$$

The equation of the ellipse is

$$\frac{(x-4)^2}{4} + \frac{(y-1)^2}{3} = 1.$$

□

Exercise Set 6.3 (Page 334)

1. $\frac{x^2}{4} + y^2 = 1$ so $a = 2, b = 1, c = \sqrt{4-1} = \sqrt{3}$

Focal points: $\left(\sqrt{3}, 0\right), \left(-\sqrt{3}, 0\right)$

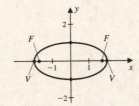

3. $x^2 + \frac{y^2}{9} = 1$ so $a = 3, b = 1, c = \sqrt{9-1} = \sqrt{8} = 2\sqrt{2}$

Focal points: $\left(0, 2\sqrt{2}\right), \left(0, -2\sqrt{2}\right)$

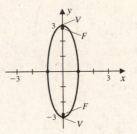

5. $16x^2 + 25y^2 = 400$ implies $\frac{x^2}{25} + \frac{y^2}{16} = 1$ so $a = 5, b = 4, c = \sqrt{25-16} = \sqrt{9} = 3$

Focal points: $(3, 0), (-3, 0)$

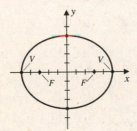

7. $3x^2 + 2y^2 = 6$ implies $\frac{x^2}{2} + \frac{y^2}{3} = 1$ so $a = \sqrt{3}, b = \sqrt{2}, c = 1$

Focal points: $(0, 1), (0, -1)$;

Vertices: $(0, \sqrt{3}), (0, -\sqrt{3})$

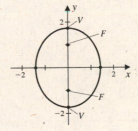

9. $4x^2 + y^2 + 16x + 12 = 0$ implies $4(x^2 + 4x + 4 - 4) + y^2 = -12$ or $4(x+2)^2 + y^2 = 4$

so $(x+2)^2 + \frac{y^2}{4} = 1$ and $a = 2, b = 1, c = \sqrt{4-1} = 3$

Focal points: $\left(-2, \sqrt{3}\right), \left(-2, -\sqrt{3}\right)$; Center: $(-2, 0)$

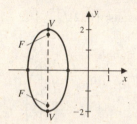

11. $x^2 + 4y^2 - 2x - 16y + 13 = 0$ implies
$x^2 - 2x + 1 - 1 + 4(y^2 - 4y + 4 - 4) = -13$
or $(x-1)^2 + 4(y-2)^2 = 4$
so $\frac{(x-1)^2}{4} + (y-2)^2 = 1$
and $a = 2, b = 1, c = \sqrt{4-1} = \sqrt{3}$
Focal points: $(1 - \sqrt{3}, 2), (1 + \sqrt{3}, 2)$;
Vertices: $(-1, 2), (3, 2)$; Center: $(1, 2)$

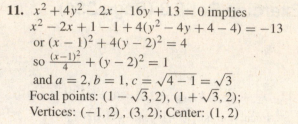

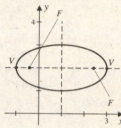

13. $2x^2 + 4y^2 + 4x - 16y + 2 = 0$

implies $2(x^2 + 2x + 1 - 1) + 4(y^2 - 4y + 4 - 4) = -2$ or $2(x+1)^2 + 4(y-2)^2 = 16$

so $\frac{(x+1)^2}{8} + \frac{(y-2)^2}{4} = 1$

and $a = \sqrt{8} = 2\sqrt{2}, b = 2, c = \sqrt{8-4} = 2$

Focal points: $(1, 2), (-3, 2)$; Center: $(-1, 2)$

Vertices: $\left(-1 + 2\sqrt{2}, 2\right), \left(-1 - 2\sqrt{2}, 2\right)$

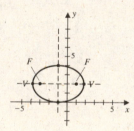

15. x-intercepts: $(\pm 4, 0)$; y-intercepts: $(0, \pm 3)$

The intercepts give the vertices of the ellipse, so the ellipse is in standard position, center at the origin, major axis along the x-axis, and $a = 4, b = 3$. The equation is $\frac{x^2}{16} + \frac{y^2}{9} = 1$.

17. Foci: $(\pm 2, 0)$; Vertices: $(\pm 3, 0)$

Since the foci are centered about the origin and on the x-axis, the ellipse is in standard position with center at the origin and major axis on the x-axis. So

$$a = 3, c = 2, \text{ and } c^2 = a^2 - b^2 \text{ implies } 4 = 9 - b^2 \text{ so } b^2 = 5.$$

The equation is $\frac{x^2}{9} + \frac{y^2}{5} = 1$.

19. Foci: $(0, \pm 1)$; x-intercepts: $(\pm 2, 0)$

Since the foci are centered about the origin and on the y-axis, the ellipse is in standard position with center at the origin and major axis on the y-axis. So

$$b = 2, c = 1, \text{ and } c^2 = a^2 - b^2 \text{ implies } 1 = a^2 - 4 \text{ so } a^2 = 5.$$

The equation is $\frac{y^2}{5} + \frac{x^2}{4} = 1$.

21. Length of major axis 5; Length of minor axis 3; Foci on the x-axis;

Since the ellipse is in standard position, $a = \frac{5}{2}$, $b = \frac{3}{2}$, and the equation is $\frac{x^2}{25/4} + \frac{y^2}{9/4} = 1$ or $\frac{4x^2}{25} + \frac{4y^2}{9} = 1$.

23. Foci: $(3, 0), (1, 0)$; Vertex: $(0, 0)$

Since the foci are on the x-axis, the major axis of the ellipse is on the x-axis. The foci are centered about $(2, 0)$, so $c = 1$. Since a vertex is $(0, 0)$, the other vertex is $(4, 0)$, and the length of the major axis 4. This gives $a = 2$. Then

$$c^2 = a^2 - b^2 \text{ implies } 1 = 4 - b^2 \text{ so } b^2 = 3.$$

The equation is $\frac{(x-2)^2}{4} + \frac{y^2}{3} = 1$.

25. Focus: $(-4, 0)$; Vertices: $(-4, -2), (-4, 8)$

Since the vertices are on the vertical line $x = -4$, the major axis of the ellipse is on the line $x = -4$. The vertices are 10 units apart, which is the length of the major axis, so $a = 5$. The center is $(-4, 3)$, so $c = 3$. Then

$$c^2 = a^2 - b^2 \text{ implies } 9 = 25 - b^2 \text{ so } b^2 = 16.$$

The equation is $\frac{(x+4)^2}{16} + \frac{(y-3)^2}{25} = 1$.

27. Vertices: $(3, 3), (3, -1)$; Passing through: $(2, 1)$

The center is midway between the two vertices, so is $(3, 1)$, and $a = 4$. The ellipse has the form

$$\frac{(x-3)^2}{b^2} + \frac{(y-1)^2}{16} = 1.$$

If the ellipse passes through $(2, 1)$, then

$$\frac{(2-3)^2}{b^2} + \frac{(1-1)^2}{16} = 1 \text{ so } b^2 = 1.$$

The equation is $(x-3)^2 + \frac{(y-1)^2}{16} = 1$.

29. Let P be the point of intersection of the latus rectum and the upper half of the ellipse. The length of the latus rectum is then $2y$, where the coordinates of $P = (c, y)$. Since the equation of the ellipse is

$$\frac{x^2}{a^2} + \frac{y^2}{b^2} = 1 \text{ implies } \frac{y^2}{b^2} = 1 - \frac{x^2}{a^2} \text{ so } y = \pm\sqrt{b^2\left(1 - \frac{x^2}{a^2}\right)}.$$

Letting $x = c$, multiplying by 2, and noting $b^2 = a^2 - c^2$, the length of the latus rectum is then

$$2\sqrt{b^2\left(1 - \frac{c^2}{a^2}\right)} = 2\sqrt{b^2\left(\frac{a^2 - c^2}{a^2}\right)} = 2\sqrt{\frac{b^4}{a^2}} = \frac{2b^2}{a}.$$

31. Since the major axis has length 480,

$$2a = 480 \quad \text{implies} \quad a = 240 \quad \text{so} \quad a^2 = 57600,$$

and since the minor axis has length 280,

$$2b = 280 \quad \text{implies} \quad b = 140 \quad \text{so} \quad b^2 = 19600.$$

The equation is $\frac{x^2}{57600} + \frac{y^2}{19600} = 1$.

33. The length of the major axis of the satellites orbit is the diameter of the earth, 12760, plus 160 plus 16000, which equals 28920. Then $a = \frac{28920}{2} = 14460$, $c = 14460 - 6540 = 7920$

so $e = \frac{c}{a} = \frac{7920}{14460} \approx 0.6$. Then $b^2 = a^2 - c^2 \approx 146365200$. Since $a^2 = 14460^2 = 209091600$, the equation of the orbit is $\frac{x^2}{209091600} + \frac{y^2}{146365200} = 1$.

· ·

6.4   Hyperbolas

A *hyperbola* is a set of points in the plane for which the magnitude of the difference between the distances from two fixed points is a given constant. The two fixed points are called the *focal points*, the line passing through the focal points is called the *axis*, and the points of intersection of the axis with the hyperbola are called the *vertices*. A hyperbola centered at the origin and with axis along one of the coordinate axes is in *standard position*.

Standard Position Hyperbolas

Equation: $a > b$	$\frac{x^2}{a^2} - \frac{y^2}{b^2} = 1$	$\frac{y^2}{a^2} - \frac{x^2}{b^2} = 1$
Axis	x-axis	y-axis
Vertices	$(-a, 0), (a, 0)$	$(0, -a), (0, a)$
Focal Points: $c^2 = a^2 + b^2$	$(-c, 0), (c, 0)$	$(0, -c), (0, c)$
Center	$(0, 0)$	$(0, 0)$
Asymptotes	$y = \pm\frac{b}{a}x$	$y = \pm\frac{a}{b}x$
Eccentricity: $c = \sqrt{a^2 + b^2}$	$e = \frac{c}{a}$	$e = \frac{c}{a}$

If the center of the hyperbola in standard position is shifted to the point (h, k), then the hyperbola will have an equation of the form

$$\frac{(x-h)^2}{a^2} - \frac{(y-k)^2}{b^2} = 1 \quad \text{or} \quad \frac{(y-k)^2}{a^2} - \frac{(x-h)^2}{b^2} = 1,$$

and the asymptotes become, respectively,

$$y - k = \pm\frac{b}{a}(x-h) \quad \text{or} \quad y - k = \pm\frac{a}{b}(x-h).$$

EXAMPLE 1
Find the vertices, focal points, eccentricity, and equations of the asymptotes of the hyperbola. Then sketch the graph.
(a) $\frac{x^2}{9} - \frac{y^2}{16} = 1$ (b) $4y^2 - 25x^2 = 100$

Solution (a) The equation describes a parabola in standard position. Since the y^2 term is negative, the axis of the hyperbola is along the x-axis with center at the origin.

$$a^2 = 9, \ a = 3$$
$$b^2 = 16, \ b = 4$$
$$c = \sqrt{a^2 + b^2} = \sqrt{9 + 16} = 5$$

- Vertices: $(-a, 0) = (-3, 0)$, $(a, 0) = (3, 0)$
- Focal points: $(-c, 0) = (-5, 0)$, $(c, 0) = (5, 0)$
- Eccentricity: $e = \frac{c}{a} = \frac{5}{3}$
- Asymptotes: $y = \pm\frac{b}{a}x = \pm\frac{4}{3}x$

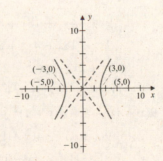

(b) First place the hyperbola in standard position by dividing both sides of the equation by 100.

$$4y^2 - 25x^2 = 100$$
$$\frac{y^2}{25} - \frac{x^2}{4} = 1$$

Since the x^2 term is negative, the axis is along the y-axis.

$$a^2 = 25, \ a = 5$$
$$b^2 = 4, \ b = 2$$
$$c = \sqrt{a^2 + b^2} = \sqrt{29}$$

- Vertices: $(0, -a) = (0, -5)$, $(0, a) = (0, 5)$
- Focal points: $(0, -c) = (0, -\sqrt{29})$, $(0, c) = (0, \sqrt{29})$
- Eccentricity: $e = \frac{c}{a} = \frac{\sqrt{29}}{5}$
- Asymptotes: $y = \pm \frac{a}{b} x = \pm \frac{5}{2} x$

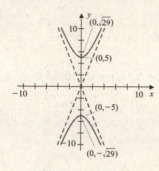

EXAMPLE 2

Find the vertices, focal points, eccentricity, and equations of the asymptotes, and sketch the graph of the hyperbola $9x^2 - 4y^2 + 18x + 8y = 31$.

Solution To compare the equation with the equation of a hyperbola in standard position, complete the square on both the x terms and on the y terms.

$$9x^2 - 4y^2 + 18x + 8y = 31$$
$$9x^2 + 18x - 4y^2 + 8y = 31$$
$$9(x^2 + 2x) - 4(y^2 - 2y) = 31$$
$$9(x^2 + 2x + 1 - 1) - 4(y^2 - 2y + 1 - 1) = 31$$
$$9(x + 1)^2 - 9 - 4(y - 1)^2 + 4 = 31$$
$$9(x + 1)^2 - 4(y - 1)^2 = 36$$
$$\frac{(x + 1)^2}{4} - \frac{(y - 1)^2}{9} = 1$$

The graph of the hyperbola is obtained from shifting the graph of the hyperbola in standard position, $\frac{x^2}{4} - \frac{y^2}{9} = 1$, to the left 1 unit and upward 1 unit. The standard position hyperbola has $a = 2$, $b = 3$, $c = \sqrt{13}$, vertices $(-2, 0)$, $(2, 0)$, focal points $(-\sqrt{13}, 0)$, $(\sqrt{13}, 0)$, and asymptotes $y = \pm \frac{b}{a} x = \pm \frac{3}{2} x$. The hyperbola in this problem has the following properties.

- Vertices: $(-a - 1, 1) = (-3, 1)$, $(a - 1, 1) = (1, 1)$
- Center: $(-1, 1)$
- Focal points: $(-c - 1, 1) = (-\sqrt{13} - 1, 1)$, $(c - 1, 1) = (\sqrt{13} - 1, 1)$
- Eccentricity: $e = \frac{c}{a} = \frac{\sqrt{13}}{2}$
- Asymptotes:

$$y - 1 = \pm\frac{b}{a}(x + 1)$$

$$y - 1 = \pm\frac{3}{2}(x + 1)$$

$$y = \frac{3}{2}x + \frac{5}{2}, \; y = -\frac{3}{2}x - \frac{1}{2}.$$

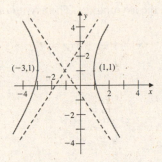

□

EXAMPLE 3

Find an equation of the hyperbola that satisfies the conditions.

(a) Foci at $(\pm7, 0)$, vertices at $(\pm3, 0)$.

(b) Foci at $(6, 1)$ and $(-2, 1)$, equations of asymptotes $y = \frac{3}{4}x - \frac{1}{2}$ and $y = -\frac{3}{4}x + \frac{5}{2}$.

Solution   (a) Since the foci and vertices are on the x axis and centered about the origin, the hyperbola is in standard position with axis on the x axis and equation of the form

$$\frac{x^2}{a^2} - \frac{y^2}{b^2} = 1.$$

The vertices are $(\pm a, 0) = (\pm3, 0)$, so $a = 3$. The foci are $(\pm c, 0) = (\pm7, 0)$, so $c = 7$. To find the equation we need the value b^2, which we obtain from the equation

$$c^2 = a^2 + b^2$$
$$b^2 = c^2 - a^2$$
$$= 49 - 9$$
$$= 40.$$

The equation of the hyperbola is

$$\frac{x^2}{9} - \frac{y^2}{40} = 1.$$

(b) Since the foci both lie on the horizontal line $y = 1$, the axis of the hyperbola is parallel to the x-axis and the equation has the form

$$\frac{(x - h)^2}{a^2} - \frac{(y - k)^2}{b^2} = 1,$$

where (h, k) is the center. The center is midway between the foci, so the center is the point

$$\left(\frac{6-2}{2}, \frac{1+1}{2}\right) = (2, 1),$$

and the equation can be written as

$$\frac{(x-2)^2}{a^2} - \frac{(y-1)^2}{b^2} = 1.$$

To find a and b we use the information from the asymptotes. The asymptotes are of the form

$$y - 1 = \pm\frac{b}{a}(x - 2),$$

so

$$\frac{b}{a} = \frac{3}{4}$$
$$b = \frac{3a}{4}.$$

Another equation is needed to find the two unknowns a and b. The value c is half the distance between the foci, so

$$c = \frac{1}{2} \cdot 8 = 4,$$

and

$$c^2 = a^2 + b^2$$
$$16 = a^2 + \left(\frac{3a}{4}\right)^2$$
$$16 = a^2 + \frac{9a^2}{16}$$
$$16 = \frac{25a^2}{16}$$
$$a^2 = \frac{256}{25}$$
$$b^2 = \frac{9a^2}{16} = \frac{9}{16} \cdot \frac{256}{25} = \frac{144}{25}.$$

The equation of the hyperbola is

$$\frac{25(x-2)^2}{256} - \frac{25(y-1)^2}{144} = 1.$$

$\square$

Exercise Set 6.4 (Page 342)

1. $\frac{x^2}{4} - \frac{y^2}{9} = 1$ implies $a = 2, b = 3$
so $c^2 = a^2 + b^2 = 13$ and $c = \sqrt{13}$;
Vertices: $(2, 0)$, $(-2, 0)$;
Foci: $\left(\sqrt{13}, 0\right)$, $\left(-\sqrt{13}, 0\right)$;
Asymptotes: $y = \pm\frac{3}{2}x$

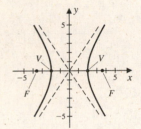

3. $\frac{y^2}{4} - \frac{x^2}{9} = 1$ implies $a = 2, b = 3$
so $c^2 = 13$ and $c = \sqrt{13}$;
Vertices: $(0, 2)$, $(0, -2)$;
Foci: $\left(0, \sqrt{13}\right)$, $\left(0, -\sqrt{13}\right)$;
Asymptotes: $y = \pm\frac{2}{3}x$

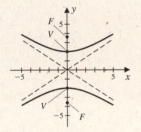

5. $x^2 - y^2 = 1$ implies $a = 1, b = 1$
so $c^2 = 2$ and $c = \sqrt{2}$;
Vertices: $(1, 0)$, $(-1, 0)$;
Foci: $\left(\sqrt{2}, 0\right)$, $\left(-\sqrt{2}, 0\right)$;
Asymptotes: $y = \pm x$

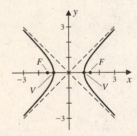

7. $9y^2 - 18y - 4x^2 = 9(y^2 - 2y + 1 - 1) - 4x^2 = 27$
implies $9(y - 1)^2 - 4x^2 = 36$ or $\frac{(y-1)^2}{4} - \frac{x^2}{9} = 1$.
So $a = 2, b = 3$ implies $c^2 = 13$ so $c = \sqrt{13}$.
Vertices: $(0, 3)$, $(0, -1)$;
Foci: $\left(0, 1 + \sqrt{13}\right)$, $\left(0, 1 - \sqrt{13}\right)$;
Asymptotes: $y - 1 = \pm\frac{2}{3}x$ or $y = \pm\frac{2}{3}x + 1$.

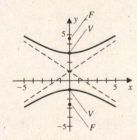

9. $3x^2 - y^2 = 6x$ implies $3(x^2 - 2x + 1 - 1) - y^2 = 0$ so $3(x - 1)^2 - y^2 = 3$

or $(x - 1)^2 - \frac{y^2}{3} = 1$.

So $a = 1, b = \sqrt{3}$ implies $c^2 = 4$ so $c = 2$.

Vertices: $(0, 0)$, $(2, 0)$; Foci: $(3, 0)$, $(-1, 0)$;

Asymptotes: $y = \pm\sqrt{3}(x - 1)$

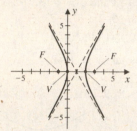

11. $9x^2 - 4y^2 - 18x - 8y = 9(x^2 - 2x + 1 - 1) - 4(y^2 + 2y + 1 - 1) = 31$

implies $9(x - 1)^2 - 4(y + 1)^2 = 36$ or $\frac{(x-1)^2}{4} - \frac{(y+1)^2}{9} = 1$.

So $a = 2, b = 3$ implies $c^2 = 13$ so $c = \sqrt{13}$.

Vertices: $(3, -1)$, $(-1, -1)$;

Foci: $\left(1 - \sqrt{13}, -1\right)$, $\left(1 + \sqrt{13}, -1\right)$;

Asymptotes: $y + 1 = \pm\frac{3}{2}(x - 1)$

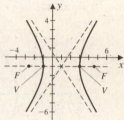

13. $9y^2 - 4x^2 - 36y + 16x - 16 = 0$ implies $9(y^2 - 4y + 4 - 4) - 4(x^2 - 4x + 4 - 4) = 16$

so $9(y - 2)^2 - 4(x - 2)^2 = 36$

or $\frac{(y-2)^2}{4} - \frac{(x-2)^2}{9} = 1$.

So $a = 2, b = 3$, and $c = \sqrt{13}$.

Vertices: $(2, 4)$, $(2, 0)$;

Foci: $\left(2, 2 + \sqrt{13}\right)$, $\left(2, 2 - \sqrt{13}\right)$

Asymptotes: $y - 2 = \pm\frac{2}{3}(x - 2)$

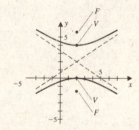

15. Foci: $(\pm 5, 0)$; Vertices: $(\pm 3, 0)$

Since the foci are centered about the origin and on the x-axis the axis is on the x-axis, $a = 3, c = 5$, and

$$c^2 = a^2 + b^2 \text{ implies } 25 = 9 + b^2 \text{ so } b^2 = 16.$$

The equation is $\frac{x^2}{9} - \frac{y^2}{16} = 1$.

17. Foci: $(0, \pm 5)$; Vertices: $(0, \pm 4)$

Since the foci are centered about the origin and on the y-axis, the axis is on the y-axis, $a = 4, c = 5$, and

$$c^2 = a^2 + b^2 \text{ implies } 25 = 16 + b^2 \text{ so } b^2 = 9.$$

The equation is $\frac{y^2}{16} - \frac{x^2}{9} = 1$.

19. Focus: $(2, 2)$; Vertices: $(2, 1), (2, -3)$

    Since the focus and the vertices are on the line $x = 2$, the axis is parallel to the y-axis. Since the vertices are centered about $(2, -1)$, $a = 2$, and since the focus $(2, 2)$ is 3 units above the center $(2, -1)$, $c = 3$. Then

    $$c^2 = a^2 + b^2 \text{ implies } 9 = 4 + b^2 \text{ so } b^2 = 5.$$

    The equation is $\frac{(y+1)^2}{4} - \frac{(x-2)^2}{5} = 1$.

21. Foci: $(-1, 4), (5, 4)$; Vertex: $(0, 4)$

    Since the foci and the one vertex are on the line $y = 4$, the axis is parallel to the x-axis. Since the foci are centered about the point $(2, 4)$, $c = 3$, and since the vertex $(0, 4)$ is 2 units to the left of the center $(2, 4)$, $a = 2$. Then

    $$c^2 = a^2 + b^2 \text{ implies } 9 = 4 + b^2 \text{ so } b^2 = 5.$$

    The equation is $\frac{(x-2)^2}{4} - \frac{(y-4)^2}{5} = 1$.

23. Vertices: $(0, \pm 2)$; Passing through: $(3, 4)$

    Since the vertices are centered about the origin and on the y-axis, the hyperbola is in standard position with axis on the y-axis and has the form $\frac{y^2}{a^2} - \frac{x^2}{b^2} = 1$. Then $a = 2$, and since the hyperbola passes through $(3, 4)$, we have

    $$\frac{(4)^2}{4} - \frac{(3)^2}{b^2} = 1 \text{ implies } \frac{9}{b^2} = 3 \text{ so } b = \sqrt{3}.$$

    The equation is $\frac{y^2}{4} - \frac{x^2}{3} = 1$.

25. Vertices: $(\pm 3, 0)$; Asymptotes: $y = \pm \frac{4}{3} x$.

    Since the vertices are centered about the origin on the x-axis, the hyperbola is in standard position with axis on the x-axis. The vertices imply $a = 3$, and the asymptotes imply

    $$\frac{b}{a} = \frac{4}{3} \text{ implies } \frac{b}{3} = \frac{4}{3} \text{ so } b = 4.$$

    The equation is $\frac{x^2}{9} - \frac{y^2}{16} = 1$.

27. Let $P(x, y)$ be the point of intersection of the latus rectum and the hyperbola as shown in the figure. The length is then $2y$, where the coordinates of $P = (c, y)$. Since the equation of the hyperbola is

    $$\frac{x^2}{a^2} - \frac{y^2}{b^2} = 1 \text{ implies } \frac{y^2}{b^2} = \frac{x^2}{a^2} - 1 \text{ so } y = \pm \sqrt{b^2 \left(\frac{x^2}{a^2} - 1 \right)}.$$

    Letting $x = c$, multiplying by 2, and noting $b^2 = c^2 - a^2$, the length of the latus rectum is

    $$2\sqrt{b^2 \left(\frac{c^2}{a^2} - 1 \right)} = 2\sqrt{b^2 \left(\frac{c^2 - a^2}{a^2} \right)} = 2\sqrt{\frac{b^4}{a^2}} = \frac{2b^2}{a}.$$

29. The axes of the hyperbolas are perpendicular to each other.

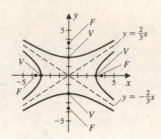

31. If the total cost of production and delivery from plants A and B to a destination C are denoted T_A and T_B, then

$$T_A - T_B = 130 + d(A, C) - d(B, C).$$

If $d(B, C) = d(A, C) + 130$, then $T_A - T_B = 0$, and it makes no difference from which plant the car is shipped. If $d(B, C) < d(A, C) + 130$, then

$$T_A - T_B > 0 \text{ so } T_A > T_B,$$

and the car should be shipped from plant B. If $d(B, C) > d(A, C) + 130$, then

$$T_A - T_B < 0 \text{ so } T_A < T_B,$$

and the car should be shipped from plant A.

· ·

6.5   Polar Coordinates

In the rectangular coordinate system, a point in the plane is specified by an ordered pair (x, y) that describes the location of the point using a rectangular grid. The first coordinate specifies the vertical distance to the x-axis along the line perpendicular to the x-axis, and the second coordinate specifies the horizontal distance to the y-axis along the line perpendicular to the y-axis.

The *polar coordinate* system represents a point in the plane as an ordered pair (r, θ). The location of the point uses a distance, r, from the point to a fixed point called the *pole*, and the angle, θ, made by the ray from the pole to the point and a fixed half ray extending from the pole. This fixed ray is called the *polar axis*.

It is also convenient to allow negative entries for the first coordinate of a point given in polar coordinates. The point $(-r, \theta)$ is obtained by reflecting the point (r, θ) through the origin, so

$$(r, \theta) \quad \text{and} \quad (-r, \theta + \pi) \quad \text{represent the same point.}$$

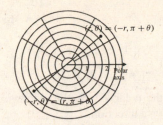

EXAMPLE 1

Plot the point with the given polar coordinates. Then give two other pairs of polar coordinates that represent the point, one with $r > 0$ and one with $r < 0$.

(a) $\left(2, \frac{\pi}{4}\right)$ (b) $\left(-3, \frac{\pi}{3}\right)$ (c) $\left(1, -\frac{\pi}{6}\right)$

Solution (a) Rotate the polar axis $\pi/4$ radians, or $45°$, in the counterclockwise direction, and place the point on the rotated ray 2 units from the pole.

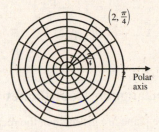

Another representation of the point is given by

$$\left(2, \frac{\pi}{4} + 2\pi\right) = \left(2, \frac{9\pi}{4}\right).$$

To find a representation of the point $\left(2, \frac{\pi}{4}\right)$ with $r < 0$, we can add π to the angle $\frac{\pi}{4}$ and change $r = 2$ to $r = -2$ to obtain $\left(-2, \frac{\pi}{4} + \pi\right) = \left(-2, \frac{5\pi}{4}\right)$, as shown in the figure.

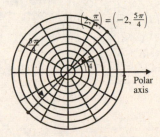

(b) The point $\left(-3, \frac{\pi}{3}\right)$ is the reflection through the pole of the point $\left(3, -\frac{\pi}{3}\right)$.

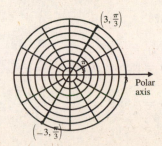

- A Second Representation, $r < 0$: $\left(-3, \frac{\pi}{3} + 2\pi\right) = \left(-3, \frac{7\pi}{3}\right)$
- A Third Representation, $r > 0$: $\left(3, \frac{\pi}{3} + \pi\right) = \left(3, \frac{4\pi}{3}\right)$

(c) The negative angle means we rotate the polar axis in the clockwise direction to find the point. Subtracting 2π makes an additional rotation in the clockwise direction.

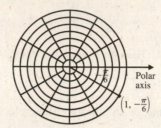

- A Second Representation, $r > 0$: $\left(1, -\frac{\pi}{6} - 2\pi\right) = \left(1, -\frac{13\pi}{6}\right)$
- A Third Representation, $r < 0$: $\left(-1, -\frac{\pi}{6} + \pi\right) = \left(-1, \frac{5\pi}{6}\right)$

$\square$

Changing Polar Coordinates To Rectangular Coordinates

If the polar coordinates (r, θ) of a point in the plane are given, then the rectangular coordinates of the point are

$$x = r\cos\theta \quad \text{and} \quad y = r\sin\theta.$$

These relationships are consequences of the definitions of the cosine and sine of the angle θ.

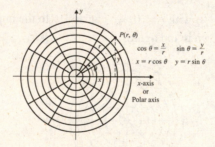

EXAMPLE 2

Convert the polar coordinates to rectangular coordinates.

(a) $\left(1, \frac{\pi}{3}\right)$ (b) $\left(-3, -\frac{7\pi}{6}\right)$

Solution (a) Since $r = 1$ and $\theta = \frac{\pi}{3}$, we have

$$x = r\cos\theta = 1\cos\frac{\pi}{3} = \frac{1}{2}$$

$$y = r\sin\theta = 1\sin\frac{\pi}{3} = \frac{\sqrt{3}}{2}.$$

(b) Since the reference angle for $-\frac{7\pi}{6}$ is $\frac{\pi}{6}$ and $-\frac{7\pi}{6}$ is in quadrant II, we have

$$\cos\left(-\frac{7\pi}{6}\right) = -\cos\frac{\pi}{6} \quad \text{and} \quad \sin\left(-\frac{7\pi}{6}\right) = \sin\frac{\pi}{6}.$$

The rectangular coordinates are

$$x = -3\cos\left(-\frac{7\pi}{6}\right) = -3\left(-\cos\frac{\pi}{6}\right) = -3\left(-\frac{\sqrt{3}}{2}\right) = \frac{3\sqrt{3}}{2}$$

$$y = -3\sin\left(-\frac{7\pi}{6}\right) = -3\sin\frac{\pi}{6} = -3\cdot\frac{1}{2} = -\frac{3}{2}.$$

$\square$

EXAMPLE 3

Convert the polar equation to a rectangular equation.
(a) $r = 4\sin\theta$ (b) $r^2 = \cos 2\theta$

Solution We know that $x = r\cos\theta$, and $y = r\sin\theta$. In addition $r^2 = x^2 + y^2$, since $(\sin x)^2 + (\cos x)^2 = 1$. If these terms are not present:

* Multiply both sides of the equation by r.
* Then make the replacements $x = r\cos\theta$, $y = r\sin\theta$, and $r^2 = x^2 + y^2$.

(a)

$$r = 4\sin\theta$$
$$r^2 = 4r\sin\theta$$
$$x^2 + y^2 = 4y$$
$$x^2 + y^2 - 4y = 0$$

This is the equation of a circle, but it is not in standard form. We need to complete the square on the y term.

$$x^2 + y^2 - 4y = 0$$
$$x^2 + y^2 - 4y + 4 = 4$$
$$x^2 + (y-2)^2 = 4$$

This is the equation of the circle with radius 2 and center at $(0, 2)$.
(b) We can make the substitution $r^2 = x^2 + y^2$.

$$r^2 = \cos 2\theta$$
$$x^2 + y^2 = \cos 2\theta$$

In the current form there is no way to replace the $\cos 2\theta$ term, but there is a trigonometric identity that is useful. If we rewrite the $\cos 2\theta$ in terms of only cosines and sines, then we can use the substitutions $x = r\cos\theta$ and $y = r\sin\theta$. The identity we need is

$$\cos 2\theta = (\cos\theta)^2 - (\sin\theta)^2.$$

Making the substitution gives

$$r^2 = \cos 2\theta$$
$$x^2 + y^2 = \cos 2\theta$$
$$x^2 + y^2 = (\cos \theta)^2 - (\sin \theta)^2$$
$$= \left(\frac{x}{r}\right)^2 - \left(\frac{y}{r}\right)^2$$
$$= \frac{x^2}{r^2} - \frac{y^2}{r^2}$$
$$= \frac{x^2 - y^2}{r^2} = \qquad\qquad \frac{x^2 - y^2}{x^2 + y^2}$$
$$\left(x^2 + y^2\right)^2 = x^2 - y^2.$$

□

Changing Rectangular Coordinates To Polar Coordinates

If the rectangular coordinates (x, y) of a point in the plane are given, then one set of polar coordinates for the point is obtained from

$$r^2 = x^2 + y^2 \text{ and } \tan \theta = \frac{y}{x}, \text{ when } x \neq 0.$$

The sign of r is chosen to ensure that the point is in the correct quadrant. When $x = 0$, one polar representation of the point is $\theta = \frac{\pi}{2}$ and $r = y$.

EXAMPLE 4
Convert the rectangular coordinates to polar coordinates.
(a) $(-2, 2)$ (b) $(3\sqrt{3}, 3)$

Solution   (a) Since $x = -2$ and $y = 2$, we have

$$\tan \theta = \frac{2}{-2} = -1$$
$$\theta = \frac{3\pi}{4}, \quad \theta = -\frac{\pi}{4}.$$

Now find r:

$$r^2 = x^2 + y^2$$
$$= (-2)^2 + (2)^2 = 8$$
$$r = \pm\sqrt{8} = \pm 2\sqrt{2}.$$

The point $(-2, 2)$ lies in quadrant II, so two different representations can be given as

$$\left(2\sqrt{2}, \frac{3\pi}{4}\right) \quad \text{and} \quad \left(-2\sqrt{2}, -\frac{\pi}{4}\right).$$

(b) We have

$$\tan \theta = \frac{3}{3\sqrt{3}} = \frac{1}{\sqrt{3}} = \frac{\sqrt{3}}{3}$$

$$\theta = \frac{\pi}{6}, \quad \frac{7\pi}{6}$$

and

$$r^2 = \left(3\sqrt{3}\right)^2 + 3^2 = 36$$

$$r = \pm 6.$$

Since the point $\left(3\sqrt{3}, 3\right)$ lies in quadrant I, two possible polar representations are

$$\left(6, \frac{\pi}{6}\right) \quad \text{and} \quad \left(-6, \frac{7\pi}{6}\right).$$

□

EXAMPLE 5

Convert the rectangular equation to polar coordinates.

(a) $x^2 + y^2 = 2x$ (b) $y = x^2$

Solution (a) Using the relations $r^2 = x^2 + y^2$ and $x = r \cos \theta$ gives

$$x^2 + y^2 = 2x$$
$$r^2 = 2r \cos \theta$$
$$r = 2 \cos \theta.$$

(b) Using the relations $x = r \cos \theta$ and $y = r \sin \theta$ gives

$$y = x^2$$
$$r \sin \theta = (r \cos \theta)^2$$
$$= r^2 (\cos \theta)^2$$
$$r = \frac{\sin \theta}{(\cos \theta)^2}$$
$$= \frac{\sin \theta}{\cos \theta} \cdot \frac{1}{\cos \theta} = \tan \theta \sec \theta.$$

□

Graphs of Polar Equations

The graph of a polar equation $r = f(\theta)$ is the collection of all points that have at least one polar representation that satisfies the equation.

$$\text{Circles:} \quad r = a, r = a \cos \theta, r = a \sin \theta$$

$$\text{Lines:} \quad \theta = a$$

$$\text{Cardioid:} \quad r = a + a\cos\theta, r = a + a\sin\theta$$

$$\text{Limacon without a loop:} \quad a > b, r = a + b\cos\theta, r = a + b\sin\theta$$

$$\text{Limacon with a loop:} \quad a < b, r = a + b\cos\theta, r = a + b\sin\theta$$

$$\text{Lemniscates:} \quad r^2 = a^2\cos 2\theta, r^2 = a^2\sin 2\theta$$

$$\text{Roses:} \quad r = a\cos n\theta, r = a\sin n\theta.$$

The number of "leaves" on the Rose is $2n$ when n is even and is n when n is odd.

EXAMPLE 6

Sketch the graph of the polar equation.
(a) $r = 3\cos\theta$ (b) $r = 1 + \sin\theta$

Solution (a) Since the period of the cosine function is 2π, the entire curve will be traced if θ varies between 0 and 2π.

θ	0	$\frac{\pi}{6}$	$\frac{\pi}{4}$	$\frac{\pi}{3}$	$\frac{\pi}{2}$	$\frac{2\pi}{3}$	$\frac{3\pi}{4}$	$\frac{5\pi}{6}$	π
	0°	30°	45°	60°	90°	120°	135°	150°	180°
$\cos\theta$	1	$\frac{\sqrt{3}}{2}$	$\frac{\sqrt{2}}{2}$	$\frac{1}{2}$	0	$-\frac{1}{2}$	$-\frac{\sqrt{2}}{2}$	$-\frac{\sqrt{3}}{2}$	-1
$r = 3\cos\theta$	3	$\frac{3\sqrt{3}}{2}$	$\frac{3\sqrt{2}}{2}$	$\frac{3}{2}$	0	$-\frac{3}{2}$	$-\frac{3\sqrt{2}}{2}$	$-\frac{3\sqrt{3}}{2}$	-3

θ	$\frac{7\pi}{6}$	$\frac{5\pi}{4}$	$\frac{4\pi}{3}$	$\frac{3\pi}{2}$	$\frac{5\pi}{3}$	$\frac{7\pi}{4}$	$\frac{11\pi}{6}$	2π
	210°	225°	240°	270°	300°	315°	330°	360°
$\cos\theta$	$-\frac{\sqrt{3}}{2}$	$-\frac{\sqrt{2}}{2}$	$-\frac{1}{2}$	0	$\frac{1}{2}$	$\frac{\sqrt{2}}{2}$	$\frac{\sqrt{3}}{2}$	1
$r = 3\cos\theta$	$-\frac{3\sqrt{3}}{2}$	$-\frac{3\sqrt{2}}{2}$	$-\frac{3}{2}$	0	$\frac{3}{2}$	$\frac{3\sqrt{2}}{2}$	$\frac{3\sqrt{3}}{2}$	3

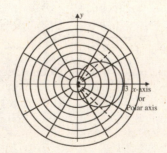

Notice that the points with $r < 0$ are reflected through the origin before they are plotted. Also, the graph is symmetric with respect to the polar axis. Since $\cos\theta = \cos(-\theta)$, it is sufficient to only plot θ ranging from 0 to π. The curve appears to be the circle with center $\left(\frac{3}{2}, 0\right)$ and radius $\frac{3}{2}$. To see this algebraically, we convert the polar equation to a rectangular equation.

$$r = 3\cos\theta$$
$$r^2 = 3r\cos\theta$$
$$x^2 + y^2 = 3x$$
$$x^2 - 3x + y^2 = 0$$
$$x^2 - 3x + \frac{9}{4} + y^2 = \frac{9}{4}$$
$$\left(x - \frac{3}{2}\right)^2 + y^2 = \left(\frac{3}{2}\right)^2$$

(b) One way to construct the graph is to plot a number of points as in part (a), but another way to construct the graph is to observe how $\sin\theta$ varies as θ varies for $0 \le \theta \le 2\pi$.

θ	$\sin\theta$	$1 + \sin\theta$
Increases from 0 to $\frac{\pi}{2}$	Increases from 0 to 1	Increases from 1 to 2
Increases from $\frac{\pi}{2}$ to π	Decreases from 1 to 0	Decreases from 2 to 1
Increases from π to $\frac{3\pi}{2}$	Decreases from 0 to -1	Decreases from 1 to 0
Increases from $\frac{3\pi}{2}$ to 2π	Increases from -1 to 0	Increases from 0 to 1

$\square$

Intersection of Polar Curves

EXAMPLE 7

Find the polar and rectangular coordinates of all points of intersection of the curves $r = 1 + \sin\theta$ and $r = -\sin\theta$.

Solution We first take an algebraic approach and find all values of θ for which the two equations yield the same values of r. Equate the two expressions for r and solve for θ.

$$1 + \sin\theta = -\sin\theta$$
$$2\sin\theta = -1$$

$$\sin\theta = -\frac{1}{2}$$

$$\theta = \frac{7\pi}{6}, \quad \theta = \frac{11\pi}{6}$$

Both of these values give $r = \frac{1}{2}$. Do we have all the values of θ? The answer is we may or we may not. The figure shows that the curves also intersect at the pole! The algebra did not yield this point.

The reason is that the equation $1 + \sin\theta = 0$ and $-\sin\theta = 0$ have solutions for different values of θ, so these values will not be found by equating the expressions. When

$$r = 1 + \sin\theta = 0,$$
$$\sin\theta = -1$$
$$\theta = \frac{3\pi}{2},$$

and when

$$r = -\sin\theta = 0,$$
$$\sin\theta = 0$$
$$\theta = 0.$$

The polar coordinates of the three points of intersection are

$$(0, 0), \quad \left(\frac{1}{2}, \frac{7\pi}{6}\right) \quad \text{and} \quad \left(\frac{1}{2}, \frac{11\pi}{6}\right).$$

Polar Coordinates	Rectangular Coordinates
$(0, 0)$:	$(0, 0)$
$\left(\frac{1}{2}, \frac{7\pi}{6}\right)$:	$x = \frac{1}{2}\cos(7\pi/6) = -\sqrt{3}/4, \ y = \frac{1}{2}\sin(7\pi/6) = -1/4$
$\left(\frac{1}{2}, \frac{11\pi}{6}\right)$:	$x = \frac{1}{2}\cos(11\pi/6) = \sqrt{3}/4, \ y = \frac{1}{2}\sin(11\pi/6) = -1/4$

$\square$

Exercise Set 6.5 (Page 353)

1. a.

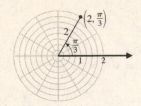

b. $x = r \cos \theta = 2 \cos \frac{\pi}{3} = 1;$
$y = r \sin \theta = 2 \sin \frac{\pi}{3} = \sqrt{3}$
c. $\left(2, \frac{7\pi}{3}\right), \left(-2, \frac{4\pi}{3}\right)$

3. a.

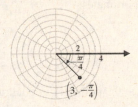

b. $x = r \cos \theta = 3 \cos \left(-\frac{\pi}{4}\right) = \frac{3\sqrt{2}}{2};$
$y = r \sin \theta = 3 \sin \left(-\frac{\pi}{4}\right) = -\frac{3\sqrt{2}}{2}$
c. $\left(3, \frac{7\pi}{4}\right), \left(-3, \frac{3\pi}{4}\right)$

5. a.

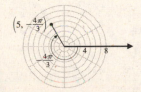

b. $x = r \cos \theta = 5 \cos \left(-\frac{4\pi}{3}\right) = -\frac{5}{2};$
$y = r \sin \theta = 5 \sin \left(-\frac{4\pi}{3}\right) = \frac{5\sqrt{3}}{2}$
c. $\left(5, \frac{2\pi}{3}\right), \left(-5, \frac{5\pi}{3}\right)$

7. a.

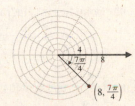

b. $x = r \cos \theta = 8 \cos \left(\frac{7\pi}{4}\right) = 4\sqrt{2};$
$y = r \sin \theta = 8 \sin \left(\frac{7\pi}{4}\right) = -4\sqrt{2}$
c. $\left(8, -\frac{\pi}{4}\right), \left(-8, \frac{3\pi}{4}\right)$

9. a.

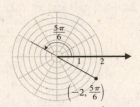

b. $x = r \cos \theta = -2 \cos \left(\frac{5\pi}{6}\right) = \sqrt{3};$
$y = r \sin \theta = -2 \sin \left(\frac{5\pi}{6}\right) = -1$
c. $\left(2, \frac{11\pi}{6}\right), \left(-2, -\frac{7\pi}{6}\right)$

11. a.

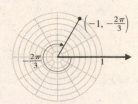

b. $x = r \cos \theta = -\cos \left(-\frac{2\pi}{3}\right) = \frac{1}{2};$
$y = r \sin \theta = -\sin \left(-\frac{2\pi}{3}\right) = \frac{\sqrt{3}}{2}$
c. $\left(1, \frac{\pi}{3}\right), \left(-1, \frac{4\pi}{3}\right)$

13. The point $(2, 0)$ lies on the positive x-axis which coincides with the ray $\theta = 0$, so the point is also $(2, 0)$ in polar coordinates.

15. The point $(0, -4)$ lies on the negative y-axis which coincides with the ray $\theta = \frac{3\pi}{2}$, so the point is $\left(4, \frac{3\pi}{2}\right)$ in polar coordinates.

17. Since
$$x = 1, y = -\sqrt{3} \text{ so } r = \sqrt{x^2 + y^2} = \sqrt{4} = 2,$$
and since the point lies in quadrant IV,
$$\tan \theta = \frac{y}{x} = -\sqrt{3} \text{ so } \theta = \frac{5\pi}{3}.$$
So the point can be represented as $\left(2, \frac{5\pi}{3}\right)$ in polar coordinates.

19. Since
$$x = -4, y = 4 \text{ so } r = \sqrt{x^2 + y^2} = \sqrt{32} = 4\sqrt{2},$$
and since the point lies in quadrant II,
$$\tan \theta = \frac{y}{x} = -1 \text{ so } \theta = \frac{3\pi}{4}.$$
So the point can be represented as $\left(4\sqrt{2}, \frac{3\pi}{4}\right)$ in polar coordinates.

21. $r = 4$ implies $r^2 = 16$ so $x^2 + y^2 = 16$

23. $\theta = \frac{3\pi}{4}$ implies $\frac{y}{x} = \tan \frac{3\pi}{4} = -1$ so $y = -x$

25. Since $r^2 = x^2 + y^2$ and $x = r \cos \theta$, we have

$r = 2 \cos \theta$ so $r^2 = 2r \cos \theta$, which implies $x^2 + y^2 = 2x$ so $x^2 - 2x + y^2 = 0$ and $x^2 - 2x + 1 - 1 + y^2 = 0$,

and $(x - 1)^2 + y^2 = 1$.

27. Since $x = r \cos \theta$ and $y = r \sin \theta$, we have

$$y = x \text{ so } r \cos \theta = r \sin \theta, \text{ which implies } \tan \theta = 1 \text{ so } \theta = \frac{\pi}{4}.$$

29. Since $r^2 = x^2 + y^2$, we have

$$x^2 + y^2 = 9, \text{ which implies } r^2 = 9 \text{ so } r = 3.$$

31. Since $r^2 = x^2 + y^2$ and $y = r \sin \theta$, we have

$$x^2 + y^2 = 2y, \text{ which implies } r^2 = 2r \sin \theta \text{ so } r = 2 \sin \theta.$$

33. $r = 3$

35. $\theta = 5\pi/3$

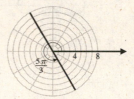

37. $r = 3 \cos \theta$

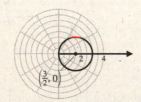

39. $r = -4 \sin \theta$

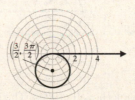

41. $r = 2 + 2 \cos \theta$

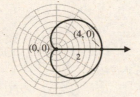

43. $r = 2 + \sin \theta$

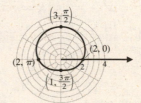

45. $r = 1 - 2\cos\theta$

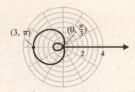

47. $r = 3\sin 3\theta$

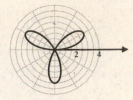

49. $r = 3\cos 2\theta$

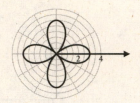

51. $r^2 = 16\cos 2\theta$

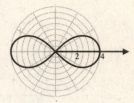

53. $r = \theta$

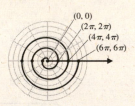

55. $r = e^\theta$

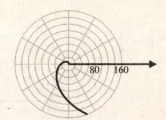

57. The graphs are all circles. The graph in (a) and (c) are symmetric about the y-axis and are reflections of each other about the x-axis. The graph in (b) and (d) are symmetric about the x-axis and are reflections of each other about the y-axis.

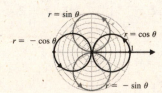

59. Graph (a) is a circle with center $\left(0, \frac{1}{2}\right)$ and radius $\frac{1}{2}$, and graphs (b)-(d) are cardioids with axis along the y-axis which approach a circle.

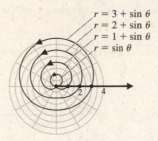

61. The graphs are all cardioids with axis along the y-axis. The graphs in (a) and (c) coincide, as do the graphs in (b) and (d). The graph in (b) is the reflection of graph (a) about the x-axis.

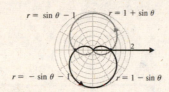

63. a. $(0, 0), \left(90, \frac{\pi}{4}\right), \left(90\sqrt{2}, \frac{\pi}{2}\right), \left(90, \frac{3\pi}{4}\right)$

   b. $(0, 0), (90, 0), \left(90\sqrt{2}, \frac{\pi}{4}\right), \left(90, \frac{\pi}{2}\right)$

65. Setting the equations for r equal, we have $1 + 2\cos\theta = 1$ implies $2\cos\theta = 0$ so $\theta = \frac{\pi}{2}$ or $\theta = \frac{3\pi}{2}$. The figure indicates the curves also intersect at the point $(1, 0)$. The polar coordinates of the points of intersection are $\left(1, \frac{\pi}{2}\right), \left(1, \frac{3\pi}{2}\right), (1, 0)$. The rectangular coordinates of the points of intersection are $(0, 1), (0, -1), (1, 0)$.

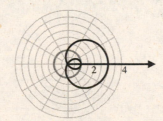

67. Setting the equations for r equal, we have $2 - 2\sin\theta = 2\sin\theta$ implies $4\sin\theta = 2$ so $\sin\theta = \frac{1}{2}$ and $\theta = \frac{\pi}{6}$ or $\theta = \frac{5\pi}{6}$. The figure indicates the curves also intersect at the pole. The polar coordinates of the points of intersection are $\left(1, \frac{\pi}{6}\right)$, $\left(1, \frac{5\pi}{6}\right)$, $(0, 0)$. The rectangular coordinates of the points of intersection are $\left(\frac{\sqrt{3}}{2}, \frac{1}{2}\right)$, $\left(-\frac{\sqrt{3}}{2}, \frac{1}{2}\right)$, $(0, 0)$.

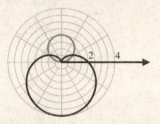

69. The graphs of $r = 1 + \sin(n\theta) + (\cos(2n\theta))^2$ for $n = 1, 2, 3, 4$ are shown below.

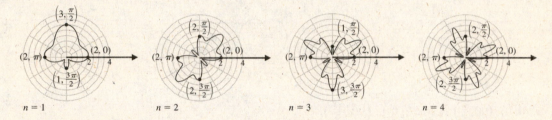

71. The graphs of $r = \sin m\theta$ and $r = |\sin m\theta|$ are shown below.

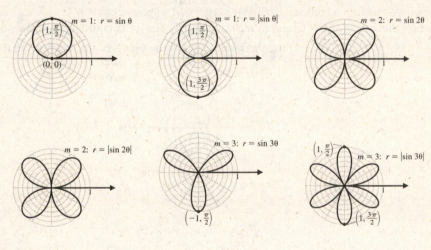

6.6   Conic Sections in Polar Coordinates

The graph of a polar equation of the form

$$r = \frac{ed}{1 \pm e \cos\theta} \quad \text{or} \quad r = \frac{ed}{1 \pm e \sin\theta}$$

is a conic section with eccentricity e.

- When $e = 1$, the graph is a parabola.

- When $e < 1$, the graph is an ellipse.

- When $e > 1$, the graph is a hyperbola.

The focus is at the pole and, the directrix is d units from the pole.

Equation	Directrix
$r = \frac{ed}{1+e\cos\theta}$	Vertical, right of pole
$r = \frac{ed}{1-e\cos\theta}$	Vertical, left of pole
$r = \frac{ed}{1+e\sin\theta}$	Horizontal, above the pole
$r = \frac{ed}{1-e\sin\theta}$	Horizontal, below the pole

EXAMPLE 1

Sketch the graph of the conic section and find a corresponding rectangular equation.

(a)

$$r = \frac{3}{1 - \cos\theta}$$

(b)

$$r = \frac{3}{3 - \sin\theta}$$

Solution   (a) First determine e and d. The polar equation is in the form

$$r = \frac{ed}{1 - e\cos\theta},$$

with

$$e = 1$$
$$ed = 3 \Rightarrow d = 3.$$

Since $e = 1$, the conic is a parabola with a vertical directrix 3 units to the left of the pole having rectangular equation $x = -3$. Since the vertex is midway between the directrix and the pole, the vertex is at the point $(-3/2, 0)$ (in rectangular coordinates). Since the parabola can not cross the directrix it opens to the right. When

$$\theta = \frac{\pi}{2}, \ \cos\theta = 0 \text{ and } r = 3;$$
$$\theta = \frac{3\pi}{2}, \ \cos\theta = 0 \text{ and } r = 3.$$

The graph is shown in the figure.

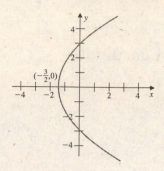

The rectangular equation of the parabola is of the form

$$x - h = \frac{1}{4c}y^2.$$

To determine the rectangular equation we convert the polar equation to the corresponding rectangular form using the relations $x = r\cos\theta$, $y = r\sin\theta$, and $r^2 = x^2 + y^2$.

$$r = \frac{3}{1 - \cos\theta}$$
$$r - r\cos\theta = 3$$
$$r = 3 + r\cos\theta = 3 + x$$
$$r^2 = (x + 3)^2$$
$$x^2 + y^2 = x^2 + 6x + 9$$
$$y^2 = 6x + 9$$
$$y^2 = 6\left(x + \frac{3}{2}\right)$$
$$x + \frac{3}{2} = \frac{1}{6}y^2$$

The vertex of the parabola in rectangular form is $(-3/2, 0)$ and $4c = 6$, so $c = 3/2$. The parabola is obtained from the parabola in standard position

$$x = \frac{1}{6}y^2,$$

shifted to the left 3/2 units.

(b) First write the polar equation in the form

$$r = \frac{ed}{1 - e\sin\theta}$$
$$= \frac{3}{3 - \sin\theta}$$
$$= \frac{3}{3(1 - \frac{1}{3}\sin\theta)}$$
$$= \frac{1}{1 - \frac{1}{3}\sin\theta}.$$

Since $e = \frac{1}{3}$ the conic is an ellipse and

$$ed = 1$$

$$\left(\frac{1}{3}\right) d = 1$$

$$d = 3.$$

The conic is an ellipse with directrix horizontal and 3 units below the pole. The graph passes through the points

$$(1, 0), \ (1, \pi), \ \left(\frac{3}{2}, \frac{\pi}{2}\right), \ \text{and} \ \left(\frac{3}{4}, \frac{3\pi}{2}\right).$$

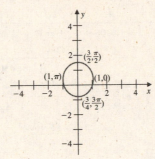

- Rectangular equation:

$$r = \frac{3}{3 - \sin\theta}$$

$$3r - r\sin\theta = 3$$

$$3r = 3 + r\sin\theta$$

$$3r = 3 + y$$

$$9r^2 = (3 + y)^2$$

$$9x^2 + 9y^2 = 9 + 6y + y^2$$

$$9x^2 + 8y^2 - 6y = 9$$

$$9x^2 + 8\left(y^2 - \frac{3}{4}y\right) = 9$$

$$9x^2 + 8\left(y^2 - \frac{3}{4}y + \frac{9}{64} - \frac{9}{64}\right) = 9$$

$$9x^2 + 8\left(y - \frac{3}{8}\right)^2 = 9 + \frac{9}{8}$$

$$9x^2 + 8\left(y - \frac{3}{8}\right)^2 = \frac{81}{8}$$

$\square$

Application

EXAMPLE 2

The earth moves in an elliptical orbit with one focal point at the sun and an eccentricity $e = 0.0167$. The major axis of the elliptical orbit is approximately 2.99×10^8 kilometers. Write a polar equation for this ellipse, assuming that the pole is at the sun.

Solution The elliptic orbit of the earth about the sun has an equation of the form

$$r = \frac{ed}{1 + e\cos\theta},$$

where the earth is closest to the sun when $\theta = 0$ and furthest from the sun when $\theta = \pi$. The length of the major axis is

$$\frac{ed}{1 + e\cos\pi} - \frac{ed}{1 + e\cos 0} = \frac{ed}{1 - e} - \frac{ed}{1 + e}$$

$$= \frac{ed(1 + e) - ed(1 - e)}{(1 - e)(1 + e)}$$

$$= \frac{ed + e^2 d - ed + e^2 d}{1 - e^2}$$

$$= \frac{2e^2 d}{1 - e^2}.$$

Since the length of the major axis is approximately 2.99×10^8 and $e = 0.0167$, we have

$$2.99 \times 10^8 = \frac{2e^2 d}{1 - e^2}$$

$$d = \frac{(2.99 \times 10^8)(1 - e^2)}{2e^2}$$

$$= \frac{(2.99 \times 10^8)(1 - 0.0167^2)}{2(0.0167)^2}$$

$$\approx 5.4 \times 10^{11}.$$

The polar equation of the earth's orbit about the sun is

$$r = \frac{0.0167\,(5.4 \times 10^{11})}{1 + 0.0167\cos\theta} = \frac{9 \times 10^9}{1 + 0.0167\cos\theta}.$$

$\square$

Exercise Set 6.6 (Page 359)

1. a. The equation $r = \frac{2}{1 + \cos\theta}$ is in the standard form $r = \frac{ed}{1 + e\cos\theta}$, where $e = 1$ implies $d = 2$, so the equation describes a parabola opening to the left with vertex $(1, 0)$, focus $(0, 0)$, and directrix along the line $x = 2$.

b. Since $r^2 = x^2 + y^2$ and $x = r\cos\theta$, we have

$$r = \frac{2}{1 + \cos\theta} \text{ implies } r + r\cos\theta = 2 \text{ so } r = 2 - r\cos\theta = 2 - x \text{ and } r^2 = 4 - 4x + x^2.$$

So

$$x^2 + y^2 = 4 - 4x + x^2 \text{ implies } x - 1 = -\frac{1}{4}y^2.$$

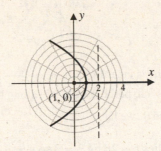

3. a. The equation

$$r = \frac{2}{2 - \sin\theta} = \frac{2}{2\left(1 - \frac{1}{2}\sin\theta\right)} = \frac{1}{1 - \frac{1}{2}\sin\theta}$$

is in the standard form $r = \frac{ed}{1 - e\sin\theta}$, where $e = \frac{1}{2}$ implies $d = 2$. Since $e < 1$, the equation describes an ellipse with major axis along the y-axis and with horizontal directrix $y = -2$.

b. Since $r^2 = x^2 + y^2$ and $y = r\sin\theta$, we have

$$r = \frac{2}{2 - \sin\theta} \text{ implies } 2r - r\sin\theta = 2 \text{ so } 2r = 2 + r\sin\theta = 2 + y \text{ and } 4r^2 = 4 + 4y + y^2.$$

So

$$4x^2 + 4y^2 = 4 + 4y + y^2 \text{ implies } 4x^2 + 3y^2 - 4y = 4.$$

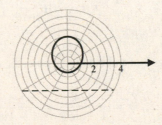

5. a. The equation $r = \frac{1}{1 + 2\cos\theta}$ is in the standard form $r = \frac{ed}{1 + e\cos\theta}$, where $e = 2$ and $ed = 1$ implies $d = \frac{1}{2}$. Since $e > 1$, the equation describes a hyperbola with axis along the x-axis and vertical directrix $x = \frac{1}{2}$.

b. Since $r^2 = x^2 + y^2$ and $x = r\cos\theta$, we have

$$r = \frac{1}{1 + 2\cos\theta} \text{ implies } r + 2r\cos\theta = 1 \text{ so } r = 1 - 2x \text{ and } r^2 = 1 - 4x + 4x^2.$$

So

$$x^2 + y^2 = 1 - 4x + 4x^2 \text{ implies } 3x^2 - 4x - y^2 = -1.$$

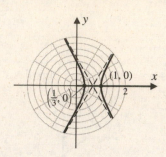

7. **a.** The equation $r = \frac{3}{1-2\sin\theta}$ is in the standard form $r = \frac{ed}{1-e\sin\theta}$, where $e = 2$ and $ed = 3$ implies $d = \frac{3}{2}$. Since $e > 1$, the equation describes a hyperbola with axis along the y-axis and horizontal directrix $y = -\frac{3}{2}$.

b. Since $r^2 = x^2 + y^2$ and $y = r\sin\theta$, we have

$$r = \frac{3}{1-2\sin\theta} \text{ implies } r - 2r\sin\theta = 3 \text{ so } r = 3 + 2y \text{ and } r^2 = 9 + 12y + 4y^2.$$

So

$$x^2 + y^2 = 9 + 12y + 4y^2 \text{ implies } 3y^2 + 12y - x^2 = -9.$$

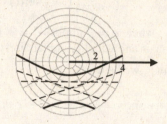

9. Since the directrix $x = -4$ is vertical and to the left of the pole, the conic has the form $r = \frac{ed}{1-e\cos\theta}$. Since $e = 2$ and $d = 4$, we have

$$r = \frac{8}{1 - 2\cos\theta}.$$

11. Since the directrix $y = -\frac{1}{4}$ is horizontal and below the pole, the conic has the form $r = \frac{ed}{1-e\sin\theta}$. Since $e = 1$ and $d = \frac{1}{4}$, we have

$$r = \frac{1/4}{1 - \sin\theta} = \frac{1}{4 - 4\sin\theta}.$$

13. Since the directrix $x = 2$ is vertical and to the right of the pole, the conic has the form $r = \frac{ed}{1+e\cos\theta}$. Since $e = 3$ and $d = 2$, we have

$$r = \frac{6}{1 + 3\cos\theta}.$$

15. Since the directrix $y = 1$ is horizontal and above the pole, the conic has the form $r = \frac{ed}{1+e\sin\theta}$. Since $e = \frac{1}{4}$ and $d = 1$, we have

$$r = \frac{1/4}{1 + \frac{1}{4}\sin\theta} = \frac{1}{4 + \sin\theta}.$$

17. The conic has equation in the form $r = \frac{ed}{1 + e\cos\theta}$, and since $(1, 0)$ and $(3, \pi)$ are on the curve,

$$1 = \frac{ed}{1 + e\cos 0} = \frac{ed}{1 + e} \text{ implies } 1 + e = ed,$$

and

$$3 = \frac{ed}{1 + e\cos\pi} = \frac{ed}{1 - e} \text{ so } 3 - 3e = ed \text{ implies } 1 + e = 3 - 3e \text{ so } e = \frac{1}{2},$$

and

$$\frac{1}{2}d = 1 + \frac{1}{2} \text{ so } d = 3.$$

So

$$r = \frac{ed}{1 + e\cos\theta} = \frac{3/2}{1 + 1/2\cos\theta} = \frac{3}{2 + \cos\theta}.$$

19. Since the orbit is elliptical if we assume the axis is horizontal, the form of the equation is

$$r = \frac{ed}{1 - e\cos\theta},$$

with $e < 1$. The satellite will be at its maximum height when $\theta = 0$, so

$$\frac{ed}{1 - e} = 3960 + 560 = 4520,$$

and at its minimum height when $\theta = \pi$, so

$$\frac{ed}{1 + e} = 3960 + 145 = 4105.$$

Then

$$4520(1 - e) = ed = 4105(1 + e) \text{ implies } 8625e = 415,$$

and

$$e = \frac{415}{8625} \approx 0.048 \text{ so } ed = 4105\left(1 + \frac{415}{8625}\right) \approx 4303.$$

The equation of the orbit is

$$r = \frac{4303}{1 - 0.048\cos\theta}.$$

· ·

6.7   Parametric Equations

When presenting curves using parametric equations the x- and y- coordinates of a point on the curve are each specified separately by functions of a third variable, called the *parameter*. When the functions are f and g, and the parameter is t,

$$x = f(t), \quad y = g(t)$$

are called *parametric equations* for the curve. As t varies, the point $(x, y) = (f(t), g(t))$ traces out the curve.

EXAMPLE 1

Sketch the graph of the curve described by the parametric equations.

$$x = 2t - 1, \; y = t^2 + 1,$$

and find a corresponding rectangular equation.

Solution First plot some representative points on the curve by making a table of values.

t	0	1	2	−1	−2
$x = 2t - 1$	−1	1	3	−3	−5
$y = t^2 + 1$	1	2	5	2	5

A smooth curve is traced between the points in the figure. The curve appears to be a parabola. Notice that as t increases the curve is traced in the direction of the arrows.

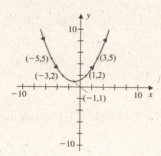

To find a corresponding rectangular equation for the curve, we eliminate the parameter. That is, solve for t in either the equation for x or the equation for y, which ever is easiest, and then substitute that value for t in the other equation.

$$x = 2t - 1$$
$$2t = x + 1$$
$$t = \frac{x + 1}{2}$$

Next substitute the value of t in the equation for y.

$$y = t^2 + 1$$
$$= \left(\frac{x + 1}{2}\right)^2 + 1$$

$$= \frac{1}{4}(x+1)^2 + 1$$

The equation describes a parabola with vertex at $(-1, 1)$ and opening upward.   □

EXAMPLE 2
For each set of parametric equations, find a corresponding rectangular equation, and sketch the graph of the curve, showing when $t = 0$ and the direction of increasing values of t.
(a) $x = t - 1,\ y = \sqrt{t}$
(b) $x = e^{2t} + 2,\ y = e^t - 3$
(c) $x = 4 + 2\cos t,\ y = 6 + 2\sin t$

Solution (a) Solve for t in the equation for x, and substitute it into the equation for y.

$$t = x + 1$$
$$y = \sqrt{x+1}$$
$$y^2 = x + 1$$

Be careful that you have the *correct* curve. Is the curve the entire parabola given by the equation $y^2 = x + 1$? In this case the answer is no! The reason is that

$$\text{since } y = \sqrt{t}, \quad y \text{ must be greater than or equal to } 0.$$

The curve consists *only* of the portion of $y^2 = x + 1$ above the x-axis. As t increases, x and y both increase and the curve is traced in the direction of the arrows in the figure.

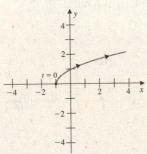

(b) In this example, we do not have to find t since finding an expression for e^t will do.

$$y = e^t - 3$$
$$e^t = y + 3$$
$$e^{2t} = \left(e^t\right)^2 = (y+3)^2$$
$$x = e^{2t} + 2 = (y+3)^2 + 2$$

Since $e^t > 0$, the restriction on y is $y = e^t - 3 > -3$, and

$$x = (y+3)^2 + 2, \quad \text{for} \quad y > -3.$$

Note that as

$$t \to -\infty,\ x = e^{2t} + 2 \to 2 \text{ and } y = e^t - 3 \to -3.$$

The point on the curve corresponding to $t = 0$ is $(3, -2)$.

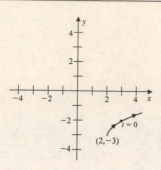

(c) In problems involving sine and cosine, always look to use the identity

$$(\sin t)^2 + (\cos t)^2 = 1.$$

When $x - 4$ and $y - 6$ are each squared and then added, the resulting expression involves this basic identity and can then be simplified.

$$
\begin{aligned}
(x-4)^2 + (y-6)^2 &= (4 + 2\cos t - 4)^2 + (6 + 2\sin t - 6)^2 \\
&= (2\cos t)^2 + (2\sin t)^2 \\
&= 4(\cos t)^2 + 4(\sin t)^2 \\
&= 4((\cos t)^2 + (\sin t)^2) \\
&= 4
\end{aligned}
$$

This is the equation of the circle with center $(4, 6)$ and radius 2.

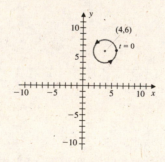

□

EXAMPLE 3

Sketch the graph described by the parametric equations, showing when $t = 0$ and the direction of increasing t.

(a) $x = t, \ y = t^3$ (b) $x = t^3, \ y = t$
(c) $x = t^2, \ y = t^6$ (d) $x = t^6, \ y = t^2$

Solution (a) $x = t, \ y = t^3$

- Eliminating the parameter: $y = x^3$
- When $t = 0$: $(x, y) = (0, 0)$
- Increasing t: As t increases from 0 toward ∞, both x and y increase, so the points trace the curve upward along the right half of $y = x^3$. As t goes toward $-\infty$, x goes to $-\infty$ and $y = x^3$ goes to $-\infty$. So as t increases from negative values towards 0, the left half of the curve is traced upward toward 0.

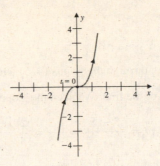

(b) $x = t^3$, $y = t$
- Eliminating the parameter: $x = y^3$

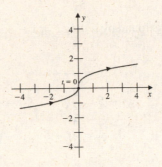

(c) $x = t^2$, $y = t^6$
- Eliminating the parameter: $y = t^6 = (t^2)^3 = x^3$

But since $x = t^2 \geq 0$, the curve is restricted to $x \geq 0$.

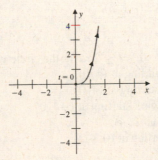

(d) $x = t^6$, $y = t^2$
- Eliminating the parameter: $x = y^3$, $x \geq 0$

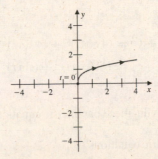

$\square$

EXAMPLE 4

Find parametric equations for the line passing through the points $(4, 5)$ and $(-1, 3)$.

Solution First find the rectangular equation of the line. We need the slope, which is

$$m = \frac{3-5}{-1-4} = \frac{-2}{-5} = \frac{2}{5}.$$

Using the point $(4, 5)$ and the point-slope equation of a line gives

$$y - 5 = \frac{2}{5}(x - 4).$$

One set of parametric equations for the line is found by setting $t = x - 4$.

$$x = t + 4$$
$$y - 5 = \frac{2}{5}t$$
$$y = \frac{2}{5}t + 5$$

A set of parametric equations is given by

$$x = t + 4, \quad y = \frac{2}{5}t + 5.$$

Note that $t = 0$ gives $(4, 5)$, and $t = -5$ gives $(-1, 3)$. $\square$

EXAMPLE 5

Find parametric equations in the parameter t, $0 \leq t \leq 2\pi$, for the circle with equation $x^2 + y^2 = 16$.
(a) Traced once around, counterclockwise, starting at $(4, 0)$.
(b) Traced twice around, counterclockwise, starting at $(4, 0)$.
(c) Traced three times around, counterclockwise, starting at $(-4, 0)$.
(d) Traced twice around, clockwise, starting at $(0, 4)$.
(e) Traced three times around, clockwise, starting at $(0, 4)$.

Solution The idea is to recognize that if

$$x = 4\cos t, \ y = 4\sin t,$$

then

$$x^2 + y^2 = (4\cos t)^2 + (4\sin t)^2$$
$$= 16((\cos t)^2 + (\sin t)^2)$$
$$= 16,$$

which is the equation of the circle. By modifying these parametric equations, the manner in which the circle is traced can be changed.
(a) No modification is required. The parametric equations

$$x = 4\cos t \quad \text{and} \quad y = 4\sin t, \quad \text{for} \ \ 0 \leq t \leq 2\pi,$$

trace the circle starting at $(4, 0)$, when $t = 0$, and ending back at $(4, 0)$, when $t = 2\pi$. To check which direction the curve is traced, select a few values of increasing t. For example,

$$t = 0, \ (x, y) = (4, 0)$$
$$t = \frac{\pi}{2}, \ (x, y) = (0, 4)$$
$$t = \pi, \ (x, y) = (-4, 0)$$
$$t = \frac{3\pi}{2}, \ (x, y) = (0, -4),$$

and the curve is being traced counterclockwise.

(b) To trace the circle twice, reduce the period of the cosine and sine functions by half so that as t varies from 0 to 2π, the points on the circle are traced twice. To halve the period, multiply the argument t by 2. Note that the period of $y = \cos 2t$ and of $y = \sin 2t$ is $\frac{2\pi}{2} = \pi$. So the parametric equations are

$$x = 4 \cos 2t \quad \text{and} \quad y = 4 \sin 2t, \quad \text{for} \quad 0 \le t \le 2\pi.$$

Notice also that when,

$$t = 0, \ (x, y) = (4, 0)$$
$$t = \frac{\pi}{4}, \ (x, y) = (0, 4)$$
$$t = \frac{\pi}{2}, \ (x, y) = (-4, 0)$$
$$t = \frac{3\pi}{4}, \ (x, y) = (0, -4)$$
$$t = \pi, \ (x, y) = (4, 0).$$

(c) To go three times around starting at $(-4, 0)$, let

$$x = -4 \cos 3t \quad \text{and} \quad y = -4 \sin 3t, \quad \text{for} \quad 0 \le t \le 2\pi.$$

Then

$$t = 0, \ (x, y) = (-4, 0)$$
$$t = \frac{\pi}{6}, \ (x, y) = (0, -4)$$
$$t = \frac{\pi}{3}, \ (x, y) = (4, 0)$$
$$t = \frac{\pi}{2}, \ (x, y) = (0, 4)$$
$$t = \frac{2\pi}{3}, \ (x, y) = (-4, 0).$$

(d) To trace the curve clockwise, use the sine function to represent x and the cosine function to represent y, as

$$x = 4 \sin 2t \quad \text{and} \quad y = 4 \cos 2t, \quad \text{for} \quad 0 \le t \le 2\pi.$$

Now for

$$t = 0, \ (x, y) = (0, 4)$$
$$t = \frac{\pi}{4}, \ (x, y) = (4, 0)$$
$$t = \frac{\pi}{2}, \ (x, y) = (0, -4)$$
$$t = \frac{3\pi}{4}, \ (x, y) = (-4, 0)$$
$$t = \pi, \ (x, y) = (0, 4).$$

(e) Finally, to go around three times clockwise, let

$$x = 4 \sin 3t \quad \text{and} \quad y = 4 \cos 3t, \quad \text{for} \quad 0 \le t \le 2\pi.$$

So for

$$t = 0, \ (x, y) = (0, 4)$$
$$t = \frac{\pi}{6}, \ (x, y) = (4, 0)$$
$$t = \frac{\pi}{3}, \ (x, y) = (0, -4)$$
$$t = \frac{\pi}{2}, \ (x, y) = (-4, 0)$$
$$t = \frac{2\pi}{3}, \ (x, y) = (0, 4).$$

$\square$

Exercise Set 6.7 (Page 364)

1. a. $x = 3t$, $y = \frac{t}{2}$ implies $t = \frac{x}{3}$ and $y = \frac{1}{2}\frac{x}{3} = \frac{x}{6}$.
b.

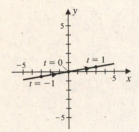

3. a. $x = \sqrt{t}$, $y = t + 1$ implies $t = x^2$ and $y = x^2 + 1$, only for $x \geq 0$.
b.

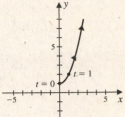

5. a. $x = \sin t$, $y = (\cos t)^2$ implies
$y = 1 - (\sin t)^2 = 1 - x^2$ for $y \geq 0$.
b.

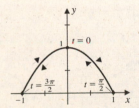

7. a. $x = \sec t$, $y = \tan t$ and $(\tan t)^2 + 1 = (\sec t)^2$.
So $y^2 + 1 = x^2$ implies $x^2 - y^2 = 1$. Since
$x = \sec t$, $|x| \geq 1$.
b.

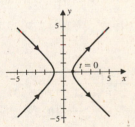

9. a. $x = 3\cos t$, $y = 2\sin t$ implies
$\frac{x^2}{9} + \frac{y^2}{4} = \frac{9(\cos t)^2}{9} + \frac{4(\sin t)^2}{4}$
or $\frac{x^2}{9} + \frac{y^2}{4} = (\cos t)^2 + (\sin t)^2 = 1$
so $\frac{x^2}{9} + \frac{y^2}{4} = 1$.
b.

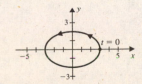

11. a. $x = e^t$, $y = e^{-t}$ implies $y = x^{-1} = \frac{1}{x}$,
for $x > 0$.
b.

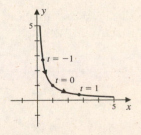

13. The graphs of the parametric equations are shown below.

a. $x = t$, $y = t^2$ implies $y = x^2$

b. $x = t^2$, $y = t$ implies $x = y^2$

c. $x = t^2$, $y = t^4$ implies $y = x^2$, $x \geq 0$

d. $x = t^4$, $y = t^2$ implies $x = y^2$, $y \geq 0$

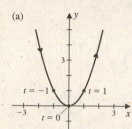 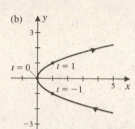 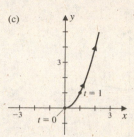 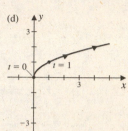

15. The graphs of the parametric equations are shown below.

a. $x = \sin t$, $y = \cos t$ implies $x^2 + y^2 = (\sin t)^2 + (\cos t)^2 = 1$

b. $x = \sin t$, $y = \cos t + 1$ implies $x^2 + (y - 1)^2 = (\sin t)^2 + (\cos t)^2 = 1$

c. $x = \cos t + 1$, $y = \sin t$ implies $(x - 1)^2 + y^2 = (\cos t)^2 + (\sin t)^2 = 1$

d. $x = \cos t$, $y = \sin t + 1$ implies $x^2 + (y - 1)^2 = (\cos t)^2 + (\sin t)^2 = 1$

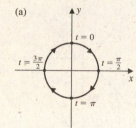 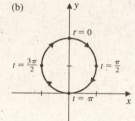 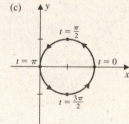 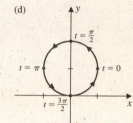

17. The graphs of the parametric equations are shown below.

a. $x = \cos t$, $y = \sin t$ implies $x^2 + y^2 = 1$

b. $x = \sin t$, $y = \cos t$ implies $x^2 + y^2 = 1$

c. $x = t$, $y = \sqrt{1 - t^2}$ implies $y = \sqrt{1 - x^2}$

d. $x = -t$, $y = \sqrt{1-t^2}$ implies $y = \sqrt{1-x^2}$

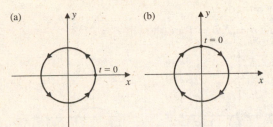

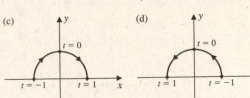

19. The graphs of the parametric equations are shown below.

   a. $x = t$, $y = \ln t$ implies $y = \ln x$, $t > 0$

   b. $x = e^t$, $y = t$ implies $\ln x = t$, $y = \ln x$

   c. $x = t^2$, $y = 2\ln t$ implies $y = \ln t^2 = \ln x$

   d. $x = \frac{1}{t}$, $y = -\ln t$ implies $y = \ln t^{-1} = \ln x$

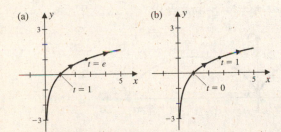

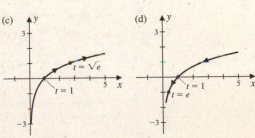

21. A rectangular equation of the line with slope $\frac{1}{3}$ and passing through $(2, -1)$ is

$$y + 1 = \frac{1}{3}(x - 2).$$

   One set of parametric equations is obtained by setting

$$t = x - 2 \text{ so } x = t + 2, \, y = \frac{1}{3}t - 1.$$

23. A parabola with vertex $(1, -2)$ and passing through the points $(0, 0)$ and $(2, 0)$ has the form
   $y = a(x - 1)^2 - 2$. Since $(0, 0)$ is on the curve, $0 = a(-1)^2 - 2$ implies $a = 2$ so $y = 2(x - 1)^2 - 2$. One set
   of parametric equations is obtained by setting

$$t = x - 1 \text{ so } x = t + 1, \, y = 2t^2 - 2.$$

25. a. $x = r \cos t$, $y = r \sin t$ **b.** $x = r \cos 2t$, $y = r \sin 2t$ **c.** $x = -r \cos 3t$, $y = -r \sin 3t$

d. $x = r \sin 2t$, $y = r \cos 2t$;   **e.** $x = r \sin 3t$, $y = r \cos 3t$

27.   The curves when $a = 1, b = 1$; $a = 1, b = 2$; and $a = 2, b = 1$ are shown.

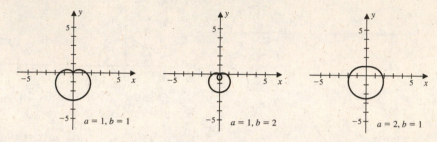

29.   The parameter a determines the maximum height of the curve, $2a$, and the period of the curve, $2\pi a$, which equals the circumference of the rolling circle. The curve touches the x-axis for those t for which the y-coordinate is 0. That is,

$$a(1 - \cos t) = 0 \text{ implies } \cos t = 1 \text{ so } t = \pm 2k\pi, \text{ for } k = 0, 1, 2, \ldots,$$

and

$$x = a(t - \sin t) = at = \pm 2ak\pi \text{ for } k = 0, 1, 2, \ldots.$$

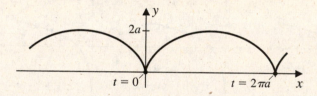

Review Exercises for Chapter 6 (Page 365)

1. $4y - x^2 = 0$ implies $y = \frac{1}{4}x^2$ so $c = 1$
Vertex: $(0, 0)$; Focus: $(0, 1)$;
Directrix, D: $y = -1$

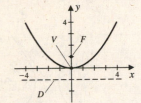

3. $4x - y^2 + 6y - 17 = 0$ implies
$-(y^2 - 6y + 9 - 9) = 17 - 4x$
so $-(y - 3)^2 = 8 - 4x$ and $x - 2 = \frac{1}{4}(y - 3)^2$ implies $c = 1$;
Vertex: $(2, 3)$; Focus: $(3, 3)$;
Directrix, D: $x = 1$

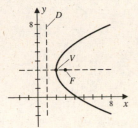

5. $x^2 + 4y^2 = 4$ implies $\frac{x^2}{4} + y^2 = 1$ so $a = 2, b = 1$ and $c^2 = 3$ implies $c = \sqrt{3}$
Foci: $\left(\sqrt{3}, 0\right), \left(-\sqrt{3}, 0\right)$; Vertices: $(2, 0), (-2, 0)$

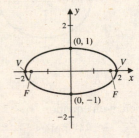

7. $4(x - 1)^2 + 9(y + 2)^2 = 36$ implies
$\frac{(x-1)^2}{9} + \frac{(y+2)^2}{4} = 1$ so $a = 3, b = 2$
and $c^2 = 5$ implies $c = \sqrt{5}$
Foci: $\left(1 - \sqrt{5}, -2\right), \left(1 + \sqrt{5}, -2\right)$;
Vertices: $(4, -2), (-2, -2)$

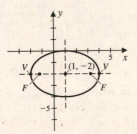

9. $x^2 - 2y^2 = 4$ implies $\frac{x^2}{4} - \frac{y^2}{2} = 1$ so $a = 2$, $b = \sqrt{2}$ and $c^2 = 6$ implies $c = \sqrt{6}$

Foci: $\left(\sqrt{6}, 0\right)$, $\left(-\sqrt{6}, 0\right)$;

Vertices: $(2, 0)$, $(-2, 0)$;

Asymptotes: $y = \pm\frac{\sqrt{2}}{2}x$

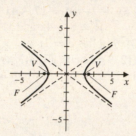

11. $2x^2 - 4x - 4y^2 + 1 = 0$ implies $2(x^2 - 2x + 1 - 1) - 4y^2 = -1$ so $2(x - 1)^2 - 4y^2 = 1$ and $a = \frac{\sqrt{2}}{2}$, $b = \frac{1}{2}$ implies $c^2 = \frac{1}{2} + \frac{1}{4} = \frac{3}{4}$ so $c = \frac{\sqrt{3}}{2}$

Foci: $\left(1 - \sqrt{3}/2, 0\right)$, $\left(1 + \sqrt{3}/2, 0\right)$;

Vertices: $\left(1 + \sqrt{2}/2, 0\right)$, $\left(1 - \sqrt{2}/2, 0\right)$;

Asymptotes: $y = \pm\frac{\sqrt{2}}{2}(x - 1)$

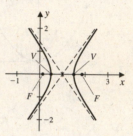

13. Parabola with equation
$x^2 = -2(y - 5)$ implies $y - 5 = -\frac{1}{2}x^2$.

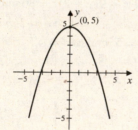

15. Ellipse with equation
$16x^2 + 25y^2 = 400$ implies $\frac{x^2}{25} + \frac{y^2}{16} = 1$.

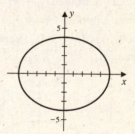

17. Parabola with equation $x^2 - 2x - 4y - 11 = 0$ implies $x^2 - 2x + 1 - 1 = 4y + 11$ so $(x - 1)^2 = 4(y + 3)$
and $y + 3 = \frac{1}{4}(x - 1)^2$.

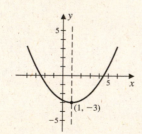

19. Hyperbola with equation
$9x^2 - 16y^2 = 144$ implies $\frac{x^2}{16} - \frac{y^2}{9} = 1$.

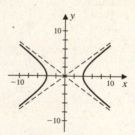

21. Hyperbola with equation $16x^2 - 64x - 25y^2 + 150y = 561$
implies $16(x^2 - 4x + 4 - 4) - 25(y^2 - 6y + 9 - 9) = 561$
so $16(x - 2)^2 - 25(y - 3)^2 = 400$ and $\frac{(x-2)^2}{25} - \frac{(y-3)^2}{16} = 1$.

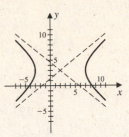

23. The vertex is midway between the focus $(0, 0)$ and the directrix $y = 2$, so it is $(0, 1)$, and the equation has the form $y - 1 = \frac{1}{4c}x^2$. Since the distance from the focus to the vertex is 1, and the directrix is above the focus, $c = -1$, $y - 1 = -\frac{1}{4}x^2$.

25. Since the foci $(\pm 3, 0)$ are centered about the origin, the hyperbola is in standard position with axis on the x-axis, and $c = 3$. Since a vertex is $(1, 0)$, $a = 1$, and

$$c^2 = a^2 + b^2 \text{ implies } 9 = 1 + b^2 \text{ so } b^2 = 8.$$

The equation is

$$x^2 - \frac{y^2}{8} = 1.$$

27. Since the foci of the ellipse are $(0, \pm 5)$, the equation has the form $\frac{y^2}{a^2} + \frac{x^2}{b^2} = 1$, and $c = 5$. Since $(4, 0)$ is on the curve,

$$\frac{16}{b^2} = 1 \text{ implies } b^2 = 16.$$

Then

$$c^2 = a^2 - b^2 \text{ implies } a^2 = 25 + 16 = 41,$$

and the equation is

$$\frac{y^2}{41} + \frac{x^2}{16} = 1.$$

29. Since the directrix $y = -2$ is horizontal and below the pole, the conic has the form $r = \frac{ed}{1 - e \sin \theta}$. Since $e = 3 > 1$, the conic is a hyperbola, and

$$r = \frac{(3)\,2}{1 - 3\sin\theta} = \frac{6}{1 - 3\sin\theta}.$$

31. $r = 4 + 4\cos\theta$

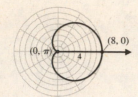

33. $r = 1 + 3\sin\theta$

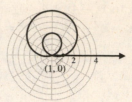

35. $r = 2\sin\theta$

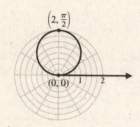

37. $r = 2\cos 2\theta$

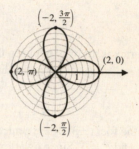

39. $\theta = 1/2$

41. Since $x^2 + y^2 = r^2$ and $x = r\cos\theta$,

$$r = \frac{3}{1 + \cos\theta} \text{ implies } r + r\cos\theta = 3 \text{ so } r = 3 - x \text{ and}$$

$$r^2 = 9 - 6x + x^2 \text{ implies } x^2 + y^2 = 9 - 6x + x^2 \text{ so}$$

$$y^2 = 9 - 6x = -6\left(x - \frac{3}{2}\right) \text{ or } x - \frac{3}{2} = -\frac{1}{6}y^2.$$

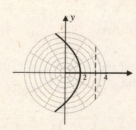

43. Since $x^2 + y^2 = r^2$ and $x = r\cos\theta$,

$$r = \frac{4}{2 - \cos\theta} \text{ implies } 2r - r\cos\theta = 4 \text{ so } 2r = 4 + x \text{ and } 4r^2 = 16 + 8x + x^2 \text{ or}$$

$$4x^2 + 4y^2 = 16 + 8x + x^2 \text{ implies } 3x^2 - 8x + 4y^2 = 16,$$

and

$$3\left(x^2 - \frac{8}{3}x + \left(\frac{4}{3}\right)^2 - \left(\frac{4}{3}\right)^2\right) + 4y^2 = 16 \text{ implies } 3\left(x - \frac{4}{3}\right)^2 + 4y^2 = \frac{64}{3}$$

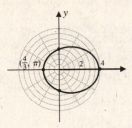

45. $x = t^2 - 1$, $y = t + 1$ implies $t = y - 1$,
$x = (y - 1)^2 - 1$ so $x + 1 = (y - 1)^2$.

47. $x = e^t$, $y = 1 + e^{-t}$ implies $y = 1 + x^{-1}$ so $y = 1 + \frac{1}{x}$ for $x > 0$,
since $x = e^t > 0$.

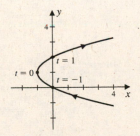

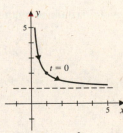

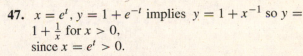

48 $x = t + 2$, $y = (t - 1)^2 + 1$ implies $t = x - 2$ and $y = (x - 2 - 1)^2 + 1 = (x - 3)^2 + 1$. PI6R40

49. $x = (\sin t)^2 + 1$, $y = (\cos t)^2$ implies
$x + y = (\sin t)^2 + (\cos t)^2 + 1 = 2$ so
$y = -x + 2$ for $1 \le x \le 2$, $0 \le y \le 1$.

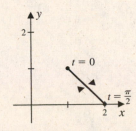

51. The number of leafs in the curves $r = 1 + \sin 2n\theta + (\cos n\theta)^2$ is $2n$.

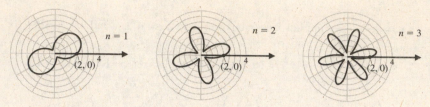

53. The curves in the figures are given by the parametric equations

$$x = (1 - a\sin t)\cos t, \; y = (1 - a\sin t)\sin t, \; \text{for } 0 \leq t \leq 2\pi.$$

If $a = 1$, the curve is a cardioid. If $a < 1$, the curve is a limacon without a loop.
If $a > 1$, the curve is a limacon with a loop.

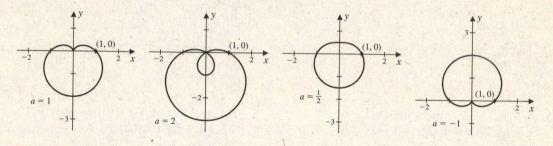

55. The epicycloids defined by the parametric equations

$$x = (a + b)\cos t - b\cos\frac{a+b}{b}t, \; y = (a + b)\sin t - b\sin\frac{a+b}{b}t, \; \text{for } 0 \leq t \leq 2\pi.$$

are shown below.

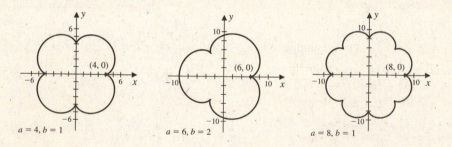

Chapter 6 Exercises for Calculus (Page 366)

1. a. The equation has the general form $y = ax^2$. Since (x_1, y_1) is on the curve, $y_1 = ax_1^2$ implies $a = \frac{y_1}{x_1^2}$, and

$$y = \frac{y_1}{x_1^2}x^2.$$

b. The equation has the general form $y - k = a(x - h)^2$. Since (x_1, y_1) is on the curve, $y_1 - k = a(x_1 - h)^2$ implies $a = \frac{y_1-k}{(x_1-h)^2}$, and

$$y - k = \frac{y_1 - k}{(x_1 - h)^2}(x - h)^2.$$

3. a. The equation has the general form $y - k = a(x - h)^2$. Since $(h + 1, k + 1)$ is on the curve, $k + 1 - k = a(h + 1 - h)^2$ implies $a = 1$, and $y - k = (x - h)^2$.

b. The equation has the general form $x - h = a(y - k)^2$. Since $(h + 1, k + 1)$ is on the curve, $h + 1 - h = a(k + 1 - k)^2$ implies $a = 1$, and

$$x - h = (y - k)^2.$$

5. The equation of the ellipse is $\frac{x^2}{4} + \frac{y^2}{9} = 1$, so $a^2 = 4, b^2 = 9$. The equation of the tangent at the point $\left(1, \frac{3\sqrt{3}}{2}\right)$ is

$$\frac{x(1)}{4} + \frac{y\left(3\sqrt{3}/2\right)}{9} = 1 \text{ implies } 9x + 6\sqrt{3}y = 36 \text{ so } y = -\frac{\sqrt{3}}{2}x + 2\sqrt{3}.$$

7. The equation of the hyperbola is $\frac{x^2}{4} - \frac{y^2}{3} = 1$, so $a^2 = 4, b^2 = 3$. The equation of the tangent at the point $\left(2\sqrt{2}, \sqrt{3}\right)$ is

$$\frac{x(2\sqrt{2})}{4} - \frac{y\left(\sqrt{3}\right)}{3} = 1 \text{ implies } 6\sqrt{2}x - 4\sqrt{3}y = 12 \text{ so } y = \frac{\sqrt{6}}{2}x - \sqrt{3}.$$

9. Since the two triangles in the figure are similar, $\frac{\sqrt{36-y^2}}{6} = \frac{x}{2}$, so the equation for the point on the ladder is

$$\sqrt{36 - y^2} = 3x \text{ implies } 9x^2 + y^2 = 36.$$

So the point moves along the portion of the ellipse $y = 3\sqrt{4 - x^2}$ from the point $(0, 6)$ to the point $(2, 0)$.

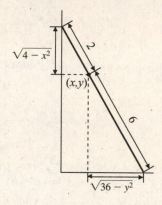

11. The light source should be placed at the focal point of the parabola. Since $y = \frac{1}{4c}x^2 = \frac{1}{4}x^2$, the axis of the parabola is the y-axis, and $c = 1$, so the focal point is $(0, 1)$. Place the light source one unit along the axis from the vertex.

13. If the parabolic cross section of the mirror is a parabola of the form $x = \frac{1}{4c}y^2$, the information implies the point $(3.75, 100) = \left(\frac{15}{4}, 100\right)$ lies on the parabola, so

$$\frac{15}{4} = \frac{1}{4c}(100)^2 \text{ implies } 4c = \frac{4(100)^2}{15} = \frac{8000}{3} \text{ so } c = \frac{2000}{3}.$$

So the focal point of the mirror is $\left(\frac{2000}{3}, 0\right)$, and the observers viewing area is $\frac{2000}{3}$ inches from the center of the mirror.

15. Since

$$r = a\cos\theta + b\sin\theta \text{ implies } r^2 = ar\cos\theta + br\sin\theta \text{ so } x^2 + y^2 = ax + by,$$

we have

$$x^2 - ax + y^2 - by = 0 \text{ implies } x^2 - ax + \frac{a^2}{4} + y^2 - by + \frac{b^2}{4} = \frac{a^2 + b^2}{4},$$

and

$$\left(x - \frac{a}{2}\right)^2 + \left(y - \frac{b}{2}\right)^2 = \frac{a^2 + b^2}{4}.$$

17. $r = \frac{ed}{1 - e\cos\theta}$

a. The aphelion, a, occurs when $\theta = 0$ and the perihelion, p, when $\theta = \pi$. So

$$a = \frac{ed}{1 - e\cos 0} = \frac{ed}{1 - e}, \qquad p = \frac{ed}{1 - e\cos\pi} = \frac{ed}{1 + e},$$

and

$$\frac{a(1 - e)}{e} = d = \frac{p(1 + e)}{e} \text{ implies } a(1 - e) = p(1 + e) \text{ so } a - ae = p + pe,$$

and

$$a - p = ae + pe \text{ implies } a - p = e(a + p) \text{ so } e = \frac{a - p}{a + p}.$$

b. The length of the major axis, $2R$, satisfies

$$2R = a + p = \frac{ed}{1 - e} + \frac{ed}{1 + e}.$$

From part (a),

$$d = \frac{a(1 - e)}{e} \text{ implies } 2R = \frac{e\left(\frac{a(1-e)}{e}\right)}{1 - e} + \frac{e\left(\frac{a(1-e)}{e}\right)}{1 + e} \text{ so } 2R = a + \frac{a(1 - e)}{1 + e},$$

and

$$2R = \frac{a(1 + e) + a(1 - e)}{1 + e} = \frac{2a}{1 + e} \text{ implies } a = R(a + e).$$

Also from part (a),

$$d = \frac{p(1 + e)}{e} \text{ implies } 2R = \frac{e\left(\frac{p(1+e)}{e}\right)}{1 - e} + \frac{e\left(\frac{p(1+e)}{e}\right)}{1 + e} \text{ so } 2R = \frac{p(1 + e)}{1 - e} + p,$$

and

$$2R = \frac{p(1 + e) + p(1 - e)}{1 - e} \text{ implies } 2R = \frac{2p}{1 - e} \text{ so } p = R(1 - e).$$

19. We have

$$\sqrt{x^2 + (y - 1)^2} + 1 = \sqrt{x^2 + (y + 1)^2}$$

$$x^2 + y^2 - 2y + 1 + 2\sqrt{x^2 + (y - 1)^2} + 1 = x^2 + y^2 + 2y + 1$$

$$2\sqrt{x^2 + (y - 1)^2} = 4y - 1$$

$$4(x^2 + y^2 - 2y + 1) = 16y^2 - 8y + 1$$

$$12y^2 - 4x^2 = 3,$$

which is the equation of a hyperbola.

....

Chapter 6 Chapter Test (Page 368)

1. True.

2. False. The equation is $x = \frac{3y^2}{4}$.

3. True.

4. True.

5. False. The vertex of the parabola is $(3, 1)$.

6. True.

7. True.

8. False. The focus is $(0, -2)$.

9. False. The equation of the parabola is $y - 1 = \frac{1}{4}(x - 3)^2$.

10. True.

11. False. The focal points of the ellipse are $\left(2\sqrt{3}, 0\right)$ and $\left(-2\sqrt{3}, 0\right)$.

12. True.

13. False. The equation is $\frac{x^2}{16} + \frac{y^2}{4} = 1$.

14. False. The major axis is on the x-axis.

15. False. The equation is $\frac{y^2}{9} - \frac{x^2}{4} = 1$.

16. False. The equation is $(x - 1)^2 - \frac{(y-1)^2}{3} = 1$.

17. False The asymptotes are $y = \pm \frac{2x}{3}$.

18. False. The asymptotes of the hyperbola are $y - 1 = \pm \frac{3}{2}(x - 3)$.

19. True.

20. False. The polar coordinates $(2, -\frac{3\pi}{4})$ and $(2, \frac{5\pi}{4})$ represent the same point, but $(2, -\frac{\pi}{4})$ does not. To be correct, it could be replaced by $(-2, \frac{\pi}{4})$.

21. True.

22. True.

23. True.

24. False. The linear equation is $x = 2y - 1$.

25. True.

26. True.

27. False. The rectangular equation is $x^2 + (y - 1)^2 = 1$.

28. True.

29. False. The curve could have polar equation $r = 2\cos\theta$.

30. False. The curve could have polar equation $r = 2\sin\theta$.

31. False. The curve could have polar equation $r = 1 + \cos\theta$.

32. False. The curve could have polar equation $r = 1 + \sin\theta$.

33. False. The curve could have polar equation $r = 1 + 2\cos\theta$.

34. False. The curve could have polar equation $r = 2 + \cos\theta$.

35. False. It is a parabola that opens to the left.

36. False. The polar equation $r = \frac{5}{5 - \cos\theta}$ describes an ellipse with vertical directrix.

37. False. The polar curves $r = 1 + 2\sin\theta$ and $r = 1$ intersect at three points. These points have rectangular coordinates $(1, 0)$, $(0, 1)$, and $(-1, 0)$.

38. True

39. True.

40. False.   The parametric equations $x = 2t - 1,$ $y = 3t + 2$ describe a line with slope $\frac{3}{2}$ and y-intercept $\frac{7}{2}$.

41. True.

42. False. The parametric equations $x = e^t + 1,$ $y = e^{2t} - 1$ describe a parabola with vertex $(1, -1)$ and opening upward, but only when $x > 1$.

43. True.

44. True.

45. False. The parametric equations $x = 2\cos t,$ $y = \sin t$ describe an ellipse with rectangular equation $x^2 + 4y^2 = 4$.

46. False. The parametric equations $x = (\cos t)^2,$ $y = (2\sin t)^2$ describe the portion of the line with equation $4x + y = 4$ when $x \geq 0$.

47. False. One set of parametric equations for the line passing through $(3, 1)$ and $(1, 2)$ is $x = 2t + 1, y = -t + 2$.

48. True.

49. False. The parametric equations $x = e^t, y = e^{-t}$ and the rectangular equation $y = 1/x$ describe the same curve only for $x > 0$.

50. True.